DESIGN OF

FLUID THERMAL SYSTEMS

THE PWS-KENT SERIES IN ENGINEERING

D E S I G N   O F
FLUID THERMAL SYSTEMS

WILLIAM S. JANNA
Memphis State University

PWS PUBLISHING COMPANY
I(T)P An International Thomson Publishing Company
Boston • Albany • Bonn • Cincinnati • Detroit • London • Madrid • Melbourne
Mexico City • New York • Paris • San Francisco • Singapore • Tokyo • Toronto • Washington

PWS PUBLISHING COMPANY
20 Park Plaza, Boston, MA 02116-4324

Sponsoring Editor: Jonathan Plant
Assistant Editor: Mary Thomas
Editorial Assistant: Cynthia Harris
Production Editor: Monique A. Calello
Manufacturing Coordinator: Marcia A. Locke
Marketing Manager: Nathan Wilbur
Cover Design: Monique A. Calello
Cover Printer: Henry N. Sawyer Company
Text Printer and Binder: Malloy Lithographing

I(T)P ™

International Thomson Publishing
The trademark ITP is used under license.

Janna, William S.
    Design of fluid thermal systems / William S. Janna.
        p. cm. — (The PWS series in engineering)
    Includes bibliographical references and index.
    ISBN 0-534-93373-4 (acid-free paper)
    1. Piping—Design and construction. 2. Heat exchangers—Design and construction. I. Title. II. Series.
    TJ930.J36 1993   92-46059
    621.402'2—dc20   CIP

Printed and bound in the United States of America.
  2 3 4 5 6 7 8 9 10 — 99 98 97 96 95

To Him who is our source of grace,

our source of love and our source of knowledge;

And to Marla whose love is a source of joy.

Preface

For many years, it seems that the objective in Engineering Education has been to teach courses and topics to engineering students as if all of them planned to go to graduate school. The emphasis has been on theoretical topics with practical applications playing only a minor part. Actual trends show, however, that the majority of engineering graduates prefer to work as practicing engineers. It is toward this group that this text is aimed.

The course for which this book is intended is a capstone type of course in the energy systems (or thermal sciences) area that corresponds to the machine design course in the mechanical systems area. This text is written for seniors in engineering who intend to practice fluid/thermal design. Fluid mechanics is a prerequisite. Heat transfer is a prerequisite or at least it should be taken with this course.

Contents

The text is divided into two major sections. The first is on piping systems blended together with economics of pipe size selection and the sizing of pumps for piping systems. The second is on heat exchangers or, more generally, devices available for the exchange of heat between two process streams. The list of topics that can be added is almost endless (including, for example: water hammer, fluid meters, etc.).

The text begins with an introductory chapter in which examples of fluid/thermal systems are provided. A pump and piping system, a household air conditioner, a baseboard heater, a water slide, and a vacuum cleaner are such examples. Also presented are dimensions and unit systems that are used in conventional engineering practice (i.e., Engineering and British Gravitational systems). Although the SI unit system is not currently being used widely in industry, it too will be presented. The student is expected to know information on unit systems, but they are presented in Chapter 1 to introduce conversion factor tables in the Appendix and to familiarize the reader with the notation in this text.

Chapter 2 is a review chapter on the properties of fluids and the equations of fluid mechanics. Chapter 2 is included so that the student will become familiar with the tables of fluid properties in the Appendix. This chapter serves as a lead-in to Chapter 3 which is about piping systems. It is expected that by the time students take this course, they have learned about piping systems in a first course in fluid mechanics.

Here, however, the subject of piping systems is covered in greater detail and depth. Specifications for pipes and tubes are discussed. Circular, square, rectangular, and annular cross sections are presented. Laminar and turbulent flow in each of these cross sections is modeled. Chapter 4, a continuation of Chapter 3, covers minor losses and economics of pipe size selection. Both piping systems chapters contain modified pipe friction diagrams useful in solving special types of problems.

Chapter 5 is about pumps. Types of machines are discussed, and testing methods for centrifugal pumps are presented. Charts from manufacturers' catalogs are given and are used to illustrate the steps in sizing a pump for a piping system. At the conclusion of studying Chapters 3, 4, and 5, an engineer should be able to start with a needed flow rate and select the most economical pipe size, pipe material, pipe fittings, pump, hangers, and hanger spacing.

Chapter 6 provides an introduction to heat transfer basics in order to present the appropriate heat transfer properties and the heat transfer tables in the Appendix. Chapter 7 is about heat exchangers. Double pipe, shell and tube, and cross flow heat exchangers are all discussed along with methods of analysis. The emphasis is on design and selection of heat exchangers, however. Chapter 8 gives a brief analysis of economics applied to heat exchanger design, selection, and configuration. Basic heat exchanger cost data are given and optimum outlet temperature analysis is presented. The main thrust of Chapters 7 and 8 is on the selection of the most economical heat exchanger for a given service.

Chapter 9 is on the design process. A design project example is given and the steps involved in completing it are presented. These steps include the bid process, project management, construction of a bar chart of project activities, written and oral reports, internal documentation, and evaluation and assessment of results.

Where appropriate, a section entitled "Show and Tell" has been added to various chapters. Students are required to provide very brief presentations on selected topics. For example, in Chapter 3, one Show and Tell requires the student to give a presentation on various types of valves that are commonly used. The valves that are available are brought to class and taken apart (or cut in half prior to class) to illustrate how each works. A Show and Tell on this and many other topics is far more effective than a photograph in a text. In addition, the student will have practice in making an oral presentation.

Chapter 10 is an introduction to the projects. The course for which this text is intended requires the students to complete term projects. Each project has associated with it a project description which begins with a few introductory comments and concludes with several tasks that are to be completed. Each project has an estimate of the number of engineers required to finish it in the given school term. The students are responsible for selecting project partners and, as a group, deciding on which projects they would like to work. Each group elects its own project manager or leader.

Projects

Where the projects are concerned, the instructor is like a general contractor who has a number of projects that need to be worked on. The student groups are like small consulting companies that must decide on who gets what project. The awarding of projects is done on a "lowest bidder" basis. All group members earn the same salary (e.g., $30,000 per year or as assigned by the instructor). All group leaders likewise earn the same salary (slightly more than the group members). Based on each group's estimate of the number of person-hours required to complete the tasks of the project, a personnel cost is calculated. Other costs include benefits, fees for experts, computer time, etc.

Each group fills out a bid sheet (see Chapter 9 for an example) for every project they are interested in—usually no less than three. The bids are sealed in envelopes and one class period is spent in a "bid opening ceremony." The lowest bidder for each project then has the option of accepting (or not accepting) that project to work on. A project for which a group is lowest bidder might not be the one that group would most like to work on. Each group will then have one project to devote the entire semester to completing.

The projects must be managed to insure that all the work is done before the last one or two weeks of classes, when quality will suffer because of the frantic, last minute pace. Each group must complete a task planning sheet. Each task is shown on the sheet along with who is to complete that task, and when it is to be completed. Each and every student is to keep a spiral ring (or equivalent) notebook in which everything, from actual design work to a mere phone call, that student does on the project and time spent is recorded. The project leaders make sure that the responsible group member is completing his/her assigned task on schedule as per the task planning sheet. The task planning sheet can be changed during the school term but an equitable division of labor must be adhered to and individual tasks must be completed. At the end of the term, students tally their hours and compare the actual cost of completing the project to the estimated cost at time of bid.

Project reports are to be given in two forms: written and oral. The written report should detail the solution to all phases of the project as outlined in the original description or as modified in discussion sessions with the instructor. The oral report should summarize the findings and give recommendations. The oral report should be limited in time.

It must be emphasized that this text does not provide a complete description in any one area. The objective here is to provide some design concepts currently used by practicing engineers in the area of fluid/thermal systems. The student should remember that actual design details of various systems can be found in textbooks, reference books, and periodicals.

Instructor's Guide and Solutions Manual

There is an *Instructor's Guide and Solutions Manual* available to accompany this text. The *Guide* provides solutions to the problems in the text, and also gives a detailed outline of the course. The outline is laid out in a twelve-week plan showing problem assignments, Show and Tell assignments, and project scheduling.

Acknowledgments

I wish to thank the many individuals, students and faculty alike, who made valuable suggestions on how to improve the text. Moreover, I am greatly indebted to the following reviewers who read over the manuscript and made helpful suggestions: Dr. Edwin P. Russo, University of New Orleans; Dr. Gerald S. Jakubowski, Loyola Marymount University; Don Dekker, Rose Hulman Institute of Technology; and, Dr. Ray W. Brown, Christian Brothers University. I extend my thanks also to the individuals at PWS who supported this project and who worked toward its completion: Jonathan Plant, Sponsoring Editor; Mary Thomas, Assistant Editor; Cynthia Harris, Editorial Assistant; and to Monique A. Calello, Production Editor, for her creative cover design. I also wish to extend gratitude to Memphis State University for providing help with various tasks associated with this project. Finally, I wish to acknowledge the encouragement and support of my lovely wife, Marla, who made many sacrifices during the writing of this text.

William S. Janna

Contents

Nomenclature

Symbol	Definition	SI	Engineering
		Unit	
A	area	m^2	ft^2
a	acceleration	m/s^2	ft/s^2
C_p	specific heat	$J/(kg \cdot K)$	$BTU/(lbm \cdot °R)$
C_o	orifice coefficient	—	—
C_v	venturi coefficient	—	—
D	diameter	m	ft
$D_h = 4A/P$	hydraulic diameter	m	ft
D_e	heat transfer characteristic dimension	m	ft
D_{eff}	effective diameter	m	ft
F	force	N	lbf
g	gravitational acceleration	m/s^2	ft/s^2
g_c	conversion factor	—	$32.17\ lbm \cdot ft/(lbf \cdot s^2)$
h	enthalpy	J/kg	BTU/lbm
h_c	convection coefficient	$W/(m^2 \cdot K)$	$BTU/(ft^2 \cdot hr \cdot °R)$
k_f	thermal conductivity	$W/(m \cdot K)$	$BTU/(ft \cdot hr \cdot °R)$
L	length	m	ft
m	mass	kg	lbm
$\dot{m}$	mass flow rate	kg/s	lbm/s
Nu	Nusselt number	—	—
NTU	number of transfer units	—	—
P_T	pitch of tube bank	m	ft
P	perimeter	m	ft
Pr	Prandtl number	—	—
p	pressure	$Pa = N/m^2$	lbf/in^2
Q	volume flow rate	m^3/s	ft^3/s
Q_{ac}	actual flow rate	m^3/s	ft^3/s
Q_{th}	theoretical flow rate	m^3/s	ft^3/s
q	heat transferred	W	BTU/hr
q''	heat transferred/area	W/m^2	$BTU/(ft^2 \cdot hr)$
R	gas constant	—	—
R_h	hydraulic radius	m	ft

Nomenclature
(concluded)

Symbol	Definition	Unit SI	Engineering
r	radius or radial coord	m	ft
Ra	Rayleigh number	—	—
Re	Reynolds number	—	—
T	temperature	K or °C	°R or °F
t	time	s	s
U	overall heat transfer coefficient	$W/(m^2 \cdot K)$	$BTU/(ft^2 \cdot hr \cdot °R)$
V	velocity	m/s	ft/s
$\mathcal{V}$	volume	m^3	ft^3
dW/dt	power	J/s	ft-lbf/s or HP

Greek Letters

Symbol	Definition	SI	Engineering
$\alpha = k_f/\rho C_p$	thermal diffusivity	m^2/s	ft^2/s
η	efficiency	—	—
μ	viscosity	$N \cdot s/m^2$	$lbf \cdot s/ft^2$
$v = \mu g_c/\rho$	kinematic viscosity	m^2/s	ft^2/s
ρ	density	kg/m^3	lbm/ft^3
σ	surface tension	N/m	lbf/ft

DESIGN OF

FLUID THERMAL SYSTEMS

CHAPTER 1   Introduction

Fluid thermal systems is a very broad term that includes many designs and devices. A pump and pipe combination is an example of a fluid system in which fluid is being conveyed. An air conditioner is a device in which a fluid is conveyed, so it is an example of a fluid system. Moreover, because heat transfer effects are important in the air conditioner, we can consider it a fluid thermal system.

For purposes of illustration, let us consider the air conditioner in more detail. Figure 1.1 is a sketch of an air conditioner, used typically as a "central unit." The fluid within the system is known as a *refrigerant*. It is compressed by the compressor and leaves as a superheated vapor. The vapor enters what is called a *heat exchanger* (like the radiator of a car). A fan moves atmospheric air over the coils or tubes of the condenser. Heat is transferred from the refrigerant within the tubes to the air outside the tubes. During this process, the refrigerant condenses. The liquid refrigerant next goes to a capillary tube, which is a long tube of very small diameter. Liquid refrigerant passing through a capillary tube experiences a significant loss of pressure and, correspondingly, a decrease in temperature. The cold liquid refrigerant is then piped to an *evaporator*, a device similar to the condenser. Air moving past the outside of the evaporator coils loses energy to the refrigerant inside. The refrigerant gains enough energy to vaporize. Once past the evaporator, the refrigerant is returned to the compressor and the cycle is repeated.

When this system is used to cool the air in a house or a refrigerator, the evaporator is located within the house or refrigerator and inside air is moved past the coils. The condenser and compressor are usually located outside and ambient air is moved past the condenser coils. Thus the refrigerant transfers energy from the evaporator within and from the compressor to the condenser.

As indicated in the above discussion, the compressor moves the fluid throughout the system. The fluid itself undergoes a change in phase at places within the system and effects an energy transfer from the evaporator to the condenser. The compressor power must be determined, the fluid conveying lines must be sized, the heat exchangers must be selected, the entire system must be housed, and the fluid itself must be

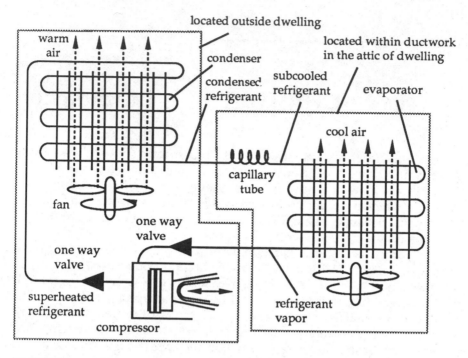

FIGURE 1.1. *Sketch of an air conditioning unit.*

picked from among many available. Moreover, the overall cost of the system must be kept to within competitive and affordable limits. Its initial cost, operating cost, and life expectancy must also be considered. Obviously, the design of this common fluid thermal system is not trivial, but instead requires careful thought and extensive planning.

Consider next a homeowner about to install a concrete driveway that is to contain tubes to convey heated water during winter so that snow and ice will not accumulate on the concrete surface. Figure 1.2 is a sketch of the tubing system over which the concrete is to be poured. The system consists of inlet and outlet manifolds connected with smaller diameter tubes that span the width of the driveway. Enough heated water must be supplied to keep the concrete surface sufficiently warm.

The manifolds and tube sizes must be selected. The hot water source must be sized and it must be determined if a pump is necessary to circulate the water. The tubing material must be selected and the entire system must be affordable. Clearly a high hot water cost can be eliminated with an inexpensive snow blower and rock salt. The life expectancy as well as initial and operating costs must be considered. Obviously, the design of such a system is not trivial but instead requires much thought and careful planning.

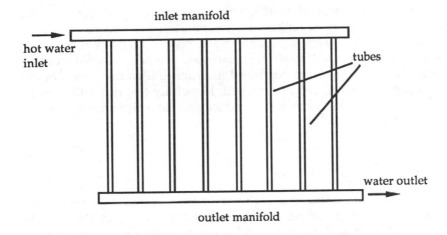

FIGURE 1.2. *Plan view of a tubing system for heating a concrete driveway.*

The air conditioner and piping system discussed above are two examples of the many fluid thermal systems that exist and that must be designed. Other examples include:

- A meter that gives an instantaneous reading of miles/gallon for an automobile.

- A system useful for measuring thrust developed by a diver who is testing swim fins (or flippers).

- A carburetor that develops a sonic shock which helps to vaporize or break up fuel droplets.

- A piping system to provide sufficient heat removal to create an ice rink.

- A funnel that signals the user to stop pouring before an overflow occurs.

- A device for measuring air pressure in truck tires and that displays the pressure even when the vehicle is moving.

- A system for testing the efficiency of ceiling fan blades.

- A ventilation system for mines.

- A sand blaster that uses ice or dry ice instead of sand, in order to minimize health hazards and ease cleanup.

The list can be expanded to include many more examples. Each system requires considerable design work, extensive refining, and inevitably some redesigning. The objective in this text is to discuss some of the theoretical concepts learned in engineering and in economics courses, and synthesize them into a coherent presentation in which practical applications are stressed. Fundamental concepts of fluid mechanics, thermodynamics, heat transfer, material science, manufacturing methods, and economics are combined in order to illustrate how devices and systems are engineered or designed. Hopefully this text will provide the engineer with ideas and design concepts that will enhance his/her future practice.

1.1 Dimensions and Units

The unit systems used in this text are primarily the British gravitational, the engineering or U.S. Customary system, and the SI unit system. Fundamental dimensions and units in each of these systems are listed in Table 1.1. Also shown are dimensions and units in other systems that have been developed, namely, the British absolute and the CGS absolute.

When using U.S. Customary units, a conversion factor between force and mass units must be used. This conversion factor is

$$g_c = 32.2 \frac{\text{lbm·ft}}{\text{lbf·s}^2} \qquad \text{(U.S. Customary)} \qquad (1.1)$$

In the other unit systems listed in Table 1.1, the conversion g_c is not necessary nor is it used. The equations of this text will contain g_c and if U.S. Customary units are not used, the reader is advised to either ignore g_c or set it equal to

$$g_c = 1 \frac{\text{mass unit·length unit}}{\text{force unit·time unit}^2} \qquad \text{(Other unit systems)} \qquad (1.2)$$

When solving problems, a unit system must be selected for use. All equations that we write are dimensionally consistent. Therefore all parameters we substitute into the equations must be in proper units. The proper units are the fundamental units in each system. Suppose for example that we are using the equation for volume flow rate,

$$Q = AV$$

Say we are given $Q = 50$ gallons per minute, an area of 17 in.^{2}, and that we wish to find velocity. We first convert to fundamental units:

$$Q = 50 \, \text{gpm} = 0.11 \, \text{ft}^3/\text{s}$$

$$A = 17 \, \text{in.}^2 = 0.118 \, \text{ft}^2$$

TABLE 1.1. *Conventional Unit Systems.*

Dimension	British Gravitational	SI	Engineering	British Absolute	CGS Absolute
Mass (M) (derived)	slug				
Mass (M) (fundamental)		kg	lbm	lbm	gram
Force (F) (derived)		N		poundal	dyne
Force (F) (fundamental)	lbf		lbf		
Length (L)	ft	m	ft	ft	cm
Time (T)	s	s	s	s	s
Temperature (t)	°R °F	K† °C	°R °F	°R °F	°K °C
Conversion factor	g_c	—	$32.2\dfrac{\text{lbm·ft}}{\text{lbf·s}^2}$	—	—

†Note that in SI, the degree Kelvin is properly written without the ° symbol.

The velocity becomes

$$V = 0.11/0.118 = 0.94 \text{ ft/s}$$

With the huge volume of parameters that have acquired specialized units (e.g., horsepower, ton of air conditioning, BTU, tablespoon, etc.), we can easily conclude that the process of converting to fundamental units is a never ending but ever present necessity. To ease this burden, a set of conversion factor tables is provided in Appendix Table A.2. Appendix Table A.1 gives prefixes that are used when working in SI units.

1.2 Summary

In this chapter we have examined some fluid thermal systems and indirectly defined them. We also briefly discussed unit systems including SI, U.S. Customary, and British gravitational systems. Mention was made of some specialized units that have arisen in industry and that fundamental units should be used when solving problems. These topics are amplified in the questions and problems that follow.

1.3 Questions for Discussion Chapter 1

The following questions should be addressed by groups of 4 or 5 individuals who spend 20 minutes on each assigned question. At the end of the discussions, the conclusions should be shared with the other groups in the form of a 3-minute or shorter oral report.

1. Discuss the properties of a plastic that is to be used for cassettes.

2. What are the expected properties of a paint that is used on streets as a lane marker?

3. Discuss the desirable properties of a tube or tubes used in a solar collector. The tubes are to convey water that is heated by the sun.

4. What are the desirable properties of a material that is to be made into an inflatable float for people to use at a swimming pool?

5. Discuss the desirable properties of a tank that is to be used to store liquid oxygen.

6. What should be the properties of a material used as brake lining in a conventional automobile?

7. What are the desirable properties of a material used as a bathtub?

8. Discuss the factors that contribute to the decision on how tight a manufacturer should make the threaded top on a jar of mayonnaise.

9. Discuss the factors that contribute to the decision on how much weight a ladder should be able to support and what material it should be made of.

1.4 Problems Chapter 1

1. Consult a dictionary to determine the meaning of the following units (remember to look up the proper pronunciation as well):

   (a) coomb   (b) scruple   (c) cord of wood   (d) ream of paper

2. Repeat Problem 1 for the following:

   (a) gill   (b) degree-day   (c) ton vs metric ton   (d) long ton vs
                                                                 short ton

3. Consult a dictionary to determine the conversion factors associated with the following units:

   (a) the number of *dashes* per teaspoon;
   (b) the number of *teaspoons* per tablespoon;
   (c) the number of *tablespoons* per cup;
   (d) the number of *cups* per quart; and,
   (e) the number of *quarts* per gallon.

4. Consult a dictionary to determine the conversion factors that apply to:

   (a) parsecs per mile   (b) stadia per mile

5. How is the height of a horse expressed? The height is measured from the ground to what part of the horse's anatomy? What is the conversion of this height unit to feet?

6. How is blood pressure expressed? Why isn't the result expressed in psig or kpa?

7. What is the relationship between an ounce (16 ounces per pound) and a fluid ounce (16 ounces equals one pint)?

8. A patient is being fed intravenously. Somewhere in the flow line is a transparent chamber where the liquid flow is controlled so that it passes through a droplet at a time. The number of droplets of liquid per unit time can be observed and measured. What is the conversion factor between droplets/time to ft^3/s. Is the droplet/time measurement convenient?

9. What is the origin of the "horsepower"? Why would anyone wish to express power in the unit of horsepower?

10. Which is heavier—a grain or a dram? Express both in the appropriate English fundamental unit.

11. Canned foods are quite common. How are can sizes measured or expressed by can manufacturers? Why aren't more commonly recognizable units used such as the pint?

12. In the plastics industry, what is a gaylord?

13. Whose foot was the model for our unit of measurement we call the "foot"?

14. How many barleycorns are in an inch?

15. Why is a mile 5,280 ft?

16. How long is the circumference of a racetrack; i.e., how many laps must be made in a one mile race?

17. What is the significance of an acre, and how many square feet are contained in one? Which is larger, an acre or an arpent?

18. How many years does a furlong designate? Furthermore, if a furlong is a linear measurement, what does this have to do with "years"? Why?

19. What is a carat?

CHAPTER 2   Fluid Properties and Basic Equations

In this chapter, we review some fundamental principles of fluid mechanics. Fluid properties are briefly defined in an effort to refresh the reader's memory and to make the reader familiar with the notation used in this text. The equations of continuity, momentum and energy, and the Bernoulli equation are stated but not derived. Review problems requiring the application of these equations are also provided.

2.1 Fluid Properties

The fluid properties we will discuss in this section include density, specific gravity, specific weight, absolute or dynamic viscosity, kinematic viscosity, specific heat, internal energy, enthalpy, and bulk modulus. We will also examine some of the common techniques used for measuring selected properties.

Density, Specific Gravity, and Specific Weight

Density of a fluid is defined as its mass per unit volume, and is denoted by the letter ρ. Density has dimensions of M/L^3 (lbm/ft^3 or kg/m^3).

Specific gravity of a fluid is the ratio of its density to the density of water at 4°C:

$$s = \frac{\rho}{\rho_w} \tag{2.1}$$

where ρ_w is the density of water. Values of specific gravity for various fluids appear in the property tables of the appendix. It is permissible here to take the water density to be

$$\rho_w = 62.4 \text{ lbm/ft}^3 = 1.94 \text{ slug/ft}^3 = 1\,000 \text{ kg/m}^3$$

Specific weight is a useful quantity related to density. While density is a *mass* per unit volume, specific weight is a *force* per unit volume. Density and specific weight are related by

$$SW = \frac{\rho g}{g_c} \qquad\qquad (2.2)$$

The dimension of specific weight is F/L^3 (lbf/ft^3 or N/m^3).

For a liquid, density can be measured directly by weighing a known volume. Specific weight can be determined by submerging an object of known volume into the liquid. The weight of the object in air minus its weight measured while it is submerged gives the buoyant force exerted on the object by the liquid. The buoyant force divided by the volume of the object is the specific weight of the liquid.

In the petroleum industry, specific gravity of a fuel oil is expressed as Sp.Gr. 60°F/60°F. This nomenclature indicates that specific gravity is the ratio of oil density at 60°F to water density at 60°F. *API gravity* is the standard used. The API gravity is related to specific gravity by

$$\text{Sp.Gr. } 60°F/60°F = \frac{141.5}{131.5 + °API} \qquad\qquad (2.3)$$

where °API is read as "degrees API."

For a gas, specific gravity can be found using any of a number of tests. One such method is the direct weighing procedure in which a volume of gas and an equal volume of air (both at standard conditions for which air properties are known) are collected and weighed. The weight differential allows for calculating the specific gravity of the gas. Other methods involve variations on the theme of measuring a differential weight or mass.

Viscosity

The viscosity of a fluid is a measure of the fluid's resistance to motion under the action of an applied shear stress. Consider the sketch of Figure 2.1. A liquid layer of thickness Δy is between two parallel plates. The lower plate is stationary while the upper plate is being pulled to the right by a force F. The area of contact between the moving plate and the liquid is A. The applied shear stress then is $\tau = F/A$. The liquid continuously deforms under the action of the applied shear stress. The continuous deformation is expressed in terms of a strain rate which physically is the slope of the resultant linear velocity distribution within the liquid. At the stationary surface, the liquid velocity is zero while at the moving plate the liquid velocity equals the plate velocity. This apparent adhering of the liquid to solid boundaries is known as the *non-slip condition*. For each applied shear stress there will correspond only one strain rate. A series of measurements (forces vs resultant strain rates) would yield data for a graph of shear stress vs strain rate. Such a plot is called a *rheological diagram*, an example of which is provided in Figure 2.2. If the liquid between the plates is water or oil, for example, then the line labelled "Newtonian" in Figure 2.2 results. The slope of that line is known as the *dynamic* or *absolute viscosity* of the liquid. For a Newtonian fluid, we have

$$\tau = \mu \frac{dV}{dy} \qquad \text{(Newtonian)} \qquad\qquad (2.4)$$

where μ is the absolute viscosity, τ is the applied shear stress, and dV/dy is the strain rate. The dimension of viscosity is $F \cdot T/L^2$ (lbf·s/ft² or N·s/m²). Other Newtonian fluids are air, oxygen, nitrogen, and glycerine, to name a few.

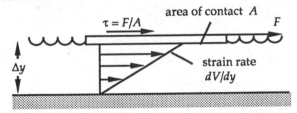

FIGURE 2.1. *Definition sketch for viscosity determination.*

Also shown in Figure 2.2 is the curve for a fluid known as a "Bingham Plastic." Such fluids behave as solids (i.e., no continuous deformation under the action of an applied shear stress) until an *initial yield stress* τ_0 is exceeded. As an example, suppose a jar of peanut butter is inverted. If the peanut butter is a Bingham plastic then it will not flow out of the jar. The force of gravity does not exert a great enough shear stress on the fluid such that the initial yield stress of the peanut butter is exceeded. If the applied shear stress exceeds the initial shear stress, the Bingham plastic behaves like a Newtonian fluid. Bingham plastics are characterized by

$$\tau = \tau_0 + \mu_0 \frac{dV}{dy} \qquad \text{(Bingham plastic)} \qquad\qquad (2.5)$$

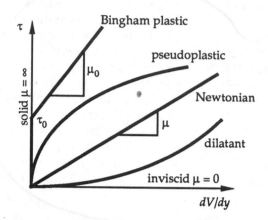

FIGURE 2.2. *A rheological diagram characterizing various fluids.*

where μ_0 is the apparent viscosity. Chocolate mixtures, some greases, paints, paper pulp, drilling muds, soap, and toothpaste are examples of Bingham plastics.

Fluids that exhibit a decrease in viscosity with increasing shear stress are known as *pseudoplastic fluids*. Some greases, mayonnaise, and starch suspensions are examples of such fluids. A power law equation, called the Ostwald-deWaele equation, describes the curve

$$\tau = K\left(\frac{dV}{dy}\right)^n \qquad (n < 1 \text{ pseudoplastic}) \qquad (2.6)$$

where K is called a *consistency index* with dimensions of $F \cdot T^n / L^2$ (lbf·s^n/ft^2 or N·s^n/m^2) and n is a dimensionless *flow behavior index*. As discernible from the figure, a pseudoplastic fluid exhibits less resistance to flow as shear stress (or in the case of pipe flow, flow rate) increases. Thus for pumping purposes, the higher the shear stress (or flow rate), the greater the frictional effects for a pseudoplastic fluid.

Fluids that exhibit an increase in viscosity with increasing shear stress are known as *dilatant fluids*. Wet beach sand, starch in water, and water solutions containing a high concentration of powder are examples of such fluids. Again, a power law equation describes the curve

$$\tau = K\left(\frac{dV}{dy}\right)^n \qquad (n > 1 \text{ dilatant}) \qquad (2.7)$$

A dilatant fluid's resistance to flow increases with flow rate or rate of shear.

Other types of fluids not shown in Figure 2.2 are *rheopectic*, *thixotropic*, and *viscoelastic* fluids. For a rheopectic fluid, the applied shear stress would have to increase with time to maintain a constant strain rate—a gypsum suspension is an example. With a thixotropic fluid, the applied shear stress would decrease with time to maintain a constant strain rate—liquid foods and shortening are examples. A viscoelastic fluid exhibits elastic and viscous properties. Such fluids partly recover from deformations caused during flow—flour dough is an example.

Kinematic Viscosity

In many equations of fluid mechanics, the term $\mu g_c/\rho$ appears frequently. This ratio is called the kinematic viscosity v which has dimensions of L^2/T (ft^2/s or m^2/s).

2.2 Measurement of Viscosity

Viscosity can be measured with a number of commercially available viscometers or viscosimeters. Each works on basically the same principle. A laminar motion of the fluid is caused and suitable measurements are taken. For the created laminar conditions, an exact solution to the equation of motion will exist that relates the viscosity to the geometry of the device and to the data obtained.

One such device consists of two concentric cylinders, one of which is free to rotate (see Figure 2.3). Liquid is placed in the annulus between the two cylinders. The outer cylinder is rotated at a known and carefully controlled angular velocity, and the momentum of the outer cylinder is transported through the liquid. In turn, a torque is exerted on the inner cylinder. The inner cylinder might be suspended by a torsion wire or a spring that measures the angular deflection caused by the fluid. The dynamic viscosity μ is proportional to the torque transmitted. If the gap between the cylinders is very narrow, no more than 0.1 of the inner cylinder radius, then the velocity distribution of the fluid in the annulus is approximately linear. Otherwise, the velocity distribution is better described as a parabola. For a linear velocity profile, we have

$$\tau = \mu \frac{dV}{dy} = \mu \frac{V}{\delta} \qquad (2.8)$$

$$\tau = \frac{\mu(R + \delta)\omega}{\delta}$$

where ω is the rotational speed. The torque exerted on the inner cylinder is

$$T = \text{shear force} \times \text{distance} = \text{shear stress} \times \text{area} \times \text{distance}$$

$$T = \tau(2\pi R L)R$$

The shear stress in terms of torque then is

$$\tau = \frac{T}{2\pi R^2 L}$$

Substitution into Equation 2.8 gives

$$\frac{T}{2\pi R^2 L} = \frac{\mu(R + \delta)\omega}{\delta}$$

Solving for viscosity, we get

$$\mu = \frac{T\delta}{2\pi R^2 (R + \delta)L\omega} \qquad (2.9)$$

All parameters on the right hand side of the above equation, except torque T and rotational speed ω, are geometric terms. Torque and rotational speed are the only dynamic measurements required.

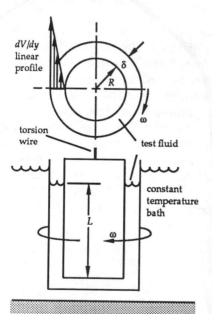

FIGURE 2.3. A narrow gap rotating cup viscometer.

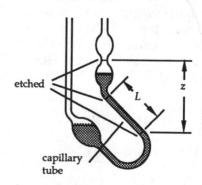

FIGURE 2.4. A capillary tube viscometer.

Another device for measuring viscosity is called the *capillary viscometer* (see Figure 2.4). This device consists of a glass tube of small diameter etched at three locations. By applying a vacuum to the right side, the liquid level is raised until it just reaches the uppermost etched line. At this point the liquid is released and allowed to flow under the action of gravity. The time required for the liquid level to fall from the middle to the lowest etched line is measured. Obviously, this problem is unsteady but if average values are used, acceptable results are obtained. The volume of liquid involved is called the *efflux volume* which is carefully measured for the experiment. The average flow rate through the capillary tube is

$$Q = \frac{\mathcal{V}}{t}$$

where $\mathcal{V}$ is the efflux volume and t is the time recorded. For laminar flow through a tube,

$$Q = \left(-\frac{dp}{dz}\right)\frac{\pi R^4}{8\mu}$$

where R is the tube radius and $(-dp/dz)$ is a positive pressure drop

$$\left(-\frac{dp}{dz}\right) = \frac{p_2 - p_1}{L}$$

The pressure drop equals the available hydrostatic head, which contains the gravity term (gravity is the driving force):

$$p_2 - p_1 = \frac{\rho g}{g_c} z$$

Substituting into the volume flow equation gives

$$\frac{\mathcal{V}}{t} = \frac{\rho g}{g_c} \frac{z}{L} \frac{\pi R^4}{8\mu}$$

Solving for viscosity, we get

$$\frac{\mu g_c}{\rho} = \nu = \left(\frac{z \pi R^4 g}{8 L \mathcal{V}}\right) t \qquad (2.10)$$

For a given viscometer, the quantity in parentheses (a geometric quantity) is a constant. So kinematic viscosity is proportional to the time.

There are other types of viscometers such as the cone and plate viscometer, the falling sphere viscometer and the wide gap concentric cylinder type. All contain geometric and dynamic terms that allow for calculating the viscosity of a fluid.

EXAMPLE 2.1. Tomato paste was tested in a viscometer and the following data were obtained. Determine if the fluid is Newtonian and its descriptive equation.

τ (N/m²)*	51	71.6	90.8	124.0	162.0
dV/dy (rad/s)	0.95	4.7	12.3	40.6	93.5

Solution: A plot of these data appears in Figure 2.5. The fluid is pseudoplastic so we assume that Equation 2.6 applies:

$$\tau = K\left(\frac{dV}{dy}\right)^n$$

Taking the natural logarithm of both sides of this equation yields a form recognized as a straight line:

$$\ln \tau = \ln K + n \ln\left(\frac{dV}{dy}\right)$$

or in simpler notation,

$$T = b_0 + b_1 V'$$

where b_0 (= $\ln K$) and b_1 (= n) are unknown constants. Using the method of least squares, the solution for the constants is found with

$$b_1 = \frac{m\Sigma[T_i V_i'] - \Sigma V_i' \Sigma T_i}{m\Sigma V_i'^2 - [\Sigma V_i']^2}$$

and $$b_0 = \frac{\Sigma T_i}{m} - b_1 \frac{\Sigma V_i'}{m}$$

---

* Data from *Fundamentals of Food Engineering* by Stanley E. Charm, 2nd ed., Avi Publishing Co., Westport, Conn., 1971, p. 62.

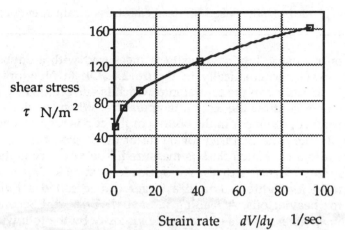

FIGURE 2.5. *Data of Example 2.1.*

The calculations are summarized in the following table:

dV/dy	$ln(dV/dy)$	τ	$ln(\tau)$	TV'
0.95	-0.0513	51	3.93	-0.202
4.7	1.55	71.6	4.27	6.62
12.3	2.51	90.8	4.51	11.3
40.6	3.70	124	4.82	17.8
93.5	4.54	162	5.09	23.1
Σ	12.2		22.62	58.6

Also, $m = 5$ and $\Sigma V^2 = 43.0$. Substituting into the above equations gives:

$$b_1 = \frac{5(58.6) - 12.2(22.62)}{5(43.0) - (12.2)^2} = 0.257$$

and $\quad b_0 = \frac{22.62}{5} - 0.257\frac{12.2}{5} = 3.90$

Also $\quad K = \exp(b_0) = 49.3$ and $n = b_1 = 0.257$

The final equation for the tomato paste is:

$$\tau = 49.3\left(\frac{dV}{dy}\right)^{0.257}$$

The fluid is thus pseudoplastic, based on its shear stress-strain rate curve.

In the oil industry, liquid viscosity is measured with a Saybolt viscometer, illustrated schematically in Figure 2.6. Oil (a Newtonian fluid) to be tested is placed in the central cup which is surrounded by an oil bath. The oil bath controls the test oil temperature. When the desired temperature is reached, a stopper in the bottom of the cup is removed and test oil flows out. The time required for 60 ml of oil to pass through a standard orifice into a calibrated flask is measured. The viscosity is then expressed in terms of *Saybolt seconds.* An orifice known as a *Universal-type orifice* is used for lighter oils and a larger orifice called a *Furol orifice* is used for heavier oils. A reading in Saybolt Universal Seconds (SUS) can be converted into units of kinematic viscosity by the following equations:

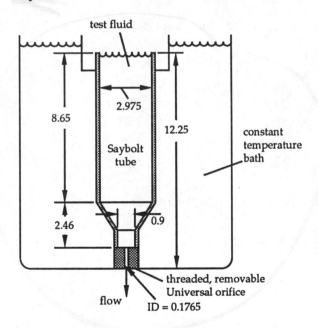

FIGURE 2.6. *Schematic of a Saybolt Viscometer. (Dimensions in cm.)*

a) 34 < SUS < 115

$$v \, (m^2/s) = 0.224 \times 10^{-6} \, (SUS) - \frac{185 \times 10^{-6}}{(SUS)} \qquad (2.11a)$$

b) 115 < SUS < 215

$$\nu \, (m^2/s) = 0.223 \times 10^{-6} \, (SUS) - \frac{155 \times 10^{-6}}{(SUS)}$$ (2.11b)

c) SUS > 215

$$\nu \, (m^2/s) = 0.2158 \times 10^{-6} \, (SUS)$$ (2.11c)

Compressibility Factor

The compressibility factor is a property that describes the change in density experienced by a liquid during a change in pressure. The compressibility factor is given by

$$\beta = -\frac{1}{\Psi} \left(\frac{\partial \Psi}{\partial p} \right)_T$$ (2.12)

where Ψ is the volume, $\partial \Psi / \partial p$ is the change in volume with respect to pressure, and the subscript indicates that the process is to occur at constant temperature. Liquids in general are incompressible. Water for example experiences only a one percent change in density for a corresponding tenfold pressure increase.

Internal Energy

Internal energy (represented by the letter U) is the energy associated with the motion of the molecules of a substance. An increase in the internal energy of a substance is manifested usually as an increase in temperature. Internal energy has dimensions of F·L. Internal energy is often expressed on a per unit mass basis ($u = U/m$) and has dimensions of F·L/M (ft·lbf/lbm or BTU/lbm or N·m/kg).

Enthalpy

Enthalpy is defined as the sum of internal energy and flow work:

$$H = U + p\Psi$$

On a per unit mass basis, we have

$$h = u + pv = u + \frac{p}{\rho}$$

The sum of internal energy and flow work appears in the energy equation and the introduction of enthalpy is a simplification. Note that enthalpy is merely a combination of known properties. Enthalpy per unit mass has the same dimensions as internal energy F·L/M (ft·lbf/lbm or BTU/lbm or N·m/kg).

Continuum

In this text we shall adopt a macroscopic approach to fluid mechanics problems wherein we consider the fluid to be a continuous medium. We are thus interested in the average effects of many molecules which means we are treating the fluid as a continuum.

Pressure

Pressure at a point is a time-averaged normal force exerted by molecules impacting a unit surface. The area must be small but large enough, compared to molecular distances, to be consistent with the continuum approach. Thus pressure is defined as

$$p = \lim_{A \to A^*} \frac{F}{A}$$

where A^* is a small area experiencing enough molecular collisions to be representative of the fluid bulk, and F is the time-averaged normal force. Note that if A^* were to shrink to zero, then it is possible that no molecules would strike it, yielding a zero normal force and a definition of pressure that has little physical significance. The dimensions of pressure are F/L^2 (lbf/ft^2 or psi in the English system and Pa = 1 N/m^2 in SI).

2.3 Measurement of Pressure

We will now examine techniques and devices available for measuring pressure: pressure gages and manometers.

A pressure gage consists of a housing with a fitting for attaching it to a pressure vessel. Inside the housing is a curved elliptical tube called a Bourdon tube. This tube is connected to the fitting at one end and to a rack and pinion assembly at its other end. When exposed to high pressure, the tube tends to straighten, in turn pulling the rack and rotating the pinion gear. The shaft of the pinion gear extends through the face of the gage. A needle is pressed or bolted onto the shaft. When the gage is exposed to atmospheric pressure the face is calibrated to read 0 (zero) because the gage itself really measures the pressure difference from inside the tube to the outside. The reading from a gage is appropriately termed *gage pressure*. A pressure gage is shown in Figure 2.7.

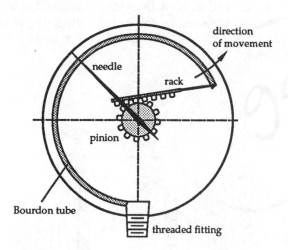

FIGURE 2.7. *Schematic of a pressure gage.*

Absolute pressure, on the other hand, would read zero only in a complete vacuum. Thus gage and absolute pressures are related by

absolute pressure = gage pressure + atmospheric pressure

In engineering units, to denote that gage pressure is being reported, the notation "psig" (pounds per square inch gage) is used. The unit "psia" (pounds per square inch absolute) is used when reporting absolute pressure. In SI, when gage pressure is being reported, the phrase "a gage pressure of . . ." is used, with a similar phrase for reporting absolute pressure.

Atmospheric pressure can be measured with a barometer which consists of a sufficiently long tube that is inverted while submerged and full of liquid. A vacuum is created above the liquid column in the tube. (See Figure 2.8.) The height of liquid above the reservoir surface is related to atmospheric pressure by the hydrostatic equation

$$p_{atm} = \frac{\rho g}{g_c} z$$

Atmospheric pressure in this text is taken to be 14.7 psia or 101.3 kPa (absolute).

Pressure differences can be measured with vertical columns of liquid. One device that can be used for this measurement is called a *manometer*. Manometers can be set up in a number of configurations depending on the application and on how tall the columns of liquid can be. Figure 2.9 illustrates a U-tube manometer configuration. One leg of the manometer is attached to a pressure vessel and the other leg is open to the atmosphere. Applying the hydrostatic equation ($p = \rho g \Delta z / g_c$) to each leg of the manometer gives

$$p_A + \frac{\rho_1 g}{g_c} z_1 = p_B$$

$$p_B = p_C$$

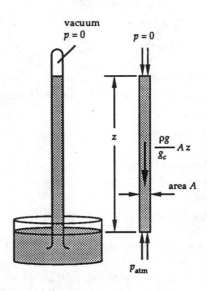

FIGURE 2.8. *A barometer.*

and $p_C = p_D + \dfrac{\rho_2 g}{g_c} z_2 = p_{atm} + \dfrac{\rho_2 g}{g_c} z_2$

Combining the above equations, we get

$$p_A - p_{atm} = (\rho_2 z_2 - \rho_1 z_1) \frac{g}{g_c} \tag{2.13}$$

     Another manometer configuration is shown in Figure 2.10. In this case a U-tube manometer is used to measure the pressure difference between two vessels. Applying the hydrostatic equation, we obtain

$$p_A + \frac{\rho_1 g}{g_c} z_1 = p_B$$

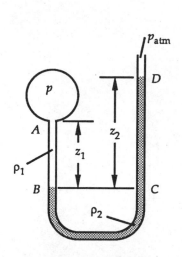

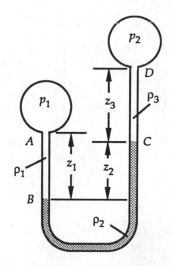

FIGURE 2.9. *U-tube manometer for measuring pressure in a vessel.*

FIGURE 2.10. *U-tube manometer for measuring pressure difference between two vessels.*

$$p_D + \frac{\rho_3 g}{g_c} z_3 + \frac{\rho_2 g}{g_c} z_2 = p_B$$

Combining, we get

$$p_A - p_D = (\rho_3 z_3 + \rho_2 z_2 - \rho_1 z_1) \frac{g}{g_c} \tag{2.14}$$

A third manometer configuration is shown in Figure 2.11. An inverted U-tube is used to measure the pressure difference between two vessels. Applying the hydrostatic equation, we write

$$p_B + \frac{\rho_1 g}{g_c} z_1 = p_A$$

and $$\quad p_B + \frac{\rho_2 g}{g_c} z_2 + \frac{\rho_3 g}{g_c} z_3 = p_D$$

Combining and simplifying, we find

$$p_A - p_D = (\rho_1 z_1 - \rho_2 z_2 - \rho_3 z_3) \frac{g}{g_c} \tag{2.15}$$

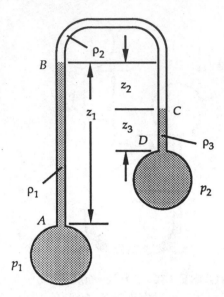

FIGURE 2.11. *An inverted U-tube differential manometer.*

EXAMPLE 2.2. Figure 2.12 shows several manometers containing different fluids and attached together. For the configuration shown, determine the pressure in the water tank at A. (All dimensions are in inches.)

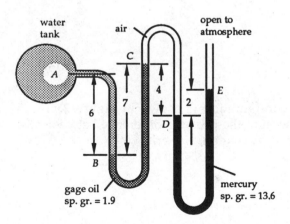

FIGURE 2.12. *The manometer of Example 2.2.*

Solution: We can apply the hydrostatic equation directly beginning at A and ending at atmospheric pressure:

$$p_A + \rho_{H_2O}g(6/12) = p_B$$

$$p_B = p_C + \rho_{oil}g(7/12)$$

$$p_D = p_C + \rho_{air}g(4/12)$$

$$p_D = p_E + \rho_{Hg}g(2/12)$$

Combining gives

$$p_A + \rho_{H_2O}g(6/12) = p_E + \rho_{Hg}g(2/12) - \rho_{air}g(4/12) + \rho_{oil}g(7/12)$$

Noting that $p_E = p_{atm} = (14.7 \text{ lbf/in}^2)\cdot(144 \text{ in}^2/\text{ft}^2)$ and that the air density ρ_{air} is negligible compared to the other fluid densities, the above equation becomes

$$p_A = 14.7(144) - \rho_{H_2O}g(6/12) + \rho_{Hg}g(2/12) + \rho_{oil}g(7/12)$$

Substituting, we get

$$p_A = 14.7(144) - 1.94(32.2)(6/12) + 13.6(1.94)(32.2)(2/12)$$
$$+ 1.9(1.94)(32.2)(7/12)$$

or $\quad p_A = 2296 \text{ lbf/ft}^2 = 16.0 \text{ psia}$

2.4 Basic Equations of Fluid Mechanics

In this section, we will discuss the definitions associated with fluid flow. We will also write the equations of fluid mechanics including continuity, momentum, energy, and the Bernoulli equation.

Flows can be characterized according to their geometries. **Closed conduit flows** are those that are completely enclosed by a restraining solid surface, such as flow in a pipe. **Open channel flows** are those that have a surface exposed to atmospheric pressure, such as flow in a river. **Unbounded flows** are those in which the fluid is not in contact with any solid surface, such as the flow that issues from a can of spray paint.

Flows can be classified according to how we mathematically describe gradients in the flow field. If the velocity of the fluid is constant at any cross section normal to the flow, or if the velocity is represented by an average value, the flow is said to be **one dimensional**. Although the velocity is constant, there will exist a driving force that changes with the flow direction. In pipe flow, for example, a pressure gradient dp/dz exists in the axial (or z) direction. The pressure gradient is what causes the fluid to flow. A pressure gradient and a constant velocity at any cross section is considered one dimensional (only one gradient). A pressure gradient dp/dz with a velocity profile that varies with only one space variable is a two dimensional flow (two gradients, $p(z)$ and $V(r)$, for example). The definition is easily extended to three dimensional flows.

Flows can be described as being **steady, unsteady,** or **quasi steady.** Steady flows have conditions that do not vary with time. Unsteady flows have conditions that do vary with time. Quasi steady flows are actually unsteady but because they proceed so slowly, they can be treated mathematically as if they were steady.

A fluid while flowing can be subjected to variations in pressure. If fluid density changes significantly as a result of pressure variations, then the fluid is considered to be **compressible.** If density practically remains unchanged with variations in pressure, then the fluid is **incompressible.** Usually, gases and vapors are compressible while liquids are incompressible. These two cases are treated differently mathematically.

We will use the **control volume approach** to model problems. In the control volume approach, we will select a region of study within the flow field and apply equations to that region. The control volume is bounded by what is called the **control surface.** Everything outside the control volume is the **surroundings.** Where to place the boundary of the control volume to best advantage is largely a matter of experience, but general guidelines will be presented where appropriate.

Continuity Equation

The continuity equation is a statement of conservation of mass. For a control volume we can write

$$\begin{pmatrix} \text{rate of} \\ \text{mass in} \end{pmatrix} = \begin{pmatrix} \text{rate of} \\ \text{mass out} \end{pmatrix} + \begin{pmatrix} \text{rate of} \\ \text{mass stored} \end{pmatrix}$$

or

$$0 = \begin{pmatrix} \text{rate of} \\ \text{mass stored} \end{pmatrix} + \begin{pmatrix} \text{net mass out} \\ \text{out minus in} \end{pmatrix}$$

In equation form, we have

$$0 = \frac{\partial m}{\partial t} \Big|_{cv} + \int\int_{cs} \rho V_n dA \tag{2.16}$$

where $\partial m/\partial t$ is the mass stored in the control volume per unit time, the integral term is to be applied at places where mass crosses the control surface, ρ is the fluid density, and V_n is the velocity normal to the control surface integrated over the area dA. For a one dimensional (V_n = a constant), steady ($\partial m / \partial t$ = 0) flow, we have

$$\sum_{\text{inlets}} \rho AV = \sum_{\text{outlets}} \rho AV \qquad (2.17)$$

The product ρAV is often called the **mass flow rate** $\dot{m}$ with dimensions of M/T (lbm/s or kg/s). Further, if the flow is incompressible, Equation 2.17 becomes

$$\sum_{\text{inlets}} AV = \sum_{\text{outlets}} AV \qquad (2.18)$$

The product AV is called the **volume flow rate** Q with dimensions of L^3/T (ft^3/s or m^3/s).

Momentum Equation

The momentum equation is a statement of conservation of linear momentum. For a control volume in Cartesian coordinates, we have

$$\Sigma F_x = \frac{1}{g_c} \frac{d(mV)_x}{dt}\bigg|_{\text{system}} = \frac{1}{g_c} \frac{\partial(mV)_x}{\partial t}\bigg|_{\text{cv}} + \frac{1}{g_c} \int\int_{\text{cs}} V_x \rho V_n dA \qquad (2.19a)$$

$$\Sigma F_y = \frac{1}{g_c} \frac{d(mV)_y}{dt}\bigg|_{\text{system}} = \frac{1}{g_c} \frac{\partial(mV)_y}{\partial t}\bigg|_{\text{cv}} + \frac{1}{g_c} \int\int_{\text{cs}} V_y \rho V_n dA \qquad (2.19b)$$

$$\Sigma F_z = \frac{1}{g_c} \frac{d(mV)_z}{dt}\bigg|_{\text{system}} = \frac{1}{g_c} \frac{\partial(mV)_z}{\partial t}\bigg|_{\text{cv}} + \frac{1}{g_c} \int\int_{\text{cs}} V_z \rho V_n dA \qquad (2.19c)$$

The ΣF term represents all external forces applied to the control volume. The first term after the second equal sign is the rate of storage of linear momentum within the control volume (CV). The last term represents the rate of linear momentum out of the control volume minus the rate of linear momentum in. For steady one dimensional flow, the equations become:

$$\Sigma F_i = \frac{1}{g_c} \int\int_{\text{cs}} V_i \rho V_n dA \qquad (2.20)$$

which can be applied to any direction i. For a steady flow system in which one fluid stream enters (*in*) and one leaves (*out*) the control volume, we have for the i direction:

$$\Sigma F_i = \frac{V_{out}(\rho AV)_{out}}{g_c}\bigg|_i - \frac{V_{in}(\rho AV)_{in}}{g_c}\bigg|_i$$

or $\Sigma F_i = \dfrac{\dot{m}}{g_c} \left(V_{out} - V_{in} \right) \Big|_i$

Energy Equation

The energy equation is known also as the First Law of Thermodynamics. It allows us to make calculations describing the transformation of energy from one form to another and includes the effects of work and heat transfer. The energy equation states that

$$\left\{ \begin{array}{c} \text{total rate of} \\ \text{change of energy} \\ \text{within system} \end{array} \right\} = \left\{ \begin{array}{c} \text{rate of} \\ \text{energy} \\ \text{stored} \end{array} \right\} + \left\{ \begin{array}{c} \text{rate of energy out} \\ \text{minus} \\ \text{rate of energy in} \end{array} \right\}$$

In equation form, we have

$$\frac{dE}{dt} \bigg|_{\text{system}} = \frac{\partial E}{\partial t} \bigg|_{\text{CV}} + \iint_{\text{cs}} e\, \rho V_n\, dA \qquad\qquad (2.21)$$

where E is the total energy of a system and e is the total energy per unit mass. The total energy traditionally is considered to consist of internal, kinetic, and potential energies:

$$E = U + KE + PE$$

and $e = \dfrac{E}{m} = u + \dfrac{V^2}{2g_c} + \dfrac{gz}{g_c}$ $\qquad\qquad (2.22)$

Experimental observations of devices and mathematical models of their behavior have led to the following relation between energy, heat transfer, and work:

$$\left\{ \begin{array}{c} \text{total rate of} \\ \text{change of energy} \\ \text{within system} \end{array} \right\} = \left\{ \begin{array}{c} \text{rate of} \\ \text{heat transferred} \\ \text{out of system} \end{array} \right\} - \left\{ \begin{array}{c} \text{rate of work} \\ \text{done by} \\ \text{system} \end{array} \right\}$$

or

$$\frac{dE}{dt} \bigg|_{\text{system}} = \frac{\partial \tilde{Q}}{\partial t} - \frac{\partial W'}{\partial t} \qquad\qquad (2.23)$$

Combining Equations 2.21, 2.22, and 2.23 gives

$$\frac{dE}{dt}\Big|_{system} = \frac{\partial \tilde{Q}}{\partial t} - \frac{\partial W'}{\partial t} = \frac{\partial E}{\partial t}\Big|_{cv} + \iint_{cs} e\,\rho V_n dA$$

or

$$\frac{\partial \tilde{Q}}{\partial t} - \frac{\partial W'}{\partial t} = \frac{\partial E}{\partial t}\Big|_{cv} + \iint_{cs}\left(u + \frac{V^2}{2g_c} + \frac{gz}{g_c}\right)\rho V_n dA \qquad (2.24)$$

The work term W' consists of all forms of work crossing the boundary of the control volume including electric, magnetic, viscous shear or friction, flow work, and shaft work. Flow work is done by or on the system when mass crosses the control surface at entrances or exits. It is customary to divide the work term W' into two components: shaft work W and flow work W_f. Thus we can write:

$$\frac{\partial W'}{\partial t} = \frac{\partial W}{\partial t} + \frac{\partial W_f}{\partial t}$$

The flow work is given by

$$\frac{\partial W_f}{\partial t} = \iint_{cs}\frac{p}{\rho}\,\rho V_n dA$$

Combining the above equations with Equation 2.24 and rearranging gives

$$\frac{\partial \tilde{Q}}{\partial t} - \frac{\partial W}{\partial t} = \frac{\partial E}{\partial t}\Big|_{cv} + \iint_{cs}\left(\frac{p}{\rho} + u + \frac{V^2}{2g_c} + \frac{gz}{g_c}\right)\rho V_n dA$$

Recall that enthalpy is defined as $h = u + p/\rho$. Substituting into the above equation gives

$$\frac{\partial \tilde{Q}}{\partial t} - \frac{\partial W}{\partial t} = \frac{\partial E}{\partial t}\Big|_{cv} + \iint_{cs}\left(h + \frac{V^2}{2g_c} + \frac{gz}{g_c}\right)\rho V_n dA \qquad (2.25)$$

For the case of steady, one dimensional flow, the above equation becomes

$$\frac{\partial \tilde{Q}}{\partial t} - \frac{\partial W}{\partial t} = \left\{\left(h + \frac{V^2}{2g_c} + \frac{gz}{g_c}\right)\Big|_{out} - \left(h + \frac{V^2}{2g_c} + \frac{gz}{g_c}\right)\Big|_{in}\right\}\rho VA$$

$$(2.26)$$

For an adiabatic process, the heat transferred is zero and the above equation reduces to

$$-\frac{\partial W}{\partial t} = \left\{\left(h + \frac{V^2}{2g_c} + \frac{gz}{g_c}\right)\Big|_{out} - \left(h + \frac{V^2}{2g_c} + \frac{gz}{g_c}\right)\Big|_{in}\right\}\rho VA \qquad (2.27)$$

The Bernoulli Equation

The Bernoulli Equation relates velocity, elevation, and pressure in a flow field. This equation results from the energy equation (2.27) for adiabatic, one dimensional flow with no work and negligible change in internal energy. The Bernoulli equation also results from applying the momentum equation to a streamline in the flow field. Thus under special conditions, the momentum and the energy equations reduce to the same equation, which is why the Bernoulli equation is often called the **mechanical energy equation**. The Bernoulli equation is written as

$$\int_1^2 \frac{dp}{\rho} + \frac{(V_2^2 - V_1^2)}{2g_c} + \frac{g(z_2 - z_1)}{g_c} = 0$$

For an incompressible fluid for which density ρ is a constant, the above equation becomes

$$\frac{p_2 - p_1}{\rho} + \frac{(V_2^2 - V_1^2)}{2g_c} + \frac{g(z_2 - z_1)}{g_c} = 0 \qquad (2.28)$$

or

$$\frac{p}{\rho} + \frac{V^2}{2g_c} + \frac{gz}{g_c} = \text{a constant} \qquad (2.29)$$

The Bernoulli equation does not account for frictional effects.

2.5 Summary

In this chapter, we examined fluid properties and wrote equations of fluid mechanics without derivation. This chapter was intended as a brief review and much detail has been omitted. The reader is referred to any text on Fluid Mechanics for more information on any of the points addressed here.

2.6 Show and Tell

Obtain a catalog from the appropriate manufacturer(s), and give an oral report on the following viscometer(s) as assigned by the instructor. In all cases present the theoretical basis for the operation of the device and, if available, demonstrate its operation.

1. The cone and plate viscometer.

2. The falling sphere viscometer.

3. The wide gap concentric cylinder viscometer.

4. Saybolt viscometer.

5. Stormer viscometer for measuring viscosity of paint.

6. Any other type of viscometer you encounter that is not mentioned in this chapter.

Obtain a catalog of the appropriate type and give an oral report on the following devices useful for measuring pressure. In all cases present the theoretical basis for the operation of the device.

7. A pitot tube and a pitot-static tube.

8. Pressure transducers.

2.7 Problems Chapter 2

1. What is the specific gravity of 40°API oil?

2. The specific gravity of manometer gage oil is 0.826. What is its density and its °API rating?

3. A popular mayonnaise is tested with a viscometer and the following data were obtained:

τ in g/cm^2	40	100	140	180
dV/dy in rev/s	0	3	7	15

Determine the fluid type and the proper descriptive equation.

4. A cod liver oil emulsion is tested with a viscometer and the following data were obtained:

τ in lbf/ft^2	0	40	60	80	120
dV/dy in rev/s	0	0.5	1.7	3	6

Graph the data and determine the fluid type. Derive the descriptive equation.

5. A rotating cup viscometer has an inner cylinder diameter of 2.00 in. and the gap between cups is 0.2 in. The inner cylinder length is 2.50 in. The viscometer is used to obtain viscosity data on a Newtonian liquid. When the inner cylinder rotates at 10 rev/min, the torque on the inner cylinder is measured to be 0.00011 in-lbf. Calculate the viscosity of the fluid. If the fluid density is 850 kg/m^3, calculate the kinematic viscosity.

6. A rotating cup viscometer has an inner cylinder whose diameter is 4 cm and whose length is 8 cm. The outer cylinder has a diameter of 5 cm. The viscometer is used to measure the viscosity of a liquid. When the outer cylinder rotates at 12 rev/min, the torque on the inner cylinder is measured to be 4×10^{-6} N·m. Determine the kinematic viscosity of the fluid if its density is 1 000 kg/m³.

7. A capillary tube viscometer is used to measure the viscosity of water (density is 62.4 lbm/ft³, viscosity is 0.89×10^{-3} N·s/m²) for calibration purposes. The capillary tube inside diameter must be selected so that laminar flow conditions (i.e., $VD/v < 2\ 100$) exist during the test. For values of $L = 3$ in. and $z = 10$ in., determine the maximum tube size permissible.

8. A Saybolt viscometer is used to measure oil viscosity and the time required for 60 ml of oil to pass through a standard orifice is 180 SUS. The specific gravity of the oil is found as 44°API. Determine the absolute viscosity of the oil.

9. A Saybolt viscometer is used to obtain oil viscosity data. The time required for 60 ml of oil to pass through the orifice is 70 SUS. Calculate the kinematic viscosity of the oil. If the oil specific gravity is 35°API, find also its absolute viscosity.

10. A manometer is used to measure pressure drop in a venturi meter as shown in Figure P2.10. Derive an equation for the pressure drop in terms of Δh. Make all measurements from the meter centerline.

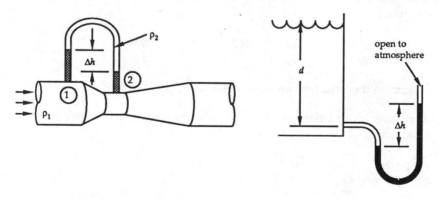

FIGURE P2.10   FIGURE P2.11

11. A mercury manometer is used to measure pressure at the bottom of a tank containing acetone as shown in Figure P2.11. The manometer is to be replaced with a gage. What is the expected gage reading in psig if Δh is measured to be 4 in. of mercury? What is the depth of the acetone?

12. An unknown fluid is in the manometer of Figure P2.12. The pressure difference between the two air chambers is 700 kPa and the manometer reading Δh is 6 cm. Determine the density and specific gravity of the unknown fluid.

13. For the system of Figure P2.13, determine the pressure of the air in the tank.

14. An inverted U-tube manometer is used to measure the pressure difference between two containers as shown in Figure P2.14. If the reading Δh is 5 ft, determine the pressure difference.

15. The glycerine within the tanks of Figure P2.14 is replaced hexane. If the reading Δh is 2 m, determine the pressure difference.

16. A manometer containing mercury as the gage oil is used to measure the pressure increase in a water pump as shown in Figure P2.16. Calculate the pressure rise if Δh is 7 in. of mercury.

17. Determine the pressure difference between the linseed and castor oils of Figure P2.17. (All dimensions in inches.)

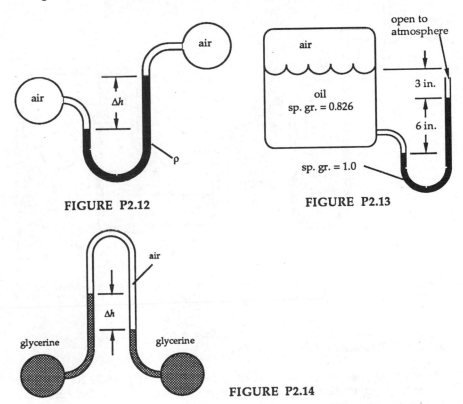

FIGURE P2.12

FIGURE P2.13

FIGURE P2.14

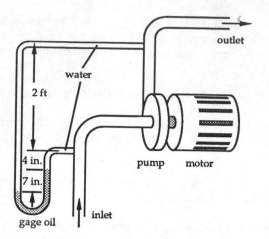

FIGURE P2.16

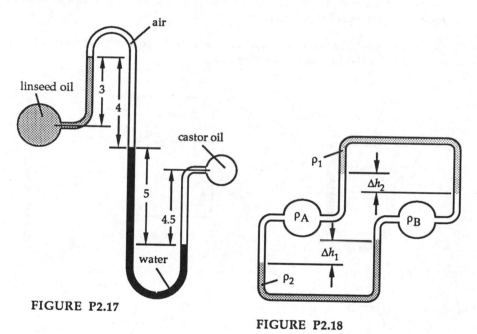

FIGURE P2.17

FIGURE P2.18

18. Two vessels are connected with two manometers as shown in Figure P2.18. What is the relationship between Δh_1 and Δh_2 if $\rho_A = \rho_B$?

19. Figure P2.19 shows a reducing bushing. Flow enters the bushing at a velocity of 0.5 m/s. Calculate the outlet velocity.

20. Three gallons per minute of water enters the tank of Figure P2.20. The inlet line is 2-1/2 in. in diameter. The air vent is 1.5 in. in diameter. Determine the air exit velocity at the instant shown.

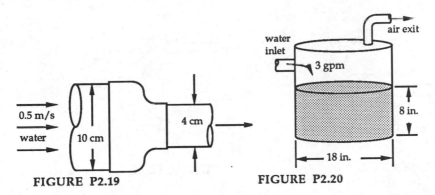

FIGURE P2.19   FIGURE P2.20

21. Figure P2.21 shows a cross flow heat exchanger used to condense Freon-12. Freon-12 vapor enters the unit at a flow rate of 0.07 kg/s. Freon-12 leaves the exchanger as a liquid at room temperature and pressure. Determine the exit velocity of the liquid.

22. Nitrogen enters a pipe at a flow rate of 0.2 lbm/s. The pipe has an inside diameter of 4 in. At the inlet, the nitrogen temperature is 540°R ($\rho = 0.073$ lbm/ft^3) and at the outlet, the nitrogen temperature is 1800°R ($\rho = 0.0213$ lbm/ft^3). Calculate the inlet and outlet velocities of the nitrogen. Are they equal? Should they be?

23. A liquid jet issuing from a faucet impacts a flat bottom sink at a right angle. Just prior to impact, the jet velocity is 1 ft/s and the jet diameter is 3/8 in. Calculate the force exerted by the jet.

24. A garden hose is used to squirt water at someone who is protecting herself with a garbage can lid. Figure P2.24 shows the jet in the vicinity of the lid. Determine the restraining force F for the conditions shown.

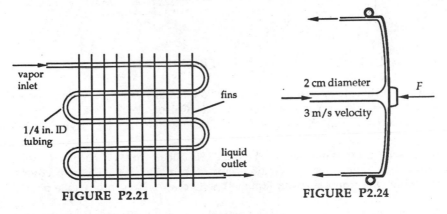

FIGURE P2.21   FIGURE P2.24

25. A two dimensional liquid jet strikes a concave semicircular object as shown in Figure P2.25. Calculate the restraining force F.

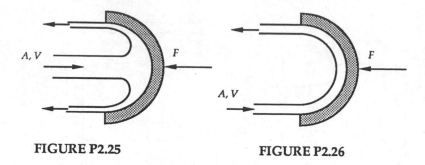

FIGURE P2.25   **FIGURE P2.26**

26. Repeat Problem 25 for the configuration of Figure P2.26.

27. A two dimensional liquid jet is turned through an angle θ ($0 < \theta < 90°$) by a curved vane as shown in Figure P2.27. The forces are related by $F_2 = 3F_1$. Determine the angle θ through which the liquid jet is turned.

28. A two dimensional liquid jet is turned through an angle θ ($0 < \theta < 90°$) by a curved vane as shown in Figure P2.28. The forces are related by $F_1 = 2F_2$. Determine the angle θ through which the liquid jet is turned.

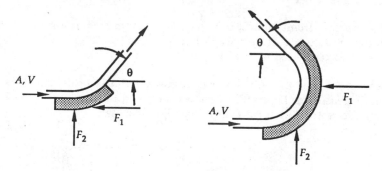

FIGURE P2.27   **FIGURE 2.28**

29. A water turbine is located in a dam. The volume flow rate through the system is 5000 gpm. The exit pipe diameter is 4 ft. Calculate the work done by (or power received from) the water as it flows through the dam. (See Figure P2.29.)

30. A kerosene pump is illustrated in Figure P2.30. The flow rate is 0.015 m^3/s; inlet and outlet gage pressure readings are -8 kPa and 200 kPa respectively. Determine the required power input to the fluid as it flows through the pump.

31. Air flows through a compressor at a mass flow rate of 0.003 slug/s. At the inlet, the air velocity is negligible. At the outlet, air leaves through an exit pipe of

diameter 2 in. The inlet properties are 14.7 psia and 75°F. The outlet pressure is 120 psia. For an isentropic (reversible and adiabatic) compression process, we have

$$\frac{T_2}{T_1} = \left\{\frac{p_2}{p_1}\right\}^{(\gamma-1)/\gamma}$$

Determine the outlet temperature of the air and the power required. Assume that air behaves as an ideal gas ($dh = c_p \, dT$, $du = c_v \, dT$ and $\rho = p/RT$).

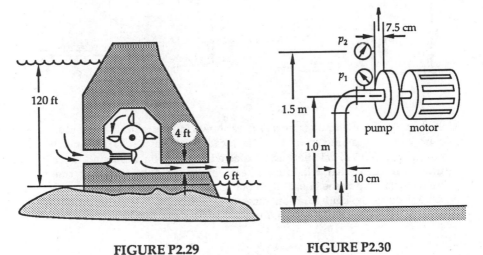

FIGURE P2.29   FIGURE P2.30

32. Figure P2.10 shows a venturi meter. Assuming that $p_2 \ll p_1$, show that the Bernoulli and continuity equations when applied combine to become:

$$Q = A_2 \sqrt{\frac{2g\Delta h}{1 - (D_2^4/D_1^4)}}$$

33. A jet of water issues from a kitchen faucet and falls vertically downward at a flow rate of 1.2 fluid ounces per second. At the faucet which is 14 inches above the sink bottom, the jet diameter is 5/8 in. Determine the diameter of the jet where it strikes the sink.

34. Calculate the force that the jet of Problem 33 exerts on the flat sink surface upon impact.

35. A garden hose is used as a siphon to drain a pool as shown in Figure P2.35. The garden hose has a 3/4 in. inside diameter. Assuming no friction, calculate the flow rate of water through the hose if the hose is 25 ft long.

36. A 2 in. ID pipe is used to drain a tank as shown in Figure P2.36. Simultaneously, a 4 in. ID inlet line fills the tank. The velocity in the inlet line is 4 ft/s. Determine the equilibrium height h of the liquid in the tank if it is ethyl alcohol. How does the height change if the liquid is octane? Assume in both cases that frictional effects are negligible.

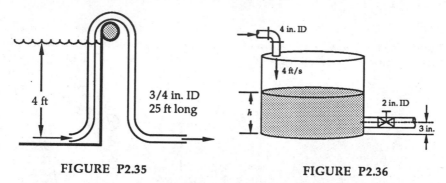

FIGURE P2.35   FIGURE P2.36

37. A pump draws castor oil from a tank as shown in Figure P2.37. A venturi meter with a throat diameter of 2 in. is located in the discharge line. For the conditions shown, calculate the expected reading on the manometer of the meter. Assume that frictional effects are negligible and that the pump delivers 0.25 HP to the liquid. If all that is available is a 6 ft tall manometer, can it be used in the configuration shown? If not, suggest an alternative way to measure pressure difference. (All measurements in inches.)

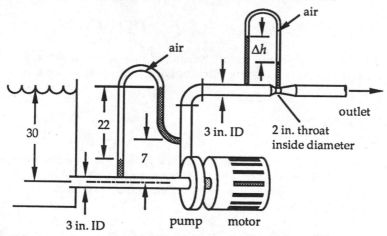

FIGURE P2.37

CHAPTER 3   Piping Systems I

In this chapter, we will review some of the basic concepts associated with piping systems. We will discuss effective and hydraulic diameters and present equations of motion for modeling flow in closed conduits. We will also examine minor losses in detail and conclude with guidelines on analyzing piping systems.

Flow in closed conduits is an extremely important area of study because it is the most common way of transporting liquids. Crude oil and its components are moved about in a refinery or across the country by pumping them through pipes. Water in the home is transported to various parts of the house through tubing. Heated and air conditioned air are distributed to all parts of a dwelling in circular and/or rectangular ducts. Examples of flow in closed conduits are everywhere.

It is important to recall that flow in a duct can be either **laminar** or **turbulent**. When laminar flow exists, the fluid flows smoothly through the duct in layers called *laminae*. A fluid particle in one layer stays in that layer. When turbulent flow exists, flowing fluid particles move about the cross section. Eddies and vortices are responsible for the mixing action; such eddies and vortices do not exist in laminar flow.

The criterion for distinguishing between laminar and turbulent flow is the observed mixing action. Experiments have shown that laminar flow exists when the **Reynolds number** is less than 2100:

$$\mathbf{Re} = \frac{\rho V D}{\mu g_c} = \frac{VD}{\nu} < 2\,100 \qquad \text{(laminar flow)} \qquad (3.1)$$

where V is the *average velocity* of the flow and D is a characteristic dimension of the duct cross section. For circular ducts, D is usually taken to be the inside diameter. For noncircular cross sections, D is usually taken to be the *hydraulic diameter* (discussed later in this chapter).

3.1 Pipe and Tubing Standards

Pipes and tubes are made of many materials. Pipes can be cast or, like tubes, can be extruded. Sizes for pipes and tubes are standardized and so are tolerances on their dimensions.

Pipe Specifications and Attachment Methods

Pipes are specified by a **nominal diameter** and a **schedule number**—for example, "2-nominal schedule 40." The nominal diameter does not necessarily equal the inside or outside diameter of the pipe. Each nominal diameter will specify one and only one outside diameter. The schedule of the pipe specifies the wall thickness such that the larger the schedule number, the thicker the pipe wall. Appendix Table D.1 gives dimensions of pipe sizes that vary from 1/8-nominal to 40-nominal in English and in SI units. Schedule 40 pipe (or the standard size) is used in common engineering applications.

Pipes can be attached together or to fittings in various ways. Pipe ends can be threaded, which is done primarily in the smaller sizes (less than 4-nominal). The number of threads per inch as well as the thread profile are standardized for each pipe size. The threaded end of the pipe is usually tapered as well. Regardless of pipe schedule, all pipe having the same nominal diameter (i.e., outside diameter OD) will have the same thread specification. Before threaded pipes are attached to fittings, the threads are coated with a viscous compound or wrapped with special tape. The thread preparation and the wedging action of tapered threads together help to ensure a fluid tight connection.

Pipes can be welded together or to fittings if the material is weldable. Welding is more commonly done with the larger pipe sizes.

Pipe ends can be threaded into or welded to flanges. Flanges are then bolted together. Usually a rubber or cork gasket is installed between the flanges to ensure that the connection is leakproof. Flanges are made in many pipe sizes, and standards have been established for their construction details including even the minimum number of bolt holes.

Plastic pipe such as polyvinyl chloride (PVC) can be attached to a fitting by threading or by using an adhesive. Plastic pipe is specified in the same manner as other pipe.

Water Tubing Specifications and Attachment Methods

Water tubing is specified by giving a **standard diameter** and a **type**—for example, 1-standard type K. The standard size does not necessarily equal the inside or outside diameter of the tube. Each standard size corresponds to one and only outside diameter. The type (K, L, or M) specifies the wall thickness. Type K is used for underground service and general plumbing. Type L is used primarily for interior plumbing while type M is made for use with soldered fittings. Appendix Table D.2 gives dimensions of copper water tubing sizes that vary from 1/4-std to 12-std in English and in SI units.

Note that copper can be a *tubing* material or a *pipe* material. If used as a pipe material, then its dimensions follow pipe specifications.

One difference between pipe and tubing is that tubing has a thinner wall and cannot sustain the high fluid pressures that pipe can.

Another type of tubing commonly used in air conditioners and heat pumps is called refrigeration tubing. It is specified in the same manner as copper water tubing and is usually made of a copper alloy. The difference is that copper water tubing is quite rigid while refrigeration tubing is rather ductile (it can be bent by hand).

Tubing can be attached to fittings in a number of ways. A tube end can be **flared** and attached to a **flared end fitting**; the tube end is flared outward uniformly with a **flaring tool**. Another attachment method involves use of a **compression fitting**, in which the tube end is inserted through a snugly fitting ring that is furnished as part of the fitting itself. When the fitting nut is tightened, it compresses the ring and causes the copper tube end to expand against the inside wall of the fitting. A third joining technique involves brazing or soldering. The tube end is inserted into a fitting which fits like a sleeve. The joint is *fluxed* to remove the oxide, and the fitting and tube are soldered or brazed (commonly referred to as **sweating**).

3.2 Equivalent Diameters for Noncircular Ducts

Noncircular ducts are found in a number of fluid conveying systems. Rectangular and square cross sections are used for air conditioning or heating ducts and for gutters and downspouts. Annular cross sections are found in double pipe heat exchangers where one tube is placed within another; the section between the tubes is annular. The question arises as to what is the characteristic dimension of a noncircular duct. Three different choices for characteristic dimension have been proposed and we will define each of them.

The **hydraulic radius** R_h is used widely for flow in open channels. The hydraulic radius is defined as

$$R_h = \frac{\text{area of flow}}{\text{wetted perimeter}} = \frac{A}{P} \qquad (3.2)$$

This definition is entirely satisfactory for an open channel but leads to an undesirable result when modeling close conduits. For a circular duct flowing full, we find

$$R_h = \frac{A}{P} = \frac{\pi D^2/4}{\pi D} = \frac{D}{4}$$

Thus the hydraulic radius for flow in a pipe is 1/4th of its diameter. Traditionally, diameter D is preferred to represent a circular duct rather than $D/4$, so we tend to not use hydraulic radius.

The **effective diameter** D_{eff} is the diameter of a circular duct that has the same area as the noncircular duct of interest. Consider, for example, a rectangular duct of dimensions $h \times w$. The effective diameter is found with

$$\frac{\pi D_{eff}^2}{4} = hw$$

or

$$D_{eff} = 2\sqrt{hw/\pi} \qquad\qquad\qquad (3.3)$$

The concept of effective diameter arises from satisfying the continuity equation for the circular and the noncircular duct. For the example above,

$$Q = AV = \frac{\pi D_{eff}^2}{4} V = hwV$$

Solving the above equation for effective diameter yields Equation 3.3.

The third equivalent diameter we will define is called the **hydraulic diameter** D_h, given as

$$D_h = \frac{4 \cdot \text{area of flow}}{\text{wetted perimeter}} = \frac{4A}{P} \qquad\qquad\qquad (3.4)$$

For a circular duct flowing full, we calculate

$$D_h = \frac{4A}{P} = \frac{4\pi D^2/4}{\pi D} = D$$

which gives us an acceptable result. For a rectangular duct of dimensions $h \times w$,

$$D_h = \frac{4hw}{2h + 2w} = \frac{2hw}{h + w}$$

This equation gives results that are entirely different from Equation 3.3 and it should be noted that the hydraulic and effective diameters cannot be made equal for any value of w given h.

As we saw in preceding paragraphs, the effective diameter satisfied the continuity equation. As we shall see, the hydraulic diameter arises when the momentum equation is applied to flow in a closed conduit. Traditionally, the hydraulic diameter is used more widely than the effective diameter. We shall use the hydraulic diameter in this text except in a few selected exercises.

3.3 Equation of Motion for Flow in a Duct

Our objective in this section is to develop an expression for pressure loss in a conduit due to frictional effects. Figure 3.1 (page 45) illustrates flow in a closed conduit. A circular cross section is illustrated but the results remain general until geometry terms for a specific cross section are introduced into the equations. The forces acting on the control volume (in this case a disk with diameter equal to the inside diameter of the duct) include pressure and friction. Gravity forces are neglected. The momentum equation is

$$\Sigma F_i = \frac{1}{g_c} \int\int_{cs} V_i \rho V_n dA$$

or $\qquad pA - \tau_w Pdz - (p + dp)A = 0$ (3.5)

where we note that the z directed velocity out of the control volume equals that into the control volume, making the right hand side of the momentum equation equal to zero. The term A is the cross sectional area and Pdz is perimeter times axial distance which equals the surface area over which the wall shear stress τ_w acts. Equation 3.5 reduces to

$$\frac{dp}{dz} = -\tau_w \frac{P}{A} = -\tau_w \frac{4P}{4A}$$

or in terms of hydraulic diameter,

$$\frac{dp}{dz} = -\frac{4\tau_w}{D_h}$$ (3.6)

The pressure change per unit length (dp/dz) is thus a function of the wall shear stress and the hydraulic diameter of the duct. Equation 3.6 applies to *any* cross section. We now introduce a **friction factor** f customarily defined as

$$f = \frac{4\tau_w g_c}{\rho V^2/2}$$ (3.7)

where V is the average velocity of the flow in the conduit. The above definition is of the **Darcy-Weisbach friction factor**. The **Fanning friction factor** is used in some texts and is defined as

$$f' = \frac{\tau_w g_c}{\rho V^2/2}$$

The two definitions for friction factor are both commonly used. The Darcy-Weisbach definition is conveniently applied when hydraulic diameter is the characteristic length. The Fanning friction factor is used in formulations where hydraulic radius is the characteristic length. We will use the Darcy-Weisbach definition of Equation 3.7.

Solving Equation 3.7 for $4\tau_w$ and substituting into Equation 3.6 gives

$$dp = -\frac{\rho V^2}{2 g_c}\frac{fdz}{D_h} \qquad\qquad (3.8a)$$

Integrating from point 1 to point 2, a distance L downstream gives

$$p_2 - p_1 = -\frac{\rho V^2}{2 g_c}\frac{fL}{D_h} \qquad\qquad (3.8b)$$

Equation 3.8 gives the pressure drop in a duct due to friction. Again, the above equation is independent of duct cross section.

To model flow in a duct, we use the Bernoulli equation. As developed in the last chapter, it is apparent that the Bernoulli equation does not account for frictional effects. For flow in a duct, friction is manifested as a loss in pressure with axial distance as shown in Equation 3.8. So to use the Bernoulli equation for flow in a duct, we must first modify it by combining it with Equation 3.8. The result is

$$\frac{p_1 g_c}{\rho g} + \frac{V_1^2}{2g} + z_1 = \frac{p_2 g_c}{\rho g} + \frac{V_2^2}{2g} + z_2 + \Sigma\frac{fL}{D_h}\frac{V^2}{2g} \qquad (3.9)$$

The above equation is actually an energy balance performed for two points a distance L apart within a duct. The equation states that

$$\left\{\begin{matrix}\text{pressure}\\ \text{head}\end{matrix} + KE + PE\right\}\Big|_1 = \left\{\begin{matrix}\text{pressure}\\ \text{head}\end{matrix} + KE + PE\right\}\Big|_2 + \left\{\begin{matrix}\text{energy loss}\\ \text{due to friction}\end{matrix}\right\}$$

The head loss is expressed as a product of a friction term (fL/D_h) and the kinetic energy of the flow. Note that Equation 3.9 can be applied to any cross section as long as the appropriate hydraulic diameter is used.

3.4 Friction Factor and Pipe Roughness

In this section, we will present methods of evaluating friction factor for a circular duct under laminar and turbulent flow conditions.

Results for noncircular ducts (rectangular and annular) will also be presented again for laminar and turbulent flow conditions.

Laminar Flow of a Newtonian Fluid in a Circular Duct

Our interest in this area is in having an equation for the velocity profile and for the average velocity. Figure 3.1 illustrates laminar flow in a duct as well as the polar coordinate system we will use in our formulation. The z directed instantaneous velocity is

$$V_z = \left(-\frac{dp}{dz}\right)\left(\frac{R^2}{4\mu}\right)\left[1 - \left(\frac{r}{R}\right)^2\right] \qquad \binom{\text{laminar flow}}{\text{circular duct}} \qquad (3.10)$$

This equation is derivable by applying the momentum equation to a control volume within the duct. (See Problem 10 for a step-by-step procedure.) Note that as axial distance z increases, the pressure p decreases. Therefore dp/dz is a negative quantity and the term $(-dp/dz)$ in Equation 3.10 is actually positive. Moreover, $(-dp/dz)$ which is the pressure drop per unit length is a constant.

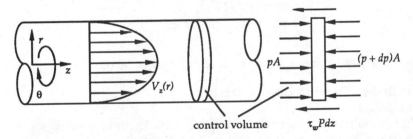

FIGURE 3.1. *Laminar flow in a circular duct.*

When Equation 3.10 is integrated over the cross sectional area (as per the continuity equation), the **volume flow rate Q** results:

$$Q = \int\int_{CS} V_n \, dA$$

$$Q = \int_0^{2\pi} \int_0^{R} \left(-\frac{dp}{dz}\right)\left(\frac{R^2}{4\mu}\right)\left[1 - \left(\frac{r}{R}\right)^2\right] r \, dr \, d\theta$$

Integrating and solving, we get

$$Q = \frac{\pi R^4}{8\mu}\left(-\frac{dp}{dz}\right) \qquad (3.11)$$

The average velocity is given by

$$V = \frac{Q}{A} = \frac{R^2}{8\mu}\left(-\frac{dp}{dz}\right) \qquad (3.12)$$

Equation 3.12 can be combined with Equation 3.8a which is

$$dp = -\frac{\rho V^2}{2g_c}\frac{f dz}{D_h} \qquad (3.8a)$$

By eliminating the pressure drop term and solving for friction factor, we find

$$f = \frac{32\mu g_c}{\rho V R} = \frac{64\mu g_c}{\rho V D}$$

or

$$f = \frac{64}{Re} \qquad \text{(laminar flow circular duct)} \qquad (3.13)$$

where the diameter D has been substituted for hydraulic diameter D_h.

Turbulent Flow in a Circular Duct

For turbulent flow, we rely on experimental methods to develop a relationship between the pertinent variables. Based on results of many tests performed using artificially roughened pipe walls, it has been determined that the friction factor is dependent upon Reynolds number **Re** and relative roughness ε/D:

$$f = f(\mathbf{Re}, \varepsilon/D) \qquad (3.14)$$

where ε is a characteristic linear dimension representing the roughness of the inside surface area of the conduit wall.

Sand particles of known dimensions or diameter (sizes separated by sieving) were attached with an adhesive to the inside surface of a pipe. The pipe was then tested; that is, pressure drop vs volume flow rate data were obtained for fluid pumped through the pipe. The test was repeated with many pipe sizes and many sand particle diameters. When tests were then performed on commercially available pipes (e.g., pressure drop vs flow rate), a comparison could be made. A commercial steel pipe exhibited the same or similar pressure drop vs flow rate behavior as a pipe coated with sand particles of size $\varepsilon = 0.00015$ ft (0.004 6 cm). Some texts call ε an "equivalent sand roughness factor." Values of ε for various materials are provided in Table 3.1 (page 51).

A graph of the data to predict the friction factor f given the Reynolds number **Re** (= $\rho VD/\mu g_c$) and the relative roughness (ε/D) is customarily known as the **Moody Diagram**. ("Friction Factors in Pipe Flow," by L. Moody; in *Transactions of ASME*, 1944, 68, 672.) Exhaustive amounts of data were compiled and consolidated into this graph.

A number of equations have been written to curve fit the Moody Diagram. The older equations are known to involve an iterative process when trying to calculate friction factor f given Reynolds number **Re** and relative roughness ε/D. Two more recently published equations, however, overcome this difficulty. The Chen Equation and the Churchill Equation both solve for f explicitly in terms of **Re** and ε/D. In other words, when **Re** and ε/D are known, the Chen and the Churchill equations allow for calculating f directly just as with the Moody Diagram. The Chen equation is valid for **Re** $\geq 2\ 100$, and is written as:

$$\frac{1}{\sqrt{f}} = -2.0 \log \left\{ \frac{\varepsilon}{3.7065D} - \frac{5.0452}{Re} \log \left(\frac{1}{2.8257} \left[\frac{\varepsilon}{D} \right]^{1.1098} + \frac{5.8506}{Re^{0.8981}} \right) \right\}$$

(3.15)

The Churchill Equation, also valid for **Re** $\geq 2\ 100$, is:

$$f = 8 \left\{ \left[\frac{8}{Re} \right]^{12} + \frac{1}{(B + C)^{1.5}} \right\}^{\frac{1}{12}}$$

(3.16)

where

$$B = \left[2.457 \ln \frac{1}{(7/Re)^{0.9} + (0.27\varepsilon/D)} \right]^{16}$$

and

$$C = \left(\frac{37\ 530}{Re} \right)^{16}$$

Figure 3.2 (see page 52) is a version of the Moody Diagram that was generated by using the Chen and the Churchill Equations. To generate the Moody diagram with these equations, a value of ε/D was selected and the Reynolds number was made to vary over the turbulent flow regime (from 2×10^3 to 10^8). Values generated were graphed by computer on computer drawn semilog paper. Both equations were used and for each **Re** and ε/D, the two calculated f's were averaged to obtain the graphed value.

The traditional form of the Moody Diagram is actually a log-log plot. The friction factor f versus Reynolds number **Re** graph of Figure 3.2, however, is a semilog graph. The friction factor axis is not logarithmic which allows the friction factor axis to extend to 0. The Reynolds number appears on the horizontal axis and varies from 2 000 to 100 000 000. The relative roughness is an independent variable and ranges from 0.008 to 0.1. The friction factor appears on the vertical axis and varies from 0 to 0.1. Note that the friction factor of Figure 3.2 is the Darcy-Weisbach friction factor.

Other forms of the Moody diagram have been developed in order to simplify calculations in problems where iterative methods (or trial and error) are required (i.e., volume flow rate Q unknown, diameter D unknown). Consider that in a piping problem, six variables can enter the problem: Δp (or Δh), Q, D, v, L, and ε. Usually in the traditional type of problem, five of these variables are known and the sixth is to be found. When pressure drop Δp (or head loss $\Delta h = \Delta p g_c / \rho g$) is unknown, then the problem can be solved in a straightforward manner using the Moody Diagram, Figure 3.2. When volume flow rate Q is unknown, use of the Moody Diagram requires a trial-and-error procedure to obtain a solution. If a graph of f vs $\mathbf{Re}\sqrt{f}$ is available, however, then the unknown Q problem can be solved in a straightforward manner. Such a graph is provided in Figure 3.3, page 53. When diameter D is unknown, use of the Moody Diagram again requires a trial-and-error procedure unless a graph of f vs $\mathbf{Re}\sqrt[5]{f}$ is available. Such a graph is given in Figure 3.4 (page 54).

Equations 3.15 and 3.16 (the Chen and the Churchill equations) were used to generate the f vs $f^{1/2}\mathbf{Re}$ graph. A value of ε/D was selected and Reynolds number was allowed to vary from 2×10^3 to 10^8. Two f's were found and averaged. As an intermediate step, $f^{1/2}\mathbf{Re}$ was calculated. Values generated were graphed on a semilog grid and the result is provided in Figure 3.3, which is a graph of f vs $f^{1/2}\mathbf{Re}$ with ε/D as an independent variable.

The trial-and-error process required when the diameter D is unknown can be eliminated only with a change of independent variable. This is due to the fact that the relative roughness term contains diameter D. In studies involving economics of pipe size selection, a new variable is introduced to rid the roughness term of diameter. The new parameter is called the Roughness Number and is defined as:

$$\mathbf{Ro} = \frac{\varepsilon/D}{\mathbf{Re}} \tag{3.17}$$

So for the diameter unknown problem, it is desirable to have a graph of f vs $f^{1/5}\mathbf{Re}$ with $\mathbf{Ro}$ [$= (\varepsilon/D)/\mathbf{Re}$] as an independent variable. This graph is provided as Figure 3.4 which was generated with Equations 3.15 and 3.16 (the Chen and the Churchill equations). A value of ε/D was selected as was a single value of **Re**. Two f's were found and averaged. The Roughness number **Ro** ($= \varepsilon/D/\text{Re}$) and $f^{1/5}\text{Re}$ were calculated. The next value of ε/D was selected in harmony with the next **Re** such that **Ro** was held constant. The objective was to generate lines of constant **Ro**. Values of $f, f^{1/5}\text{Re}$ and **Ro** were then graphed on a semilog grid and the result is provided in Figure 3.4.

The justification for using new parameters (e.g., the Roughness number) arises from dimensional analysis. The Moody Diagram is a compilation of results that are correlated according to:

$$f = f\,(\text{Re}, \varepsilon/D) \tag{3.14}$$

Any two of these three variables (f, **Re**, ε/D) raised to any power can be combined to obtain a fourth variable, and the fourth variable can be used to replace either of its constituents. Thus $f^m\text{Re}^n$ could be a new variable used to replace either f or **Re** in Equation 3.14 to obtain a new correlation. Selecting $m = 1/2$ and $n = 1$, we can therefore write

$$f = f\,(f^{1/2}\text{Re}, \varepsilon/D) \tag{3.18}$$

Equation 3.18 is the correlation used in producing Figure 3.3.

Following the same lines of reasoning, $(\varepsilon/D)^a/\text{Re}^b$ could be a new variable used to replace either **Re** or ε/D in Equation 3.14. Selecting $a = b = 1$, we define $(\varepsilon/D)/\text{Re} = \text{Ro}$. Selecting in addition $m = 1/5$ and $n = 1$, Equation 3.14 becomes

$$f = f\,(f^{1/5}\text{Re}, \text{Ro}) \tag{3.19}$$

Equation 3.19 is the correlation used in producing Figure 3.4.

For the graphs of Figures 3.3 and 3.4, m is selected as being $1/2$ or $1/5$, making the exponent for f in Figures 3.3 and 3.4, respectively, $1/2$ and $1/5$. The reason for choosing these values arises from the solution of the equations for specific problems. How the graphs are used to solve the traditional pipe flow problems will be illustrated by example.

Example 3.1. Chloroform flows at a rate of 0.05 m³/s through a 4-nominal schedule 40 wrought iron pipe. The pipe is laid out horizontally and is 250 m long. Calculate the pressure drop of the chloroform.

Solution: For chloroform, we find from Table B.1

$$\rho = 1.47(1\,000)\;\text{kg/m}^3 \qquad\qquad \mu = 0.53 \times 10^{-3}\,\text{N·s/m}^2$$

For 4-nominal schedule 40 pipe, Table D.1 shows

$$ID = D = 10.23\;\text{cm} \qquad\qquad A = 82.19\;\text{cm}^2$$

Table 3.1 for wrought iron gives $\varepsilon = 0.004\,6$ cm. The continuity equation for incompressible steady flow through the pipe is

$$A_1 V_1 = A_2 V_2$$

Because $A_1 = A_2$, then $V_1 = V_2$. The Bernoulli equation applies:

$$\frac{p_1 g_c}{\rho g} + \frac{V_1^2}{2g} + z_1 = \frac{p_2 g_c}{\rho g} + \frac{V_2^2}{2g} + z_2 + \Sigma \frac{fL}{D_h}\frac{V^2}{2g}$$

where points 1 and 2 are $L = 250$ m apart, $z_1 = z_2$ for a horizontal pipe and $p_1 - p_2$ is sought. The above equation reduces to

$$p_1 - p_2 = \frac{fL}{D_h}\frac{\rho V^2}{2g_c}$$

The flow velocity is

$$V = \frac{Q}{A} = \frac{0.05}{82.19 \times 10^{-4}} = 6.08\;\text{m/s}$$

The Reynolds number is calculated to be

$$\text{Re} = \frac{\rho V D}{\mu g_c} = \frac{1.47(1\,000)(6.08)(0.1023)}{0.53 \times 10^{-3}}$$

or

$$\text{Re} = 1.73 \times 10^6$$

$$\left.\begin{array}{c}\\\\\end{array}\right\}\qquad f = 0.016\,5 \quad\text{(Figure 3.2)}$$

Also $\dfrac{\varepsilon}{D} = \dfrac{0.004\,6}{10.23} = 0.000\,45$

The pressure loss is

$$p_1 - p_2 = \frac{fL}{D_h}\frac{\rho V^2}{2g_c} = \frac{1.47(1\,000)(6.08)^2}{2}\frac{0.016\,5(250)}{0.102\,3}$$

$$p_1 - p_2 = 110\,000\;\text{N/m}^2 = 0.11\;\text{MPa}$$

TABLE 3.1. *Roughness factor for various pipe materials.*

Pipe Material	ε, ft	ε, cm
Steel		
Commercial	0.00015	0.004 6
Corrugated	0.003–0.03	0.09–0.9
Riveted	0.003–0.03	0.09–0.9
Mineral		
Brick sewer ⎤		
Cement–asbestos ⎱	0.001–0.01	0.03–0.3
Clays ⎰		
Concrete ⎦		
Wood stave	0.0006–0.003	0.018–0.09
Cast iron	0.00085	0.025
Asphalt coated	0.0004	0.012
Bituminous lined	0.000008	0.000 25
Cement lined	0.000008	0.000 25
Centrifugally spun	0.00001	0.000 31
Drawn tubing	0.000005	0.000 15
Miscellaneous		
Brass ⎤		
Copper ⎮		
Glass ⎬	0.000005	0.000 15
Lead ⎮		
Plastic ⎮		
Tin ⎦		
Galvanized	0.0002–0.0008	0.006–0.025
Wrought iron	0.00015	0.004 6
PVC	Smooth	Smooth

FIGURE 3.2. *Moody Diagram constructed with Chen and Churchill equations.*

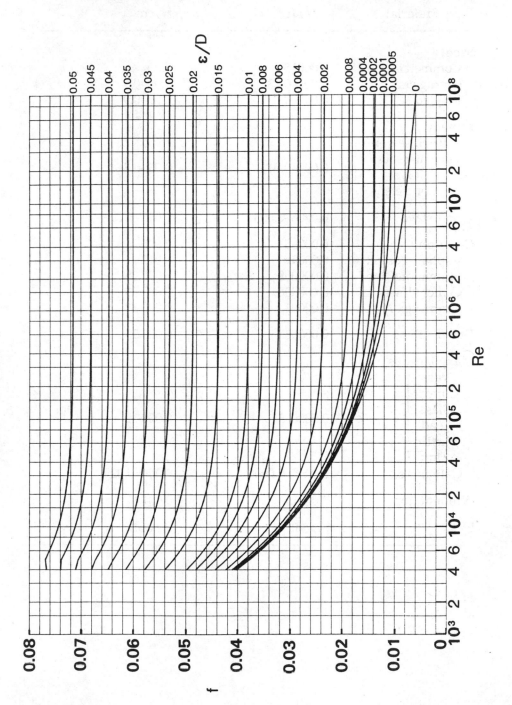

FIGURE 3.3. *Modified pipe friction diagram for solving volume flow rate unknown problems.*

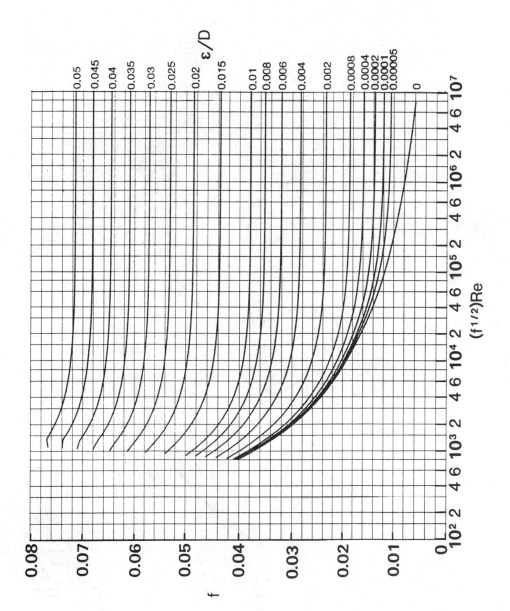

FIGURE 3.4. *Modified pipe friction diagram for solving diameter unknown problems.*

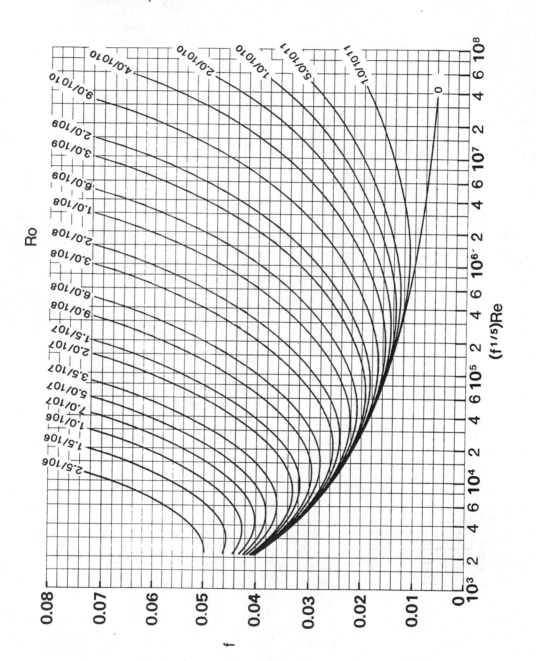

Example 3.2. A 6-nominal schedule 80 cast iron pipe is 11,270 ft long. It is to convey octane. The available pump can provide a pressure drop of 25 psi. Determine the expected flow rate of octane in the pipe.

Solution: From Table B.1, we read for octane

$$\rho = 0.701(62.4) \text{ lbm/ft}^3 \quad \text{and} \quad \mu = 1.07 \times 10^{-5} \text{ lbf·s/ft}^2$$

From Table D.1, we read for 6-nominal schedule 80 pipe,

$$ID = 0.4801 \text{ ft} \quad \text{and} \quad A = 0.1810 \text{ ft}^2$$

Table 3.1 shows that for cast iron, $\varepsilon = 0.00085$ ft. The Reynolds number of the flow is

$$\text{Re} = \frac{\rho V D}{\mu g_c} = \frac{0.701(62.4)V(0.4801)}{1.07 \times 10^{-5}(32.2)}$$

or

$$\text{Re} = 6.1 \times 10^4 V \qquad\qquad \text{(i)}$$

$$\frac{\varepsilon}{D} = \frac{0.00085}{0.4801} = 0.0018 \qquad\qquad \text{(ii)}$$

The continuity equation for incompressible, steady flow is

$$A_1 V_1 = A_2 V_2$$

Because $A_1 = A_2$, then $V_1 = V_2$. The Bernoulli equation (3.9) is

$$\frac{p_1 g_c}{\rho g} + \frac{V_1^2}{2g} + z_1 = \frac{p_2 g_c}{\rho g} + \frac{V_2^2}{2g} + z_2 + \Sigma \frac{fL}{D_h} \frac{V^2}{2g}$$

where length L and $(p_1 - p_2)$ are both given. With $z_1 = z_2$, the above equation becomes:

$$p_1 - p_2 = \Delta p = \frac{fL}{D_h} \frac{\rho V^2}{2g_c}$$

Rearranging and solving for velocity, we obtain

$$V = \sqrt{\frac{2D \Delta p g_c}{\rho f L}}$$

Substituting,

$$V = \sqrt{\frac{2(0.4801)(25)(144)32.2}{0.701(62.4)f(11270)}} = \frac{0.475}{\sqrt{f}} \qquad (iii)$$

We will use Figure 3.2 to illustrate the method. We begin by assuming that the friction factor f will correspond to the fully turbulent value of $\varepsilon/D = 0.0018$. We read

$$f = 0.023$$

Then from Equation iii above,

$$V = 0.475/\sqrt{0.023} = 3.13 \text{ ft/s}$$

The Reynolds number then is (from Equation i):

$$\text{Re} = 6.14 \times 10^4(3.13) = 1.9 \times 10^5$$

$$\frac{\varepsilon}{D} = 0.0018$$

$$\left.\right\} \quad f = 0.024 \quad \text{(Figure 3.2)}$$

Repeating the calculations with this new value of f gives

$$f = 0.024; \quad V = 3.06; \quad \text{Re} = 1.87 \times 10^5; \quad f \approx 0.024 \text{ close enough}$$

The velocity is thus 3.06 ft/s. The volume flow rate is calculated as

$$Q = AV = 0.1810(3.06) = 0.55 \text{ ft}^3/\text{s}$$

or

$$Q = 247 \text{ gpm}$$

———

Suppose we wish to use Figure 3.3 and avoid the trial-and-error procedure. We would set up the calculations and again arrive at Equations i, ii and iii:

$$\text{Re} = 6.1 \times 10^4 V \qquad (i)$$

$$\frac{\varepsilon}{D} = \frac{0.00085}{0.4801} = 0.0018 \qquad\qquad\qquad\text{(ii)}$$

$$V = \frac{0.475}{\sqrt{f}} \qquad\qquad\qquad\qquad\text{(iii)}$$

We now combine Equations i and ii to eliminate velocity V and obtain

$$\text{Re} = 6.1 \times 10^4 (0.475/\sqrt{f})$$

or

$$\left.\begin{array}{l} f^{1/2}\text{Re} = 2.9 \times 10^4 \\[2em] \dfrac{\varepsilon}{D} = 0.0018 \end{array}\right\} \qquad f = 0.024 \quad \text{(Figure 3.3)}$$

Substituting into Equation iii, we get

$$V = 0.475/\sqrt{0.024} = 3.06 \text{ ft/s}$$

yielding the same result as before.

Example 3.3. A PVC plastic pipeline is to convey fifty liters per second of ethylene glycol over a distance of 2 000 m. The available pump can overcome a frictional loss of 200 kPa. Select a suitable size for the pipe.

Solution: When diameter is unknown, it is convenient to modify the equations into a slightly different form. The continuity equation for steady incompressible flow is

$$Q = A_1 V_1 = A_2 V_2$$

Because $A_1 = A_2$, then $V_1 = V_2$. Also, we can write for a circular duct,

$$V = \frac{4Q}{\pi D^2}$$

The Bernoulli equation (3.9) is

$$\frac{p_1 g_c}{\rho g} + \frac{V_1^2}{2g} + z_1 = \frac{p_2 g_c}{\rho g} + \frac{V_2^2}{2g} + z_2 + \Sigma \frac{fL}{D_h} \frac{V^2}{2g}$$

With $z_1 = z_2$ and $V_1 = V_2$, the above equation reduces to

$$p_1 - p_2 = \Sigma \frac{fL}{D_h} \frac{\rho V^2}{2g_c}$$

Substituting for velocity in terms of flow rate, the above equation becomes

$$p_1 - p_2 = \Delta p = \frac{fL}{D} \frac{\rho\, 16Q^2}{2\pi^2 D^4 g_c}$$

Rearranging and solving for diameter D gives

$$D = \sqrt[5]{\frac{8\rho Q^2 fL}{\pi^2 \Delta p g_c}} \tag{i}$$

In terms of flow rate, the Reynolds number becomes

$$Re = \frac{\rho VD}{\mu g_c} = \frac{4\rho Q}{\pi D \mu g_c} \tag{ii}$$

For plastic tubing, Table 3.1 indicates that it is correct to use the "smooth" line of the Moody Diagram (Figure 3.2). For ethylene glycol, Table B.1 shows

$$\rho = 1.1(1\ 000)\ \text{kg/m}^3 \qquad\qquad \mu = 16.2 \times 10^{-3}\ \text{N·s/m}^2$$

Substituting into Equation i, we have

$$D = \sqrt[5]{\frac{8(1.1)(1\ 000)(50 \times 10^{-3})^2 f(2\ 000)}{\pi^2(200\ 000)}}$$

or

$$D = 0.467 f^{1/5} \tag{iii}$$

Also, $$Re = \frac{4(1.1)(1\ 000)(50 \times 10^{-3})}{\pi D(16.2 \times 10^{-3})} = 4.3 \times 10^3/D \tag{iv}$$

In order to use Figure 3.2, we assume a friction factor to initiate the trial-and-error method. Assume

$$f = 0.025 \quad \text{(randomly selected)}$$
then
$$D = 0.467(0.025)^{1/5} = 0.223\ \text{m}$$

and

$$\text{Re} = 4.3 \times 10^3 / 0.223 = 1.93 \times 10^4$$

$$\left.\begin{array}{l} \\ \\ \dfrac{\varepsilon}{D} = \text{"smooth"} \end{array}\right\} \qquad f = 0.026 \quad \text{(Figure 3.2)}$$

For the second trial,

$$f = 0.026; \quad D = 0.225; \quad \text{Re} = 1.9 \times 10^4; \quad f \approx 0.026 \text{ close enough}$$

The diameter we select then is 0.225 m. Referring to Appendix Table D.1, we find that 10 nominal schedule 120 is too large and 10 nominal schedule 140 is too small. The latter size will not deliver the required flow rate so we specify

$$D = 10 \text{ nominal schedule 120 pipe}$$

Suppose we elect to avoid the trial-and-error procedure and use Figure 3.4. We arrive at Equations iii and iv in the usual way, and then combine them to eliminate D; we obtain

$$\text{Re} = 4.3 \times 10^3 / 0.467 f^{1/5}$$

or

$$\left.\begin{array}{l} f^{1/5}\text{Re} = 9.2 \times 10^3 \\ \\ \dfrac{\varepsilon}{D} = \text{"smooth"} \end{array}\right\} \qquad f = 0.026 \quad \text{(Figure 3.4)}$$

The diameter is calculated as

$$D = 0.467(0.026)^{1/5} = 0.225 \text{ m}$$

which is the same result obtained before.

Laminar Flow of a Newtonian Fluid in an Annulus

Flow through an annulus is illustrated in Figure 3.5. The annular flow area is bounded by the outside diameter of the inner duct (OD_p) and the inside diameter of the outer duct (ID_a). Also shown in the figure is one half of the control volume we use for study. The forces acting on the control volume are due to pressure and viscosity. Gravity is neglected. We are seeking an equation for the velocity distribution V_z that we can integrate over the cross section to obtain average velocity. We will then combine the result with Equation 3.8 to find an equation for the friction factor just as is done with the circular duct. The momentum equation applied to the control volume of Figure 3.5 is

$$\Sigma F_z = \frac{1}{g_c} \int\int_{cs} V_z \, \rho V_n \, dA$$

Noting that the velocity does not change through the system, we have

$$pA + (\tau + d\tau)dA_1 - \tau dA_2 - (p + dp)A = 0$$

Simplifying gives

$$(\tau + d\tau)dA_1 - \tau dA_2 - Adp = 0 \qquad\qquad (3.20)$$

The surface areas over which the shear stresses act are evaluated as

$$dA_1 = 2\pi(r + dr)dz$$

and

$$dA_2 = 2\pi r dz$$

The cross sectional area is

$$A = \pi(r + dr)^2 - \pi r^2 = 2\pi r dr$$

Substituting these areas into Equation 3.20, we get

$$(\tau + d\tau)\, 2\pi(r + dr)dz - \tau\, 2\pi r dz - 2\pi r dr\, dp = 0$$

The quantity 2π appears in each term. Simplifying and neglecting the $drd\tau$ term as being small compared to the others, we obtain

$$\frac{\tau}{r} + \frac{d\tau}{dr} = \frac{dp}{dz}$$

which becomes

$$\frac{1}{r}\frac{d}{dr}(r\tau) = \frac{dp}{dz} \qquad\qquad (3.21)$$

For a Newtonian fluid,

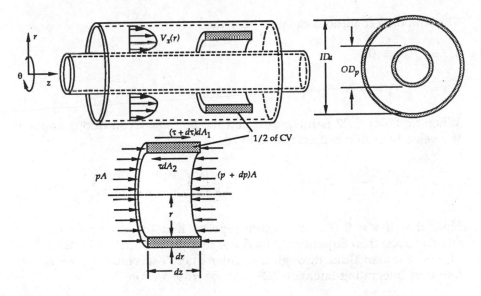

FIGURE 3.5. *Laminar flow in an annulus.*

$$\tau = \mu \frac{dV_z}{dr}$$

Combining with Equation 3.21 and simplifying yields

$$\frac{d}{dr}\left(r \, \frac{dV_z}{dr} \right) = \frac{r}{\mu} \frac{dp}{dz} \qquad (3.22)$$

with boundary conditions

1. $r = OD_p/2$, $V_z = 0$

2. $r = ID_a/2$, $V_z = 0$

The boundary conditions can be expressed in a slightly different and ultimately more convenient way. We define

$$R = ID_a/2$$

and

$$\kappa = \frac{OD_p/2}{ID_a/2} = \frac{OD_p}{ID_a}$$

The boundary conditions can now be written as

1. $r = R$, $V_z = 0$

2. $r = \kappa R$, $V_z = 0$

When Equation 3.22 is integrated and the boundary conditions applied, the velocity profile is determined to be

$$V_z = \left(-\frac{dp}{dz}\right)\left(\frac{R^2}{4\mu}\right)\left[1 - \left(\frac{r}{R}\right)^2 - \frac{1 - \kappa^2}{\ln(\kappa)} \ln\frac{r}{R}\right] \quad \left(\begin{array}{l}\text{laminar flow} \\ \text{annular duct}\end{array}\right) \quad (3.23)$$

Note that if $\kappa = 0$ (i.e., no inside pipe so that the flow passage is a circular duct) then Equation 3.23 reduces to Equation 3.10 for laminar flow of a Newtonian fluid through a circular duct. The volume flow rate is found by integrating Equation 3.23 over the cross section:

$$Q = \int_0^{2\pi} \int_{\kappa R}^{R} V_z \, r dr d\theta$$

which becomes

$$Q = \int_0^{2\pi} \int_{\kappa R}^{R} \left(-\frac{dp}{dz}\right)\left(\frac{R^2}{4\mu}\right)\left[1 - \left(\frac{r}{R}\right)^2 - \frac{1 - \kappa^2}{\ln(\kappa)} \ln\frac{r}{R}\right] r dr d\theta$$

Integrating gives the volume flow rate as

$$Q = \left(-\frac{dp}{dz}\right)\left(\frac{\pi R^4(1 - \kappa^2)}{8\mu}\right)\left(1 + \kappa^2 + \frac{1 - \kappa^2}{\ln(\kappa)}\right) \quad (3.24)$$

The average velocity is

$$V = \frac{Q}{A} = \frac{Q}{\pi(R^2 - \kappa^2 R^2)} = \frac{Q}{\pi R^2(1 - \kappa^2)}$$

or

$$V = \left(-\frac{dp}{dz}\right)\left(\frac{R^2}{8\mu}\right)\left(1 + \kappa^2 + \frac{1 - \kappa^2}{\ln(\kappa)}\right) \quad (3.25)$$

The hydraulic diameter for the annular flow section is

$$D_h = \frac{4A}{P} = \frac{4\left(\pi(ID_a/2)^2 - \pi(OD_p/2)^2\right)}{\pi(ID_a) + \pi(OD_p)}$$

which simplifies to

$$D_h = ID_a - OD_p \qquad\qquad (3.26a)$$

or $$D_h = 2R(1 - \kappa) \qquad\qquad (3.26b)$$

Equation 3.8a relates the pressure drop to the friction factor:

$$dp = - \frac{\rho V^2}{2g_c} \frac{f dz}{D_h} \qquad\qquad (3.8a)$$

Combining Equations 3.8a, 3.25 and 3.26b, we get after considerable simplification:

$$\frac{1}{f} = \frac{Re}{64} \left(\frac{1 + \kappa^2}{(1 - \kappa)^2} + \frac{1 + \kappa}{(1 - \kappa)ln(\kappa)} \right) \qquad\qquad (3.27)$$

where
$$Re = \frac{VD}{v} = \frac{2RV(1 - \kappa)}{v}$$

Note again that when $\kappa = 0$, the equations for a circular duct result.

Turbulent Flow Through an Annulus

For turbulent flow through an annulus, we cannot derive an equation for the velocity profile and continue as we did for the laminar case. Instead we rely on experimental results. When $\kappa (= OD_p/ID_a)$ is less than 0.75, the Moody Diagram can be used with little error to find the friction factor for flow in an annulus. It must be remembered, however, that the characteristic dimension $D_h [= ID_a - OD_p = 2R(1 - \kappa)]$ must be used in the Reynolds number equation ($Re = VD_h/v$) and in the relative roughness (ε/D_h).

Example 3.4. A vertically oriented, double pipe heat exchanger is made of copper water tubing and is 6 ft long. The tubes are 4 std and 2 std, both type M. Acetone flows downward through the annulus and is subjected to a 2.0 psi pressure drop opposite to the flow direction. Determine the volume flow rate of acetone through the exchanger.

Solution: From Appendix Table B.1 for acetone,

$$\rho = 0.787(62.4 \text{ lbm/ft}^3) \qquad\qquad \mu = 0.659 \times 10^{-5} \text{ lbf·s/ft}^2$$

From Appendix Table D.2 for 4 std and 2 std type M copper tubing, we write

$$ID_a = 0.3279 \text{ ft} \qquad\qquad \text{(4 std type M inside diameter)}$$

$$OD_p = 2.125/12 = 0.177 \text{ ft} \qquad\qquad \text{(2 std type M outside diameter)}$$

We calculate the annular flow area and hydraulic diameter:

$$A = \pi(ID_a{}^2 - OD_p{}^2)/4 = 0.0598 \text{ ft}^2$$

$$D_h = ID_a - OD_p = 0.151 \text{ ft}$$

The continuity equation is written as

$$Q = A_1 V_1 = A_2 V_2$$

With $A_1 = A_2$, we conclude that $V_1 = V_2$. The Bernoulli Equation applied to this system is

$$\frac{p_1 g_c}{\rho g} + \frac{V_1^2}{2g} + z_1 = \frac{p_2 g_c}{\rho g} + \frac{V_2^2}{2g} + z_2 + \Sigma \frac{fL}{D_h} \frac{V^2}{2g} \qquad\qquad (3.9)$$

The pressure drop opposite to the flow direction is given by

$$p_1 - p_2 = -2.0(144) = -288 \text{ lbf/ft}^2$$

If the outlet height z_2 is the reference datum, then

$$z_1 = 6 \text{ ft} \qquad\qquad \text{and} \qquad z_2 = 0$$

Simplifying the Bernoulli equation and substituting gives

$$\frac{(p_1 - p_2)g_c}{\rho g} + z_1 = \frac{fL}{D_h}\frac{V^2}{2g}$$

$$-\frac{288(32.2)}{0.787(62.4)(32.2)} + 6 = \frac{6f}{0.151}\frac{V^2}{2(32.2)}$$

or

$$V = \frac{0.469}{\sqrt{f}}$$

The Reynolds number is calculated as

$$Re = \frac{\rho V D_h}{\mu g_c} = \frac{0.787(62.4)(V)(0.151)}{0.659 \times 10^{-5} (32.2)}$$

or

$$Re = 3.49 \times 10^4 \, V$$

For drawn copper tubing, $\varepsilon = 0.000005$ ft. The relative roughness is found to be

$$\frac{\varepsilon}{D_h} = \frac{0.000005}{0.151} = 0.000033$$

A trial and error process is required in order to solve this problem using Figure 3.2. We assume as our first trial the fully turbulent value of friction factor corresponding to the relative roughness calculated above. Thus,

1st trial: $f = 0.0095$ (corresponding to $\frac{\varepsilon}{D_h} = 0.000033$)

Then

$$V = 0.469/\sqrt{f} = 0.469/\sqrt{0.0095} = 4.81 \text{ ft/s}$$

$$Re = 3.49 \times 10^4 \, (4.81)$$

or

$$\left.\begin{array}{l} Re = 1.68 \times 10^5 \\ \\ \frac{\varepsilon}{D} = 0.000033 \end{array}\right\} \qquad f = 0.0165 \quad \text{(Figure 3.2)}$$

2nd trial: $f = 0.0165$, $V = 3.65$, $Re = 1.27 \times 10^5$, and $f = 0.0175$

3rd trial: $f = 0.0175$, $V = 3.55$, $Re = 1.23 \times 10^5$, and $f \approx 0.0175$ (close enough)

Therefore

$$V = 3.55 \text{ ft/s}$$

and

$$Q = AV = 0.0598(3.55)$$

$$Q = 0.212 \text{ ft}^3/\text{s} = 95.5 \text{ gpm}$$

Alternatively, we could use Figure 3.3. By combining the velocity and Reynolds number equations above, we get

$$\mathbf{Re} = 3.49 \times 10^4 \, V = 3.49 \times 10^4 \, (0.469/\sqrt{f})$$

or

$$\left. \begin{array}{l} f^{1/2}\mathbf{Re} = 1.63 \times 10^4 \\[2mm] \dfrac{\varepsilon}{D} = 0.000033 \end{array} \right\} \qquad f = 0.0175 \quad \text{(Figure 3.3)}$$

which is the same result obtained with Figure 3.2.

Laminar Flow of a Newtonian Fluid in a Rectangular Duct

Flow through a rectangular duct is illustrated in Figure 3.6. The cross section is assumed to be very wide compared to its height. Flow is in the z-direction and the control volume we are working with does not extend to the wall surfaces. Applying the momentum equation to the control volume gives

$$\Sigma F_z = \frac{1}{g_c} \int\int_{cs} V_z \, \rho V_n \, dA$$

The forces we consider are due to pressure and friction. The above equation when applied to Figure 3.6 becomes

$$pA + \tau P dz - (p + dp)A = 0 \qquad\qquad (3.28)$$

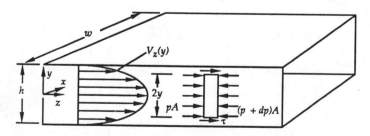

FIGURE 3.6. *Laminar flow through a rectangular duct.*

For a rectangular duct, the area and perimeter are

$$A = 2xy$$

$$P = 2y + x + 2y + x = 2x + 4y$$

The x dimension ($\sim w$) is much larger than the y dimension ($\sim 2h$) of the duct. The perimeter term can therefore be reduced to

$$P \approx 2x$$

Substituting into Equation 3.28 yields

$$2pyx + 2\tau xdz - 2(p + dp)yx = 0$$

Rearranging and simplifying, we get

$$\frac{dp}{dz} = \frac{\tau}{y} \tag{3.29}$$

For a Newtonian fluid,

$$\tau = \mu \frac{dV_z}{dy}$$

Combining with Equation 3.29 and rearranging, we obtain

$$\frac{dV_z}{dy} = \frac{y}{\mu} \frac{dp}{dz} \tag{3.30}$$

with boundary conditions

1. $y = \pm h/2$ $V_z = 0$

2. $y = 0$ $\dfrac{\partial V_z}{\partial y} = 0$

Integrating Equation 3.30, applying the boundary conditions, and simplifying yield

$$V_z = \left(-\frac{dp}{dz}\right)\left(\frac{h^2}{2\mu}\right)\left(\frac{1}{4} - \frac{y^2}{h^2}\right) \qquad \left(\genfrac{}{}{0pt}{}{\text{laminar flow}}{\text{2-D rectangular duct}}\right) \tag{3.31}$$

The volume flow rate is found by integrating the velocity V_z over the cross sectional area

$$Q = \int\limits_{0}^{w} \int\limits_{-h/2}^{+h/2} \left(-\frac{dp}{dz}\right)\left(\frac{h^2}{2\mu}\right)\left(\frac{1}{4} - \frac{y^2}{h^2}\right) dy\, dx$$

Integrating and simplifying, we find

$$Q = \frac{h^3 w}{12\mu}\left(-\frac{dp}{dz}\right) \tag{3.32}$$

The average velocity then becomes

$$V = \frac{Q}{A} = \frac{h^2}{12\mu}\left(-\frac{dp}{dz}\right) \tag{3.33}$$

Equation 3.8a relates the pressure loss to the average velocity in a duct for any cross section

$$dp = -\frac{\rho V^2}{2g_c}\frac{f dz}{D_h} \tag{3.8a}$$

Also, for this 2-dimensional duct, the hydraulic diameter is

$$D_h = \frac{4A}{P} = \frac{4h\,w}{2h + 2w} \approx \frac{4hw}{2w}$$

or $D_h = 2h$ $\hphantom{xxxxxxxxxxxxxxxxxxxxxxxxxxxxxxxxx}$ (3.34)

Combining Equations 3.8a, 3.33, and 3.34 gives

$$f = \frac{96\mu g_c}{\rho V D_h} = \frac{96\mu g_c}{\rho V(2h)}$$

or $f = \dfrac{96}{Re}$ $\hphantom{xxxxxxxxxxxxxxxxxxxxxxxxxxxxxxxxx}$ (3.35)

where

$$Re = \frac{\rho V(2h)}{\mu g_c}$$

Thus for laminar flow of a Newtonian fluid in a two dimensional duct, the friction factor x Reynolds number is 96. For other rectangular ducts, the friction factor x Reynolds number product is found as a function of the height/width ratio h/w. Results are provided in Figure 3.7.

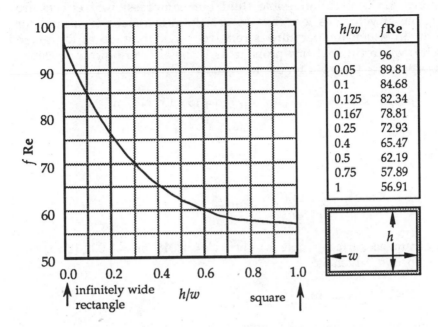

h/w	fRe
0	96
0.05	89.81
0.1	84.68
0.125	82.34
0.167	78.81
0.25	72.93
0.4	65.47
0.5	62.19
0.75	57.89
1	56.91

Curve fit Equation: $f\,Re = 96 - 95(h/w) + 56(h/w)^2$

FIGURE 3.7. *Friction factor–Reynolds number product for laminar flow of a Newtonian fluid in a rectangular duct.*

Turbulent Flow in a Rectangular Duct

For turbulent flow through a rectangular duct, we cannot develop an equation for velocity and proceed as we did for the laminar case. Experience has shown, however, that we can use the Moody Diagram. The only restriction is that the characteristic dimension to be used is the hydraulic diameter

$$D_h = \frac{4A}{P} = \frac{2hw}{h+w} \tag{3.36}$$

The characteristic dimension is substituted into the Reynolds number equation $Re = VD_h/v$ and into the relative roughness ε/D_h.

Example 3.5. Air flows through a horizontal duct that is 3 m long. The duct is rectangular (30 cm x 15 cm) and is made of galvanized sheet metal. Air flows through the duct at a velocity of 20 ft/s. Determine the pressure drop in the duct.

Solution: Air is a compressible fluid but compressibility effects are negligible at velocities less than several hundred feet per second for systems in which temperature is relatively moderate. So at the given velocity, we can model this problem as if the air were incompressible. From Appendix Table C.1 for air, we read

$$\rho = 1.17 \, \text{kg/m}^3 \qquad\qquad \mu = 18 \times 10^{-6} \, \text{N·s/m}^2$$

For the rectangular duct,

$$A = hw = 0.15(0.3) = 0.045 \, \text{m}^2$$

$$D_h = \frac{2hw}{h + w} = \frac{2(0.15)(0.3)}{0.15 + 0.3} = 0.2 \, \text{m}$$

For galvanized surface, Table 3.1 shows $\varepsilon = 0.015$ cm $= 0.000 \, 15$ m. The flow velocity is given as

$$V = 20 \, \text{ft/s} = 6.1 \, \text{m/s}$$

The continuity equation applied over the length of duct is

$$Q = A_1 V_1 = A_2 V_2$$

With $A_1 = A_2$, we conclude that $V_1 = V_2$. The Bernoulli Equation applied to this system is

$$\frac{p_1 g_c}{\rho g} + \frac{V_1^2}{2g} + z_1 = \frac{p_2 g_c}{\rho g} + \frac{V_2^2}{2g} + z_2 + \Sigma \frac{fL}{D_h} \frac{V^2}{2g} \qquad\qquad (3.9)$$

Setting $V_1 = V_2$ and $z_1 = z_2$ in the above equation and simplifying gives

$$\frac{(p_1 - p_2) g_c}{\rho g} = \frac{fL}{D_h} \frac{V^2}{2g} \qquad\qquad (i)$$

All parameters are known except the friction factor f which we now determine. The Reynolds number is

$$\text{Re} = \frac{\rho V D_h}{\mu g_c} = \frac{1.17(6.1)(0.2)}{18 \times 10^{-6}}$$

or

$$\left. \begin{array}{l} \mathbf{Re} = 8.0 \times 10^4 \\[2ex] \text{Also } \dfrac{\varepsilon}{D_h} = \dfrac{0.000\ 15}{0.2} = 0.000\ 75 \end{array} \right\} \qquad f = 0.022 \quad \text{(Figure 3.2)}$$

Substituting into the right hand side of Equation i above gives

$$\frac{(p_1 - p_2)g_c}{\rho g} = \frac{0.022(3)}{0.2} \frac{(6.1)^2}{2(9.81)} = 0.625 \text{ m} \qquad\qquad \text{(ii)}$$

The pressure drop then is

$$p_1 - p_2 = 1.17(9.81)(0.625)$$

$$p_1 - p_2 = 7.19 \text{ N/m}^2 = 7.19 \text{ Pa}$$

The problem asks for the pressure drop; however, it is customary when dealing with air flows to express the pressure loss in terms of a head of water. Thus

$$\Delta h_{H_2O} = \frac{(p_1 - p_2)g_c}{\rho_{H_2O}g}$$

Equation ii gives the pressure drop in m of air. The density term of the left hand side of ii then is of air. Consequently we write

$$\Delta h_{H_2O} = \frac{(p_1 - p_2)g_c}{\rho_{H_2O}g} = \frac{(p_1 - p_2)g_c}{\rho_{air}g} \frac{\rho_{air}}{\rho_{H_2O}}$$

Substituting, we get

$$\Delta h_{H_2O} = 0.625 \frac{1.17}{1\ 000}$$

$$\Delta h_{H_2O} = 0.000\ 73 \text{ m of } H_2O = 0.073 \text{ cm of } H_2O$$

We have thus considered laminar and turbulent flow of a Newtonian fluid in a circular, an annular, and a rectangular duct. We have a source for friction factor in each of these cases and the results are summarized in Table 3.2.

TABLE 3.2. *Summary of equations for three cross sections.*

Section	Hydraulic Diameter	Laminar flow friction factor f
	$D_h = D$	$f = \dfrac{64}{Re}$
	$D_h = ID_a - OD_p$ $\kappa = \dfrac{OD_p}{ID_a}$	$\dfrac{1}{f} = \dfrac{Re}{64}\left(\dfrac{(1 + \kappa^2)}{(1 - \kappa)^2} + \dfrac{1 + \kappa}{(1 - \kappa)ln\,(\kappa)}\right)$
	$D_h = \dfrac{2hw}{h+w}$	$f\,Re = 96 - 95\left(\dfrac{h}{w}\right) + 56\left(\dfrac{h}{w}\right)^2$

$$\text{Reynolds number } Re = \frac{VD_h}{v} = \frac{\rho V D_h}{\mu g_c}$$

Friction factor for turbulent flow: Moody Diagram
Chen Equation, Churchill Equation

3.5 Minor Losses

Minor losses is a term that refers to pressure losses encountered by a fluid as it flows through a fitting or a valve in a piping system. Fittings and valves are used to direct the flow, to connect conduits together, to re-route the fluid, and to control the flow rate. Fittings are an integral part of any piping system and how their presence affects the fluid is the subject of this section.

As fluid flows through a fitting, the fluid may undergo an abrupt change in area (increase or decrease). The fluid may also have to negotiate a sharp curve and might do so by forming a separation region within the fitting. The fluid will encounter a loss in pressure. We treat this loss mathematically by assigning to each fitting a loss factor K. The pressure loss is then expressed as a multiple of the kinetic energy of the flow:

$$p_1 - p_2 = \Sigma K \frac{\rho V^2}{2g_c} \qquad (3.37)$$

The Bernoulli equation when written to include the effects of friction and of minor losses becomes

$$\frac{p_1 g_c}{\rho g} + \frac{V_1^2}{2g} + z_1 = \frac{p_2 g_c}{\rho g} + \frac{V_2^2}{2g} + z_2 + \Sigma \frac{fL}{D_h} \frac{V^2}{2g} + \Sigma K \frac{V^2}{2g} \qquad (3.38)$$

We refer to Equation 3.38 as the **Modified Bernoulli Equation**.

Loss coefficients for a number of fittings are provided in Table 3.3. Most of the information in that table is a result of measurements made on fittings. A number of the fittings in the table have a constant value of the loss coefficient K and a corresponding equation. When performing calculations by hand, it is convenient to use a constant value which is why they are provided. When using a computer, on the other hand, it is easy to use an equation for the loss coefficient, also provided in the table. Note that loss coefficient varies with pipe diameter for many of the fittings listed.

Table 3.3 also gives minor loss coefficients for several types of valves. Valves are available in a variety of types and sizes, and selecting the right valve for the job should receive due attention. A wrong valve can have disastrous consequences and so it is desirable to have information on valve selection. Table 3.4 gives general guidelines on selecting the proper valve for a given application. Included are valve characteristics, advantages, and disadvantages.

Before proceeding to solve piping problems, it is worthwhile to review the concept of the control volume and how to apply the Modified Bernoulli Equation correctly. The first step in formulating the solution to a problem is to determine where the control volume is to be located. Next, we identify the cross sections where mass crosses the control surface. The pressure p, velocity V, and height z terms of the Modified Bernoulli Equation apply only to cross sections where mass crosses the boundary. These terms are to be applied to nothing outside or inside the control volume. The friction term fL/D_h and the minor loss coefficient K apply to what is happening within the piping system. These terms do not apply to anything that happens outside the control volume.

TABLE 3.3. *Loss coefficients for pipe fittings; inlets, exits, and elbows.*

	Square edged inlet $K = 0.5$		Basket strainer $K = 1.3$
	Re-entrant inlet or inward projecting pipe $K = 1.0$		Well rounded inlet or a bell mouth inlet $K = 0.05$
	Foot valve $K = 0.8$		Exit $K = 1.0$
	Convergent outlet or nozzle $K = 0.1(1 - D_2/D_1)$ D_2/D_1 from 0.5 to 0.9		

	threaded	flanged or glued
90 Elbow	regular $K = 1.4$ $K = 1.4(ID)^{-0.53}$ ID from 0.3 to 4 in	regular $K = 0.31$ $K = 0.44(ID)^{-0.23}$ ID from 1 to 25 in
	long radius $K = 0.75$ $K = 0.75(ID)^{-0.81}$ ID from 0.3 to 4 in	long radius $K = 0.22$ $K = 0.51(ID)^{-0.58}$ ID from 1 to 23 in
45 Elbow	regular $K = 0.35$ $K = 0.35(ID)^{-0.14}$ ID from 0.3 to 4 in	
		long radius $K = 0.17$ $K = 0.22(ID)^{-0.14}$ ID from 1 to 23 in

TABLE 3.3 continued. *Loss coefficients for pipe fittings; elbows, T-joints, and couplings*

	threaded	flanged or glued
Return bend	regular $K = 1.5$ $K = 1.5(ID)^{-0.57}$ ID from 0.3 to 4 in	regular $K = 0.3$ $K = 0.43(ID)^{-0.26}$ ID from 1 to 23 in long radius $K = 0.2$ $K = 0.43(ID)^{-0.53}$ ID from 1 to 23 in
T joint	line flow $K = 0.9$ all sizes ID from 0.3 to 4 in branch flow $K = 1.9$ $K = 1.9(ID)^{-0.38}$ ID from 0.3 to 4 in	line flow $K = 0.14$ $K = 0.27(ID)^{-0.46}$ ID from 1 to 20 in branch flow $K = 0.69$ $K = 1.0(ID)^{-0.29}$ ID from 1 to 20 in
Coupling	$K = 0.08$ $K = 0.083(ID)^{-0.69}$ ID from 0.4 to 4 in	$K = 0.08$ ID from 0.3 to 23 in
Reducing bushing (D_1, D_2)	$K = 0.5 - 0.167(D_2/D_1) - 0.125(D_2/D_1)^2 - 0.208(D_2/D_1)^3$ $0.25 < D_2/D_1 < 1$	
Sudden expansion (D_1, D_2)	$K = 0.93 + 0.592(D_1/D_2) - 3.625(D_1/D_2)^2 + 2.803(D_1/D_2)^3$ $0.2 < D_1/D_2 < 0.9$	

TABLE 3.3 continued. *Loss coefficients for pipe fittings; valves.*

	threaded	flanged or glued
Globe valve	fully open $K = 10$ $K = \exp\{2.158 - 0.459 \ln(ID)$ $+ 0.259[\ln(ID)]^2$ $- 0.123[\ln(ID)]^3\}$ ID from 0.3 to 4 in	fully open $K = 10$ $K = \exp\{2.565 - 0.916 \ln(ID)$ $+ 0.339[\ln(ID)]^2$ $- 0.01416[\ln(ID)]^3\}$ ID from 0.3 to 4 in

Gate Valve	fully open $K = 0.15$ $K = 0.24(ID)^{-0.47}$ ID from 0.3 to 4 in	fully open $K = 0.15$ $K = 0.78(ID)^{-1.14}$ ID from 1 to 20 in

All sizes

Fraction closed	0	1/2	3/8	1/2	5/8	3/4	7/8
$K =$	0.15	0.26	0.81	2.06	5.52	17.0	97.8

	fully open $K = 2.0$ $K = 4.5(ID)^{-1.08}$ ID from 0.6 to 4 in	fully open $K = 2.0$ $K = \exp\{1.569 - 1.43 \ln(ID)$ $+ 0.8[\ln(ID)]^2$ $- 0.137[\ln(ID)]^3\}$ ID from 1 to 20 in
Angle Valve		

All sizes

$\alpha^\circ =$	0	10	20	30	40	50	60	70	80
$K =$	0.05	0.29	1.56	5.47	17.3	25.6	206	485	∞

Ball Valve

Check Valves Swing Type Ball Type Lift Type	$K = 2.5$ $K = 70.0$ $K = 12.0$	$K = 2.5$ $K = 70.0$ $K = 12.0$

TABLE 3.4. *Valve selection guide. (Information from "Selecting the Proper Valve—Parts 1 and 2" by J. L. Lyons and C. Askland, Design News, December 1974, pg. 56.)*

Valve Type	Description	Applications	Advantages	Disadvantages
Ball Valve	Ported sphere within housing Rotation of sphere by 90° changes from fully open to fully closed Variety of sizes	Flow control Pressure control Shutoff Can be used at high pressures and temperatures Fluids: common, corrosive, cryogenic, viscous, and slurries	Low pressure drop Low leakage rate Small size/weight ratio Rapid opening Insensitive to contamination	Seats of ball subject to wear if used as throttle Fluid trapped in ball when closed Quick opening may cause surges or water hammer
Butterfly Valve	Disc within housing Disc rotates about a shaft Disc closes against ring seal	Low pressure systems Leakage unimportant Large diameter lines Fluids: common	Low pressure drop Lightweight Small face-to-face dimension	Leakage fairly high Seals often damaged by high velocity Require high actuation forces Limited to low pressure systems
Gate Valve	Sliding disc or gate Moves perpendicular to flow Not used as a throttle High temperature High pressures	Stop valves—fully open or fully closed Tight seal when fully closed Insensitive to contamination Fluids: common	Low pressure drop when fully open	Prone to vibration Subject to disc and seat wear Slow response characteristics Require high actuation forces Not suited for steam
Globe Valve	Closure member travels in direction perpendicular to seat	Throttling purposes Power & process piping General purpose control Fluids: common	Faster to open than gate valve Seating surface less subject to wear Pressure control	High pressure drop Require considerable power to operate Often heavier than other valves

TABLE 3.4. *Valve selection guide continued.*

Valve Type	Description	Applications	Advantages	Disadvantages
Pinch Valve	Has one or more flexible elements that can be pinched to close off flow	Small tendency to build up contamination Fluids: common, slurries	Low in cost Insensitive to contamination Low pressure drop Tight closing	Subject to wear Need periodic re-placement Limited to low pressure Limited to low tem-perature Require high actuation forces
Taper Plug Valve	Similar to ball valve Closure member is a tapered plug	High temperature Low pressure Same functions as ball, globe, and gate valves Fluids: common	Small in size Fairly low cost Tight seal Leakproof	Subject to binding and galling Not suited for steam Require lubrication
Poppet Valve	Closure member moves parallel to fluid flow and perpendicular to sealing surface Closure element can be flat, conical, or spherical	Safety & relief functions Pressure control Check valve Fluids: common	Excellent leakage control Low pressure drop Some seat surfaces subject to contamination	Subject to pressure imbalances that may cause chattering
Swing Valve	Similar to butterfly valves Disc hinged at one end	Check valve Unidirectional flow control	Low pressure drop Lightweight Low cost	May have high leakage Seal may erode Introduces turbulence at low flow rates

Before proceeding to solve piping problems, it is worthwhile to review the concept of the control volume and how to apply the Modified Bernoulli Equation correctly. The first step in formulating the solution to a problem is to determine where the control volume is to be located. Next, we identify the cross sections where mass crosses the control surface. The pressure p, velocity V, and height z terms of the Modified Bernoulli Equation apply only to cross sections where mass crosses the boundary. These terms are to be applied to nothing outside or inside the control volume. The friction term fL/D_h and the minor loss coefficient K apply to what is happening within the piping system. These terms do not apply to anything that happens outside the control volume.

As an example, consider the system shown in Figure 3.8. A piping system is connected to two tanks. We will examine five different control volumes applied to this same piping system and write the modified Bernoulli Equation for each case. In Figure 3.8a, the control volume includes all the fluid in the piping system and the fluid in both tanks. Section 1 is the free surface of the tank on the left and section 2 is the free surface of the tank on the right. We now evaluate each property at both sections:

$$p_1 = p_2 = p_{atm} = 0$$

V_1 = surface velocity ≈ 0 (compared to velocity in pipe)

$$V_2 \approx 0$$

z_1 = height at section 1; z_2 = height at section 2

$\dfrac{fL}{D}$ = friction term applied to piping system

ΣK = minor losses encountered by a fluid particle in traveling
        from section 1 to section 2; inlet, 2–90° elbows, valve, and
        exit

The modified Bernoulli Equation (3.38) is

$$\frac{p_1 g_c}{\rho g} + \frac{V_1^2}{2g} + z_1 = \frac{p_2 g_c}{\rho g} + \frac{V_2^2}{2g} + z_2 + \Sigma \frac{fL}{D_h} \frac{V^2}{2g} + \Sigma K \frac{V^2}{2g} \qquad (3.38)$$

For this application, we get for Figure 3.8a

$$z_1 = z_2 + \frac{fL}{D_h} \frac{V^2}{2g} + (K_{inlet} + 2K_{90° \, elbow} + K_{valve} + K_{exit}) \frac{V^2}{2g} \qquad (3.39)$$

This result accompanies the figure.

In Figure 3.8b, the control volume includes the fluid in the tank at the left and all the fluid in the pipe. Section 1 is the free surface of the liquid in the tank and section 2 is just at the end of the pipe. After section 2, the liquid could be discharged to the atmosphere or to another tank or to another pump. Its destination after section 2 is of no concern with regard to the analysis we formulate. The properties are:

$p_1 = p_{atm} = 0$; $p_2 = $ pressure at section 2 $\neq p_{atm}$

$V_1 = $ surface velocity ≈ 0 (compared to velocity in pipe)

$V_2 = $ velocity in the pipe

$z_1 = $ height at section 1; $z_2 = $ height at section 2

$\frac{fL}{D} = $ friction term applied to piping system

$\Sigma K = $ minor losses encountered by a fluid particle in traveling
          from section 1 to section 2; inlet, 2–90° elbows, and a valve

The exit loss is not accounted for in Figure 3.8b because the pressure loss in a fitting is realized by the fluid only *after* it passes through it. The Modified Bernoulli equation (for Figure 3.8b) reduces to:

$$z_1 = \frac{p_2 g_c}{\rho g} + \frac{V_2^2}{2g} + z_2 + \frac{fL}{D_h} \frac{V^2}{2g} + (K_{inlet} + 2K_{90° \, elbow} + K_{valve}) \frac{V^2}{2g}$$

$$(3.40)$$

where the exit velocity V_2 equals the velocity in the pipe V. This result accompanies the figure.

Figure 3.8c shows the right tank removed as does Figure 3.8b. In Figure 3.8c, however, our control volume does not end abruptly with the end of the pipe. Instead, we use a large surface area as section 2. The pressure at the exit of the pipe is usually not equal to atmospheric pressure so we allow the fluid to expand until its pressure does equal p_{atm}.

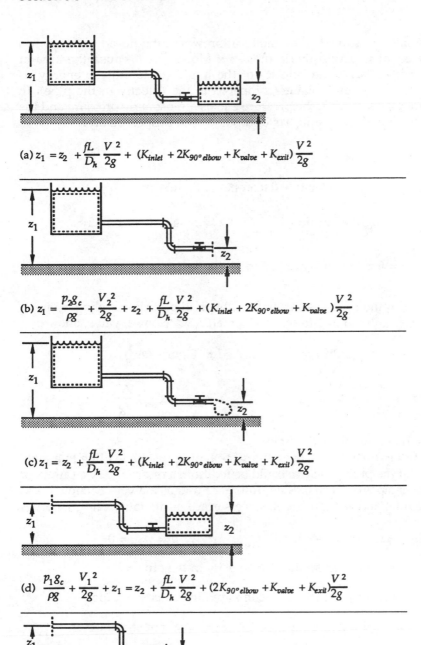

(a) $z_1 = z_2 + \dfrac{fL}{D_h} \dfrac{V^2}{2g} + (K_{inlet} + 2K_{90°\,elbow} + K_{valve} + K_{exit}) \dfrac{V^2}{2g}$

(b) $z_1 = \dfrac{p_2 g_c}{\rho g} + \dfrac{V_2^2}{2g} + z_2 + \dfrac{fL}{D_h} \dfrac{V^2}{2g} + (K_{inlet} + 2K_{90°\,elbow} + K_{valve}) \dfrac{V^2}{2g}$

(c) $z_1 = z_2 + \dfrac{fL}{D_h} \dfrac{V^2}{2g} + (K_{inlet} + 2K_{90°\,elbow} + K_{valve} + K_{exit}) \dfrac{V^2}{2g}$

(d) $\dfrac{p_1 g_c}{\rho g} + \dfrac{V_1^2}{2g} + z_1 = z_2 + \dfrac{fL}{D_h} \dfrac{V^2}{2g} + (2K_{90°\,elbow} + K_{valve} + K_{exit}) \dfrac{V^2}{2g}$

(e) $\dfrac{p_1 g_c}{\rho g} + z_1 = \dfrac{p_2 g_c}{\rho g} + z_2 + \dfrac{fL}{D_h} \dfrac{V^2}{2g} + (2K_{90°\,elbow} + K_{valve}) \dfrac{V^2}{2g}$

FIGURE 3.8. *Modified Bernoulli Equation written for various systems.*

So section 2 is assumed to be the location where the liquid pressure has become equal to atmospheric pressure. Moreover, because the area at section 2 is so large, the velocity of the liquid (or its kinetic energy) is reduced to a negligible value (compared to the velocity in the pipe). In other words at section 2, the pressure equals atmospheric pressure and the kinetic energy of the liquid has dissipated. The properties are:

$$p_1 = p_2 = p_{atm} = 0 \qquad\qquad\qquad\qquad\qquad V_2 \approx 0$$

V_1 = surface velocity ≈ 0 (compared to velocity in pipe)

z_1 = height at section 1; $\qquad\qquad$ z_2 = height at section 2

$\dfrac{fL}{D}$ = friction term applied to piping system

ΣK = minor losses encountered by fluid particle traveling from
$\qquad\qquad$ section 1 to section 2; inlet, 2–90° elbows, valve, and exit

The Modified Bernoulli Equation applied to Figure 3.8c is

$$z_1 = z_2 + \frac{fL}{D_h}\frac{V^2}{2g} + (K_{inlet} + 2K_{90° elbow} + K_{valve} + K_{exit})\frac{V^2}{2g} \qquad (3.41)$$

This result accompanies the figure.

In Figure 3.8d, we have the same piping system leading to a tank. The inlet of the piping system could be fed from a reservoir, by a pump, or by another pipeline. It makes no difference in our analysis. Section 1 is at the pipe inlet and section 2 is the free surface of the tank. The properties are:

$\qquad p_1$ = pressure at section 1; $\qquad\qquad\qquad$ $p_2 = p_{atm} = 0$

$\qquad V_1$ = velocity at section 1 = velocity in the pipe = V

$\qquad V_2$ = surface velocity ≈ 0 (compared to velocity in pipe)

$\qquad z_1$ = height at section 1; $\qquad\qquad$ z_2 = height at section 2

$\qquad \dfrac{fL}{D}$ = friction term applied to piping system

$\qquad \Sigma K$ = minor losses encountered by a fluid particle in traveling
$\qquad\qquad\qquad$ from section 1 to section 2; 2–90° elbows, valve, and
$\qquad\qquad\qquad$ exit

For Figure 3.8d, the Modified Bernoulli Equation reduces to

$$\frac{p_1 g_c}{\rho g} + \frac{V_1^2}{2g} + z_1 = z_2 + \frac{fL}{D_h}\frac{V^2}{2g} + (2K_{90°\,elbow} + K_{valve} + K_{exit})\frac{V^2}{2g} \quad (3.42)$$

in which $V_1 = V$. This result accompanies the figure.

Figure 3.8e shows the piping system without tanks attached. The liquid source or its ultimate destination do not affect our analysis. The locations of section 1 and section 2 are shown. The properties are:

p_1 = pressure at section 1; $\quad\quad\quad\quad\quad\quad$ p_2 = pressure at section 2

V_1 = velocity at section 1 = velocity in the pipe = V

V_2 = velocity at section 2 = velocity in the pipe = V

z_1 = height at section 1; $\quad\quad\quad\quad$ z_2 = height at section 2

$\dfrac{fL}{D}$ = friction term applied to piping system

ΣK = minor losses encountered by a fluid particle in traveling from section 1 to section 2; 2–90° elbows and valve

For Figure 3.8e, the Modified Bernoulli Equation reduces to

$$(e) \quad \frac{p_1 g_c}{\rho g} + z_1 = \frac{p_2 g_c}{\rho g} + z_2 + \frac{fL}{D_h}\frac{V^2}{2g} + (2K_{90°\,elbow} + K_{valve})\frac{V^2}{2g} \quad (3.43)$$

This result is shown in the figure. As indicated in the previous discussion, it is extremely important to clearly define the boundary of the control volume.

We are now equipped to model piping problems. Again we examine problems in which pressure drop Δp (or Δh), volume flow rate Q, or diameter D is unknown.

Example 3.6. Figure 3.9 shows a portion of a piping system used to convey 750 gpm of ethyl alcohol. The system contains 180 ft of 12-nominal schedule 40 commercial steel pipe. All fittings are of the long radius type and are flanged. Calculate the pressure drop over this portion of the pipeline.

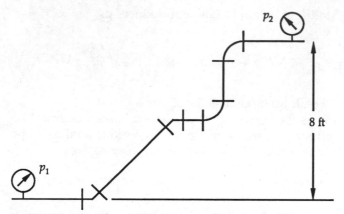

FIGURE 3.9. *The piping system of Example 3.6.*

Solution: The control volume we select includes all the liquid in the pipe and extends to each pressure gage. The calculation procedure is as follows.

Liquid Properties (Ethyl alcohol, Appendix Table B.1):

$$\rho = 0.787(62.4) \text{ lbm/ft}^3 \qquad \mu = 2.29 \times 10^{-5} \text{ lbf·s/ft}^2$$

Conduit Dimensions (12-nom sch 40, Appendix Table D.1):

$$D = 0.9948 \text{ ft} \qquad A = 0.773 \text{ ft}^2$$

Flow velocity $V = Q/A$:

$$Q = 750 \text{ gpm} = 1.67 \text{ ft}^3/\text{s}$$

$$V = \frac{1.67}{0.773} = 2.16 \text{ ft/s}$$

Reynolds number $\mathbf{Re} = \rho VD/\mu g_c$:

$$\mathbf{Re} = \frac{0.787(62.4)(2.16)(0.9948)}{2.29 \times 10^{-5} (32.2)} = 1.43 \times 10^5$$

Relative roughness (commercial steel, Table 3.1):

$$\varepsilon = 0.00015 \text{ ft}$$

$$\frac{\varepsilon}{D} = \frac{0.00015}{0.9948} = 0.00015$$

Friction factor (Figure 3.2):

$$Re = 1.43 \times 10^5$$
$$\frac{\varepsilon}{D} = 0.00015$$

$$f = 0.018$$

Modified Bernoulli Equation (3.38):

$$\frac{p_1 g_c}{\rho g} + \frac{V_1^2}{2g} + z_1 = \frac{p_2 g_c}{\rho g} + \frac{V_2^2}{2g} + z_2 + \sum \frac{fL}{D_h} \frac{V^2}{2g} + \sum K \frac{V^2}{2g} \qquad (3.38)$$

Property evaluation:

$$V_1 = V_2; \qquad z_1 = 0; \qquad z_2 = 8 \text{ ft}; \qquad L = 180 \text{ ft}$$

$$\sum K = 2K_{45° \, elbow} + 2K_{90° \, elbow} = 2(0.17) + 2(0.22) = 0.78$$

Equation of motion:

$$\frac{p_1 g_c}{\rho g} = \frac{p_2 g_c}{\rho g} + 8 + \left(\frac{0.018(180)}{0.9948} + 0.78 \right) \frac{(2.16)^2}{2(32.2)}$$

or

$$\frac{(p_1 - p_2) g_c}{\rho g} = 8 + 0.29 = 8.29 \text{ ft of ethyl alcohol}$$

Thus if we attached an air–over–ethyl alcohol, inverted U–tube manometer from section 1 to 2, it would read $\Delta h = 8.29$ ft. The pressure drop is now calculated as

$$p_1 - p_2 = \frac{8.29(0.787)(62.4)(32.2)}{32.2} = 407 \text{ lbf/ft}^2$$

or

$$p_1 - p_2 = 2.83 \text{ psi}$$

Example 3.7. A huge water reservoir is drained with a 4–nominal schedule 80 copper pipe. The piping system is shown in Figure 3.10. The fittings are regular and threaded. Determine the volume flow rate through the system.

Solution: The control volume we select includes the water in the reservoir and in the piping system. Section 1 is the free surface of the water in the reservoir and section 2 is located such that $p_2 = p_{atm}$. The calculations are:

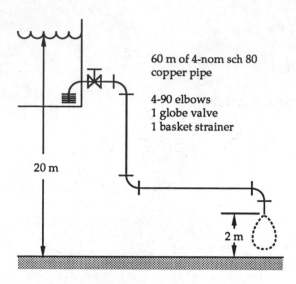

60 m of 4-nom sch 80
copper pipe

4-90 elbows
1 globe valve
1 basket strainer

20 m

2 m

FIGURE 3.10. *The piping system of Example 3.7.*

Liquid Properties (Water, Appendix Table B.1):

$$\rho = 1\,000 \text{ kg/m}^3 \qquad\qquad \mu = 0.89 \times 10^{-3} \text{ N·s/m}^2$$

Conduit Dimensions (4-nom sch 80, Appendix Table D.1):

$$D = 9.718 \text{ cm} \qquad\qquad A = 74.17 \text{ cm}^2$$

Flow velocity $V = Q/A$:

$$V = \frac{4Q}{\pi D^2}$$

Reynolds number $\mathbf{Re} = \rho V D/\mu g_c$:

$$\mathbf{Re} = \frac{1\,000(V)(0.09718)}{0.89 \times 10^{-3}} = 1.09 \times 10^5 V \qquad\qquad \text{(i)}$$

Relative roughness (copper pipe, Table 3.1):

$$\varepsilon = 0.000\,15 \text{ cm}$$

$$\frac{\varepsilon}{D} = \frac{0.000\,15}{9.718} = 0.000\,015 \qquad\qquad \text{(ii)}$$

Modified Bernoulli Equation (3.38):

$$\frac{p_1 g_c}{\rho g} + \frac{V_1^2}{2g} + z_1 = \frac{p_2 g_c}{\rho g} + \frac{V_2^2}{2g} + z_2 + \Sigma \frac{fL}{D_h} \frac{V^2}{2g} + \Sigma K \frac{V^2}{2g} \qquad (3.38)$$

Property evaluation:

$$p_1 = p_2 = p_{atm} \qquad V_1 \approx 0 \quad V_2 = 0 \quad z_1 = 20 \text{ m} \qquad z_2 = 2 \text{ m}$$

$$L = 60 \text{ m}$$

$$\Sigma K = K_{\substack{basket \\ strainer}} + 4K_{90° \, elbow} + K_{\substack{globe \\ valve}} + K_{exit}$$

$$\Sigma K = 1.3 + 4(1.4) + 10 + 1.0 = 17.9$$

Equation of motion:

$$z_1 = z_2 + \Sigma \frac{fL}{D_h} \frac{V^2}{2g} + \Sigma K \frac{V^2}{2g}$$

or

$$20 = 2 + \frac{60f}{0.09718} \frac{V^2}{2(9.81)} + 17.9 \frac{V^2}{2(9.81)}$$

Rearranging and solving for velocity give

$$18 = (31.5f + 0.912)V^2$$

or

$$V = \sqrt{\frac{18}{31.5f + 0.912}} \qquad \qquad (iii)$$

A trial and error process involving equations i, ii, and iii is required. First we assume a value of the friction factor corresponding to the fully developed turbulent flow value for which $\varepsilon/D = 0.000\,015$:

1st trial: $f = 0.008;$ $V = 3.93;$ $Re = 4.28 \times 10^5$
2nd trial: $f = 0.015;$ $V = 3.61;$ $Re = 3.93 \times 10^5$
              $f = 0.015$ close enough

The velocity is thus 3.61 m/s. The volume flow rate then is

$$Q = AV = 74.17 \times 10^{-4} (3.61)$$

or $Q = 0.026\,8 \text{ m}^3/\text{s} = 27 \text{ l/s}$

Example 3.8. Figure 3.11 shows a piping system that consists of a line connected to two branches. When the bypass branch is closed off with its valve, the flow line is to deliver 0.3 ft³/s of benzene with a pressure drop of $p_1 - p_2$ of 8.5 psi. Select a suitable size for the pipe if it is made of uncoated cast iron and has regular threaded fittings. The length of pipe required is 700 ft. Due to cost considerations, it is desirable to use schedule 40 or schedule 80 pipe.

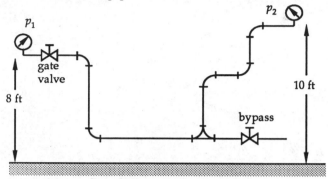

FIGURE 3.11. *The piping system of Example 3.8.*

Solution: The control volume we select includes all the fluid in the pipe from the gage at section 1 to the gage at section 2, excluding the bypass. The method of solution is:

Liquid Properties (Benzene, Appendix Table B.1):

$$\rho = 0.876(62.4) \text{ lbm/ft}^3 \qquad \qquad \mu = 1.26 \times 10^{-5} \text{ lbf·s/ft}^2$$

Flow velocity $V = Q/A$:

$$V = \frac{4Q}{\pi D^2} = \frac{4(0.3)}{\pi D^2} = \frac{0.382}{D^2}$$

Reynolds number $\mathbf{Re} = \rho V D / \mu g_c$:

$$\mathbf{Re} = \frac{0.876(62.4)(0.382/D^2)D}{1.26 \times 10^{-5}(32.2)}$$

$$\mathbf{Re} = \frac{5.15 \times 10^4}{D} \qquad\qquad\qquad\qquad\qquad \text{(i)}$$

Relative roughness (uncoated cast iron pipe, Table 3.1):

$$\varepsilon = 0.00085 \text{ ft}; \qquad\qquad \frac{\varepsilon}{D} = 0.00085/D \qquad\qquad \text{(ii)}$$

Continuity Equation:

$$Q = A_1 V_1 = A_2 V_2 \qquad A_1 = A_2, \text{ therefore } V_1 = V_2.$$

Modified Bernoulli Equation (3.38)

$$\frac{p_1 g_c}{\rho g} + \frac{V_1^2}{2g} + z_1 = \frac{p_2 g_c}{\rho g} + \frac{V_2^2}{2g} + z_2 + \Sigma \frac{fL}{D_h} \frac{V^2}{2g} + \Sigma K \frac{V^2}{2g} \qquad (3.38)$$

Property evaluation:

$$\frac{(p_1 - p_2)g_c}{\rho g} = \frac{8.5(144)(32.2)}{0.876(62.4)(32.2)} = 22.4 \text{ ft of benzene}$$

$$V_1 = V_2 \qquad\qquad z_1 = 8 \text{ ft} \qquad\qquad z_2 = 10 \text{ ft} \qquad\qquad L = 700 \text{ ft}$$

$$\Sigma K = K_{gate \atop valve} + 5K_{90° \, elbow} + K_{T\text{-}joint}$$

$$\Sigma K = 0.15 + 5(1.4) + 1.9 = 9.05$$

Equation of motion:

$$\frac{p_1 g_c}{\rho g} - \frac{p_2 g_c}{\rho g} + z_1 - z_2 = \left(\Sigma \frac{fL}{D_h} + \Sigma K \right) \frac{V^2}{2g}$$

Substituting,

$$22.4 + 8 - 10 = \left(\frac{700f}{D} + 9.05 \right) \frac{(0.382)^2}{2D^4 (32.2)}$$

or

$$9003 = \frac{700f}{D^5} + \frac{9.05}{D^4}$$

Rearranging and simplifying gives

$$f = 12.86 D^5 - 0.01293 D \qquad\qquad (iii)$$

The solution method involves a trial and error procedure which begins by assuming a diameter:

1st trial: $D = 0.25$ ft; $f = 0.00933$ (Eq. iii)   not likely
2nd trial: $D = 0.333$ ft; $f = 0.0484$ then

$$\left.\begin{array}{l} \mathbf{Re} = 5.15 \times 10^4/D = 1.55 \times 10^5 \\[2ex] \dfrac{\varepsilon}{D} = \dfrac{0.00085}{0.333} = 0.00255 \end{array}\right\} \qquad f = 0.026 \quad (\text{Figure 3.2})$$

Therefore $D = 0.333$ ft is too large. Rather than continue by using randomly selected values for diameter, we refer to sizes listed specifically in Appendix Table D.1.

3rd trial: $3^1/_2$ nom sch 40 $D = 0.2957$ ft; $f = 0.02525$ (Eq. iii)

$$\left.\begin{array}{l} \mathbf{Re} = 5.15 \times 10^4/D = 1.74 \times 10^5 \\[2ex] \dfrac{\varepsilon}{D} = \dfrac{0.00085}{0.2957} = 0.0029 \end{array}\right\} \qquad f = 0.026 \quad (\text{Figure 3.2})$$

which is close enough. The area of $3^1/_2$ nom sch 40 is 0.06867 ft². The average velocity is

$$V = \frac{Q}{A} = \frac{0.3}{0.06867} = 4.37 \text{ ft/s}$$

and $\mathbf{Re} = 1.74 \times 10^5$, $f = 0.025$.

It is important to note that the friction factor calculated with Equation iii ($f = 0.02525$) is an upper limit which fits the conditions exactly. If the actual friction factor ($f = 0.026$) is less than the maximum, then the actual pressure drop will be somewhat less than the allowable. This is due to the fact that we are using a larger pipe than is necessary. The actual pressure drop can be calculated with the equation of motion:

$$\frac{p_1 g_c}{\rho g} - \frac{p_2 g_c}{\rho g} + z_1 - z_2 = \left(\Sigma \frac{fL}{D_h} + \Sigma K \right) \frac{V^2}{2g}$$

Substituting,

$$\frac{(p_1 - p_2)(32.2)}{0.876(62.4)(32.2)} + 8 - 10 = \left(\frac{0.026(700)}{0.2957} + 9.05\right)\frac{(4.37)^2}{2(32.2)}$$

Solving, we get

$$p_1 - p_2 = 8.706 \text{ psi}$$

and we were allowed 8.5 psi.

3.6 Summary

In this chapter, we have examined pipe and tubing standards, and discussed the current specifications that apply to them. We have stated three definitions of characteristic dimensions used to represent noncircular cross sections. Equations for velocity, flow rate, Reynolds number, and friction factor were provided for circular, annular, and rectangular cross sections. The Moody Diagram was discussed and two modified versions of it were also provided. Sample problems were given to illustrate the use of these charts. Minor losses were discussed and recommended procedures for accounting for them were given.

3.7 Show and Tell

1. Prepare a demonstration that illustrates the following (dependent upon availability of tools and equipment):
   a. the taper of pipe threads;
   b. the application of "pipe dope" or tape prior to attaching a fitting to a pipe; and
   c. the cutting and threading of a pipe using a hand die and/or a machine;

2. Prepare a demonstration that illustrates the following (dependent upon availability of tools and equipment):
   a. the use of a flaring tool to flare the end of a tube;
   b. the attachment of the tube to a flared end fitting;
   c. the installation of a compression fitting; and
   d. the sweating of a tubing joint.

3. Prepare a demonstration that illustrates how the valves mentioned in this chapter operate, as assigned by the instructor and according to availability.

3.8 Problems Chapter 3

1. Five gallons per minute of methyl alcohol flows in a 2-nominal schedule 40 pipe. Is the flow laminar or turbulent?

2. Fifteen liters per second of propane flows through 4 standard type M copper tubing. Is the flow laminar or turbulent?

3. Turpentine flows through a 12 nominal schedule 40 pipe. What is the maximum flow rate permissible for laminar conditions to exist?

4. Determine the hydraulic radius of a two dimensional rectangular duct in which the width is much greater than the height ($w \gg h$).

5. Determine the effective diameter of a two dimensional rectangular duct in which the width is much greater than the height ($w \gg h$).

6. A rectangular duct has dimensions of h x w. The height h is 4 cm. Determine the width w if the hydraulic and effective diameters are equal, if possible.

7. An annular duct consists of two tubes. The inner tube is 2 std and the outer tube is 4 std, both type K. Calculate the hydraulic radius, hydraulic diameter, and effective diameter of the flow passage.

8. A flow passage is bounded by the outside area of a 1 std type L copper tube and the inside of a 4 in. x 4 in. square duct. Calculate the hydraulic radius, hydraulic diameter, and effective diameter of the flow passage.

9. Equation 3.8b was derived from Equation 3.5 using the Darcy-Weisbach definition of friction factor and the hydraulic diameter. Begin again with Equation 3.5 but use the Fanning friction factor and the hydraulic radius to derive an equation analogous to Equation 3.8b.

10. Refer to Figure P3.10 and derive the equation of velocity for laminar flow in a circular duct by following the steps outlined below:

   a. Perform a force balance on the control volume in the figure and verify that

   $$pA + \tau dA_p - (p + dp)A = 0$$

   where A is cross sectional area and dA_p is perimeter times axial length.

   b. Substitute $dA_p = 2\pi r dz$ and $A = \pi r^2$ and show that

$$\frac{dp}{dz} = \frac{2\tau}{r}$$

c. Assuming a Newtonian fluid with constant properties, let

$$\tau = \mu \frac{dV_z}{dr}$$

and verify that

$$\frac{dV_z}{dr} = \frac{r}{2\mu}\frac{dp}{dz}$$

d. Verify that the boundary condition $r = R$, $V_z = 0$ is correct. Integrate the equation for dVz/dr above and apply the boundary condition. Show that

$$V_z = \left(-\frac{dp}{dz}\right)\left(\frac{R^2}{4\mu}\right)\left[1 - \left(\frac{r}{R}\right)^2\right] \qquad \left(\begin{array}{c}\text{laminar flow} \\ \text{circular duct}\end{array}\right)$$

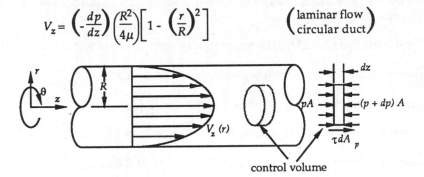

FIGURE P3.10

11. Verify that Equation 3.12 is correct.

12. Derive Equation 3.13 in detail.

13. Benzene flows at a rate of 0.012 m³/s through a 4-nominal schedule 40 cast iron pipe that is 45 m long. Calculate the pressure drop in the pipe.

14. A 12-nominal schedule 80 commercial steel pipe conveys 300 gpm of decane. The pipe length is 200 ft. Calculate the pressure drop in the pipe.

15. Heptane flows through an 8-nominal schedule 120 pipe that is 1/2 mile long. The volume flow rate is 1 ft³/s. Calculate the pressure drop if the pipe is made of wrought iron.

16. A 10-nominal schedule 40 galvanized iron pipe is 300 ft long. It conveys carbon disulfide. The pressure drop in the pipe is 3 psi. Calculate the volume flow rate of liquid.

17. Glycerine flows through 6 std type M copper water tube. The pressure loss over a length of 250 m is measured to 110 kPa. Determine the volume flow rate through the pipe.

18. Hexane flows through an 8-std type K copper tube that is 250 ft long. The pressure drop in the tube is 85 kPa. Calculate the volume flow rate.

19. A fuel line is to convey octane over a distance of 35 ft. The required flow rate is $0.3 \text{ ft}^3/s$ and the allowable pressure drop is 75 kPa. Select an appropriate line size if drawn copper is used.

20. Linseed oil is to be pumped at a flow rate of 12 gpm over a distance of 18 m. The allowable pressure drop is 15 psi. Centrifugally spun cast iron is to be used as a pipe material. Determine the appropriate line size.

21. A fuel line is to convey 80 l/s of propane with an allowable pressure drop of 13 psi over a 10 m length. Select a suitable diameter if drawn copper tubing is to be used.

22. Verify that Equation 3.11 is correct, beginning with Equation 3.10.

23. Derive Equation 3.23 in detail beginning with Equation 3.20.

24. Integrate Equation 3.23 appropriately and derive Equation 3.24.

25. Combine Equations 3.8a, 3.25, and 3.26b to derive Equation 3.27.

26. Start with Equation 3.28 and verify that Equation 3.30 is correct.

27. Derive Equation 3.31 beginning with Equation 3.30.

28. Integrate Equation 3.31 appropriately to obtain Equation 3.32.

29. Combine Equations 3.8a, 3.33, and 3.34 and verify that Equation 3.35 is correct.

30. An annular flow passage is 5 m long and is formed by placing a 2-nominal pipe within a 4-nominal pipe (both schedule 40 and made of uncoated cast iron). The flow passage is to convey 8.5 l/s of carbon disulfide. Calculate the pressure drop over the 5 m length.

31. Solve Problem 30 using effective diameter instead of hydraulic diameter, and compare the results of the two methods.

32. An annular flow area is formed by 3 std type M copper tube and 1 std type L copper tube, both 6 ft long. Glycerine is pumped through the annulus. Attached pressure gages show that the pressure drop is 3 psi. Determine the volume flow rate of glycerine.

33. Solve Problem 32 using effective diameter instead of hydraulic diameter, and compare the results of the two methods.

34. As part of a heat exchanger, an annulus is to convey ethyl alcohol at a flow rate of 10 *l*/s. The inner tube is 3/4-std type M copper tubing and the size of the outer tube is to be determined. The available pump can overcome a pressure drop of 5 psi and the annular flow passage is 8 ft long. Select a suitable outer tube size.

35. An asphalt coated, 6 m long rectangular duct has internal dimensions of 0.5 m x 1.5 m. It conveys air at a flow rate of 1.5 m^3/s. Calculate the pressure drop.

36. Solve Problem 35 using effective diameter instead of hydraulic diameter, and compare the results of the two methods.

37. A rectangular duct made of galvanized sheet metal is 3 ft wide by 6 ft tall. It delivers cooled air ($T \approx 60°F$) to a basement. The duct is 25 ft long and the allowable pressure drop is 0.01 in. of water. Calculate the flow rate of air assuming it to be an ideal gas.

38. Solve Problem 37 using effective diameter instead of hydraulic diameter, and compare the results of the two methods.

39. Figure P3.39 shows a piping system used to convey 100 gpm of turpentine. The pipe is made of 2-nom sch 40 uncoated cast iron that is 100 ft long. All fittings are regular and threaded and include 1 globe valve, 7-90° elbows, and 2-45° elbows. If the pump delivers 2 hp to the turpentine, what is the change in pressure from the pipe inlet upstream of the pump to the free surface in the tank downstream?

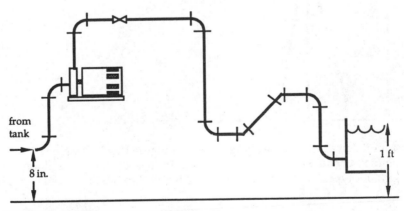

from
tank

8 in.

1 ft

FIGURE P3.39

40. A piping system is shown in Figure P3.40. It consists of 6 m of 6-std type K and 12 m of 4 std type K, both drawn copper tubing. The system conveys ethylene glycol at a rate of 0.013 m^3/s. The pressure drop from section 1 to section 2 is to be calculated. All fittings are soldered (same minor loss as flanged) and regular.

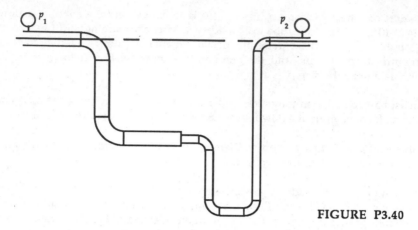

FIGURE P3.40

41. A garden hose is used to siphon water as shown in Figure P3.41. The hose is made of a rubber material ("smooth") and is 50 ft long. For the configuration shown, determine the volume flow rate through the hose if (a) frictional effects are neglected; and (b) if friction is accounted for. The inside diameter of the hose is 5/8 in. Neglect minor losses.

42. A pressurized tank and piping system are shown in Figure P3.42. The tank pressure is maintained at 175 kPa. The line is made of 12 m of 1-std type M copper tubing and it conveys gasoline (octane). What is the expected flow rate through the line? All fittings are soldered (same as flanged) and regular.

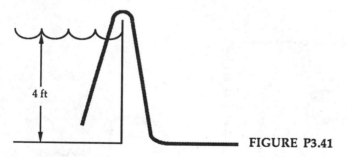

4 ft

FIGURE P3.41

43. Castor oil flows through the piping system of Figure P3.43. The pipe is made of 6-nominal schedule 40 galvanized steel. All fittings are flanged and are of the long radius type. Calculate the flow rate of liquid through the pipe if it is 250 ft long.

44. Suppose the receiver tank and discharge end of the pipe in Problem 43 are changed to the configuration shown in Figure P3.44. Rework the problem with this new setup and compare the following details between the two problems (noting of course the differences in control volume selection):
    a. Continuity Equation
    b. Modified Bernoulli Equation after simplification and before substitution of numbers

c. Minor losses
d. Reynolds numbers and friction factors
e. Solution
Assume that the pipe is 5 ft longer in this problem than in Problem 43.

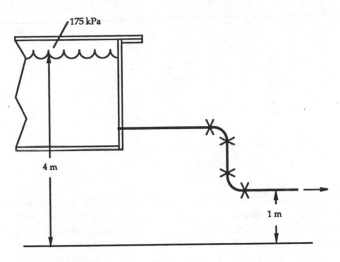

FIGURE P3.42

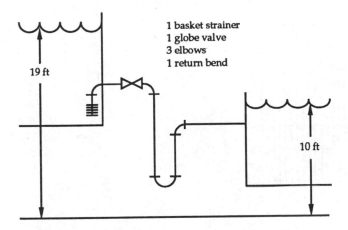

1 basket strainer
1 globe valve
3 elbows
1 return bend

FIGURE P3.43

45. The tubing arrangement of a cross flow heat exchanger is given schematically in Figure P3.45. It consists of type M drawn copper tubing with fins attached. There are 2 elbows and 7 return bends, all regular. The tube is to convey 0.25 l/s of propylene glycol. The allowable pressure drop in the system (inlet to outlet) is 85 kPa, and the tube length is 10 m. Select a suitable line diameter. All fittings are regular and soldered (same minor loss as flanged).

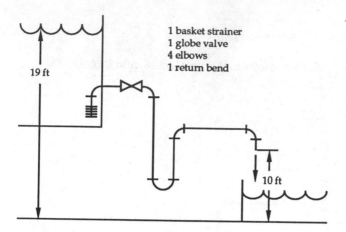

1 basket strainer
1 globe valve
4 elbows
1 return bend

19 ft

10 ft

FIGURE P3.44

46. A commercial steel pipeline is 200 ft long and is to convey 400 gpm of water. The system will contain 3 couplings and 8-90° elbows. The flow will be controlled by a gate valve. The inlet is fed by a pump and the pressure there is 125 psia. The outlet height is exactly the same height as the inlet, and water is discharged to the atmosphere. Select a suitable line size. It is desirable to use threaded fittings if the diameter is 2-nominal or smaller, and flanged fittings if the diameter is larger than 2-nominal. All fittings are to be regular. Schedule 40 is preferable.

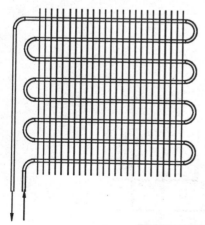

FIGURE P3.45

CHAPTER 4   Piping Systems II

In the last chapter, a study of piping systems was introduced and most of the material was reproduced from elementary fluid mechanics. In this chapter, we continue our study of piping systems by combining results of the last chapter with an economic analysis to develop a new method of pipe sizing.

Equations for the least annual cost method of economic pipe diameter selection are derived. The equations include economic and pipe friction parameters. The derivations lead to a new format for the traditional pipe friction diagram (the Moody Diagram). Three new graphs are presented that aid in determining the economic diameter when economic parameters and power costs are known. A new dimensionless group has been developed by combining the relative roughness and the Reynolds number. An example problem is provided to illustrate use of the graphs and the method in general. The method can be used successfully to select the economic diameter and satisfy least cost (first plus operating) requirements.

Next, the concept of equivalent length of minor losses is discussed. The equivalent length is defined and calculated for a fitting to illustrate the definition. Methods of graphically representing piping systems are also discussed. ANSI piping symbols are given as well.

The behavior of a system is also described in this chapter. A system curve is defined to show how frictional effects influence the volume flow rate. The chapter concludes with a section about conventional hardware available for physically supporting a piping system.

4.1 Economic Pipe Diameter

Engineers typically learn about piping systems in a first course in fluid mechanics. Three types of pipe flow problems are usually discussed: pressure drop Δp unknown, volume flow rate Q unknown, or inside diameter D unknown. In all cases, six variables enter the problem (L = pipe length, ε = surface roughness, v = kinematic viscosity of the fluid, Δp = pressure drop, Q = volume flow rate, and D = inside diameter). In any of the three types of problems, five variables are known and the sixth one is solved for.

In a real design problem, however, the value of five variables is usually not known. Suppose a tank contains liquid, for example, that is to be pumped to a bottling machine of given capacity (flow rate specified). The length of pipe, the surface roughness, and kinematic viscosity would be known. The pressure drop allowable must be determined and the size of the pipe must be selected. With only four parameters known, additional criteria must be used to solve the problem.

It is reasonable to use cost figures in the above cited problem as a selection guide. The larger the pipe diameter, the greater the cost, which suggests that a small diameter should be selected. On the other hand, fluid flowing through a small diameter pipe undergoes a large friction loss and thus a larger pump is required. A larger pump means greater initial and operating costs. In general, there exists a diameter that minimizes the initial cost of the pump, the pipe, and the fittings. This diameter is called the optimum economic diameter, D_{opt}.

Here, we present what is traditionally known as the **least annual cost method** of economic pipe diameter selection and derive appropriate dimensionless groups. Results are applicable to gravity flow situations and can be used whether pumps are present or not. An equation for the optimum economic diameter D_{opt} is derived but solving it using the classical Moody Diagram (Figure 3.6) requires a trial and error procedure. To avert trial and error, three graphs are presented of the Darcy–Weisbach friction factor f vs $f^x Re^y$ with $(\varepsilon/D)/Re$ as an independent variable. The values used for x and y will become evident in later sections.

Analysis

The optimum economic diameter is the diameter that minimizes the *total* cost of a piping system. The total cost will consist of fixed plus operating costs. The fixed costs include those for the pipe, the fittings, the hangers or supports, the pump, and the installation. The fixed costs are a function of the size of the pipe. The operating costs include those associated with pumping power requirements which could be in the form of electricity or engine fuel. The power is that needed to overcome friction losses, changes in elevation, and changes in pressure, if any. We will formulate an equation for the initial and operating costs of the pipe, fittings, installation, and pump, and express the result in a cost per year basis. Next we differentiate the expression with respect to diameter to obtain the desired result. Exactly what costs to use can vary from one formulation to another but the method is still the same. To illustrate pipe cost figures, refer to Table 4.1 and Figure 4.1.

TABLE 4.1. *Installed pipeline costs for various sizes in 1980 dollars/ft. (Data taken from "Direct Determination of Optimum Economic Pipe Diameter for Non–Newtonian Fluids" by R. Darby and J. D. Melson, J of Pipelines, v. 2, (1982), pp. 11–21.)*

Nominal Diameter in Inches	ANSI Designation				
	300#	400#	600#	900#	1500#
4	7	7	8	11	14
6	10	10	11	14	19
8	13	13	16	20	28
10	17	18	21	28	40
12	22	23	28	36	53
14	26	27	32	39	58
16	30	31	38	48	74
18	35	36	45	59	90
20	39	40	55	69	102
22	43	45	65	80	129
24	47	49	74	95	147
26	54	61	81	112	163
30	66	74	97	135	
32	70	78	108	156	
34	79	87	122	162	
36	86	96	129		

Table 4.1 lists pipe costs for various grades. These are installed costs and are expressed in dollars per feet. Figure 4.1 is a graph of the Table 4.1 cost vs pipe size data. Figure 4.2 is of the same data plotted on log-log paper and the result is approximately linear. The equation of a graph such as that shown in Figure 4.2 is

$$C_p = C_1 D^n \qquad (4.1)$$

where C_p is the pipe cost in monetary units per length = MU/L ($/ft or $/m), C_1 is the cost of a reference size (MU/L^{n+1}) and n is the (dimensionless) exponent. Referring to Figure 4.2, we select C_1 to be the cost of 12-nominal pipe, and n is the slope of the curve. (The value to use for C_1 is customarily taken to be that of the 12-nominal size.) Equation 4.1 is thus a curve fit equation of actual cost data. The value of C_1 typically varies from $22/ft^{n+1} to about $55/ft^{n+1} while the exponent n varies from 1.0 to about 1.4.

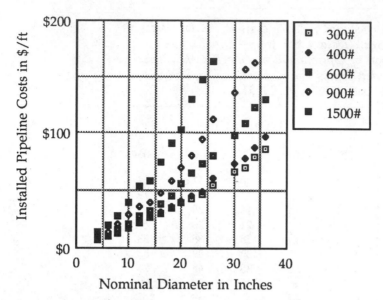

FIGURE 4.1. *Installed pipeline costs as a function of nominal pipe size (see Table 4.1 for data source).*

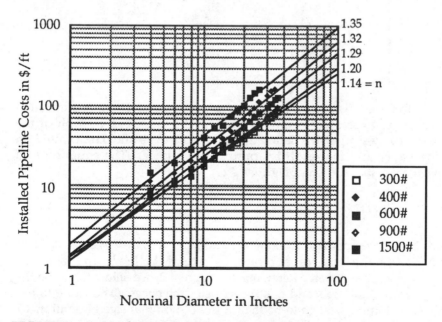

FIGURE 4.2. *Log-log graph of the data of Table 4.1.*

Now suppose we express the cost of fittings, valves, supports, pump(s), and installation as a multiplier F of the pipe costs. We get

$$C_f = FC_P = FC_1 D^n \tag{4.2}$$

where C_F is the cost of fittings, etc. in MU/L, and F is a multiplier that ranges typically from 6 to 7 (see Table 4.1 for reference source). The total cost (pipe, fittings, supports, installation) is the sum of Equations 4.1 and 4.2:

$$C_{PF} = C_P + C_F = (1 + F)C_1 D^n \tag{4.3}$$

where C_{PF} has dimensions of MU/L. The annualized amortization rate of the cost of the system is a fraction of the cost of the pipe and fittings. The fraction will be the reciprocal of the expected life in years of the system. Amortization refers to the gradual reduction of the balance in an account according to a specified schedule of amounts and schedule of time. Usually, amortization is the provision for extinguishing a debt, including the interest charge, by using a sinking fund or some similar form of payment. In this case we expect that the debt incurred by constructing the piping system will be "paid off" in, say, A years. The amortization rate then is $a = 1/(A$ years).

The annual maintenance cost is a fraction of the cost of the pipe and fittings, denoted as b. Thus the total amortized, installed cost of the piping system and its maintenance is

$$C_{PT} = (a + b)(1 + F)C_1 D^n \tag{4.4}$$

in which C_{PT} is the total annualized cost of the piping system in MU/(L·T) [\$/(ft·yr) or \$/(m·yr)]. Thus the total cost of the piping system (including a pump) is now expressed on a yearly basis.

The second factor in the total annual cost analysis is the cost of moving fluid through the pipe. This cost can reflect that associated with overcoming friction and/or changes in kinetic and potential energies. In the most general case, we elect to include all these factors in our model. The energy required per unit mass of fluid to pump the fluid through the pipeline is found with the energy equation written for a general system:

$$\frac{p_1 g_c}{\rho g} + \frac{V_1^2}{2g} + z_1 = \frac{p_2 g_c}{\rho g} + \frac{V_2^2}{2g} + z_2 + \Sigma \frac{fL}{D}\frac{V^2}{2g} + \Sigma K \frac{V^2}{2g} + \frac{g_c}{\dot{m}\,g}\frac{dW}{dt} \tag{4.5}$$

In terms of head $H = \left(\dfrac{p g_c}{\rho g} + \dfrac{V^2}{2g} + z\right)$, we have

$$H_1 = H_2 + \Sigma \frac{fL}{D}\frac{V^2}{2g} + \Sigma K \frac{V^2}{2g} + \frac{g_c}{\dot{m}\,g}\frac{dW}{dt}$$

Rearranging and solving for power, we obtain

$$\frac{d\,W}{d\,t} = -\dot{m}\left((H_2 - H_1)\frac{g}{g_c} + \Sigma\,\frac{fL}{D}\frac{V^2}{2g_c} + \Sigma\,K\,\frac{V^2}{2g_c}\right)$$

The above equation can be simplified in a number of ways. For this analysis, we assume that the minor losses are either negligible or that they can be expressed in terms of an equivalent length. In addition, we further assume that the entire pipeline consists of only one size of pipe. We therefore obtain

$$\frac{d\,W}{d\,t} = -\dot{m}\left((H_2 - H_1)\frac{g}{g_c} + \Sigma\,\frac{fL}{D}\frac{V^2}{2g_c}\right) \qquad (4.6)$$

It is convenient to rewrite the velocity in the above equation in terms of the mass flow rate using continuity:

$$V = \frac{Q}{A} = \frac{\dot{m}}{\rho A} = \frac{4\dot{m}}{\rho \pi D^2} \qquad (4.7)$$

Substituting Equation 4.7 into 4.6 and simplifying gives

$$-\frac{d\,W}{d\,t} = \dot{m}\left((H_2 - H_1)\frac{g}{g_c} + \frac{8fL\,\dot{m}^2}{\pi^2\rho^2 D^5 g_c}\right) \qquad (4.8)$$

The cost of operating the pump on a yearly basis is given by

$$C_{OP} = \frac{C_2\,t(-d\,W/d\,t)}{\eta} \qquad (4.9)$$

in which C_{OP} is the annualized cost in MU/T ($/yr), C_2 is the cost of energy in MU/(F·L) ($/(kW·hr)), t is the time during which the system operates per year (hrs/yr) and η is the efficiency of the pump (dimensionless).

   The initial cost of the pump varies with size. The larger the pump, the greater the cost. For pumping stations placed in remote locations and used to pump fluids over many miles, initial costs can vary to 2×10^6 for an installation of 4000 h.p. On the other hand, for a small installation of a 100 h.p. or less, the cost is a few thousand dollars. The initial pump cost can be accounted for in this analysis in one of two ways: separate term(s) where cost is expressed as a function of diameter; or include it in the pipe cost equation (4.4) as a part of F. As mentioned earlier, here we elect to use the latter method. The total annual cost associated with the piping system (initial + maintenance + operating + pumping) with an amortization rate of a is given by the sum of Equations 4.4 and 4.9:

$$C_T = LC_{PT} + C_{OP} = (a + b)(1 + F)C_1 D^n L + C_2 t(-dW/dt)/\eta$$

Substituting Equation 4.8 for dW/dt, the cost becomes

$$C_T = LC_{PT} + C_{OP}$$

$$= (a + b)(1 + F)C_1 D^n L + \frac{\dot{m} C_2 t}{\eta}(H_2 - H_1)\frac{g}{g_c} + \frac{8fL \dot{m}^3}{\pi^2 \rho^2 D^5 g_c} \frac{C_2 t}{\eta}$$

$$(4.10)$$

The optimum economic diameter is the one that minimizes the above equation for total cost. The minimum is found by differentiating Equation 4.10 with respect to diameter (holding all other variables constant) and setting the result equal to zero:

$$\frac{\partial C_T}{\partial D} = n(a + b)(1 + F)C_1 D^{(n-1)} L - 5\left(\frac{8fL \dot{m}^3}{\pi^2 \rho^2 D^6 g_c} \frac{C_2 t}{\eta}\right) = 0$$

Rearranging and solving for diameter gives

$$D^{n+5} = \frac{40 f \dot{m}^3 C_2 t}{n(a + b)(1 + F)C_1 \eta \pi^2 \rho^2 g_c} \tag{4.11}$$

or

$$D_{opt} = \left[\frac{40 f \dot{m}^3 C_2 t}{n(a + b)(1 + F)C_1 \eta \pi^2 \rho^2 g_c}\right]^{\frac{1}{n+5}} \tag{4.12}$$

where the parameters in the above equation are defined in Table 4.2 which also gives some typical values. Several features of importance are noticeable in Equation 4.12. First, the pipe length does not appear in the equation. Second, the viscosity of the fluid does not appear but the density does. Third, we have a problem in which diameter is unknown and so a trial and error solution will be required if the Moody Diagram (Figure 3.2) is used, because diameter is given in terms of friction factor f. Fourth, head loss (ΔH) does not appear in the equation. Finally, if there were no frictional effects, an optimum diameter would not be calculated. Equation 4.12 is dimensionally homogeneous.

TABLE 4.2. *Factors in the optimum economic diameter analysis.*

Symbol	Definition	Dimensions (Units)	Typical Values
D_{opt}	the optimum economic economic diameter	L (ft or m)	—
$\dot{m}$	mass flow rate	M/T (lbm/s or kg/s)	—
f	friction factor	—	—
C_2	cost of energy	MU/(F·L) [$/(kW·hr)]	$0.04/(kW·hr) or $0.04 s/(738 ft·lbf·hr)
t	time during which system operates per year	(hr/yr)	7880 hr/yr (10% downtime)
n	exponent of D in curve fit of pipe cost data	—	1.0 to 1.4
a	amortization rate	1/T (1/yr)	1/7 to 1/20
b	yearly maintenance cost fraction	1/T (1/yr)	0.01
F	multiplier of pipe cost representing cost of pump, fittings, installation	—	6 to 7
C_1	constant in curve fit of pipe cost data	MU/L^{n+1} ($/ft^{n+1} or $/m^{n+1})	$22/ft^{n+1} to $55/ft^{n+1}
η	efficiency of pump	—	0.6 to 0.9
ρ	density of liquid	M/L^3 (lbm/ft^3 or kg/m^3)	—

$$D_{opt} = \left[\frac{40\, f\, \dot{m}^3\, C_2\, t}{n(a + b)(1 + F)C_1 \eta \pi^2 \rho^2 g_c} \right]^{\frac{1}{n+5}}$$

$$Ro = \frac{\pi \varepsilon \mu g_c}{4\dot{m}}$$

$$(f(Re)^{n+5})^{1/6} = \left[\frac{128}{5\pi^3 g_c{}^4} \frac{\dot{m}^2}{\mu^5} \left(\frac{4\dot{m}}{\pi \mu g_c} \right)^n \left(\frac{n\,(a + b)(1 + F)C_1 \eta \rho^2}{C_2 t} \right) \right]^{1/6}$$

By appropriate manipulation of Equation 4.12, dimensionless groups can be derived to obtain a new correlation which can then be used as a graphical scaling parameter. The objective is to rid the right hand side of Equation 4.12 of the friction factor f. The reciprocal of Equation 4.12 is

$$\frac{1}{D_{opt}} = \left[\frac{n(a+b)(1+F)C_1\eta\pi^2\rho^2 g_c}{40f\dot{m}^3 C_2 t}\right]^{\frac{1}{n+5}}$$

Multiplying both sides by $4\dot{m}/\pi\mu g_c$ gives

$$\frac{4\dot{m}}{\pi\mu g_c D_{opt}} = \left[\frac{n(a+b)(1+F)C_1\eta\pi^2\rho^2 g_c}{40f\dot{m}^3 C_2 t}\frac{4^{n+5}\dot{m}^{n+5}}{\pi^{n+5}\mu^{n+5}g_c^{n+5}}\right]^{\frac{1}{n+5}}$$

or

$$\left(\frac{4\dot{m}}{\pi\mu g_c D_{opt}}\right)^{n+5} = \frac{256}{10\pi^3 g_c^4}\frac{\dot{m}^2}{\mu^5}\left(\frac{4\dot{m}}{\pi\mu g_c}\right)^n \left(\frac{n(a+b)(1+F)C_1\eta\rho^2}{f C_2 t}\right)$$

$$(4.13)$$

The term in parentheses in the left hand side is recognized as the Reynolds number. Multiplying both sides by the friction factor f and taking the sixth root rids the right hand side of friction factor f and gives

$$(f(\text{Re})^{n+5})^{1/6} = \left[\frac{128}{5\pi^3 g_c^4}\frac{\dot{m}^2}{\mu^5}\left(\frac{4\dot{m}}{\pi\mu g_c}\right)^n \left(\frac{n(a+b)(1+F)C_1\eta\rho^2}{C_2 t}\right)\right]^{1/6}$$

$$(4.14)$$

Because the friction factor f and the Reynolds number are dimensionless groups, their product is also dimensionless and can be used as a scaling parameter. Referring to a Moody diagram, we know that ε/D is a significant group, but before it can be evaluated, diameter must be known. This difficulty can be overcome by introducing another new group called the roughness number:

$$\text{Ro} = \frac{\varepsilon/D}{\text{Re}} = \frac{\varepsilon}{D}\left(\frac{\pi D\mu g_c}{4\dot{m}}\right)$$

or

$$\text{Ro} = \frac{\pi\varepsilon\mu g_c}{4\dot{m}}$$

$$(4.15)$$

For the optimum economic diameter problem, it is convenient to have a graph f vs $(f(\mathbf{Re})^{n+5})^{1/6}$ with $\mathbf{Ro}$ as an independent parameter. As indicated in Equation 4.1, n is the exponent of diameter in the pipe cost expression. The exponent n varies from 1.0 to 1.4. Three graphs for the results of this formulation have therefore been developed:

1. f vs $(f(\mathbf{Re})^6)^{1/6}$ with $\mathbf{Ro}$ as an independent variable (n = 1.0);
2. f vs $(f(\mathbf{Re})^{6.2})^{1/6}$ with $\mathbf{Ro}$ as an independent variable (n = 1.2); and,
3. f vs $(f(\mathbf{Re})^{6.4})^{1/6}$ with $\mathbf{Ro}$ as an independent variable (n=1.4).

The three graphs listed above were constructed by using two curve fit equations of the Moody Diagram: the Chen Equation and the Churchill Equation (presented in Chapter 3). Both equations solve for friction factor f in terms of Reynolds number $\mathbf{Re}$ and ε/D, and are valid in the transition and turbulent regimes. The Chen Equation is

$$\frac{1}{\sqrt{f}} = -2.0 \log\left\{\frac{\varepsilon}{3.7065D} - \frac{5.0452}{\mathbf{Re}} \log\left[\frac{1}{2.8257}\left(\frac{\varepsilon}{D}\right)^{1.1098} + \frac{5.8506}{\mathbf{Re}^{0.8981}}\right]\right\}$$

(4.16)

The Churchill Equation is

$$f = 8\left[\left(\frac{8}{\mathbf{Re}}\right)^{12} + \frac{1}{(B + C)^{1.5}}\right]^{1/12}$$

(4.17)

where

$$B = \left[2.457 ln \frac{1}{(7/\mathbf{Re})^{0.9} + (0.27\varepsilon/D)}\right]^{16}$$

and

$$C = \left(\frac{37\,530}{\mathbf{Re}}\right)^{16}$$

Reynolds number $\mathbf{Re}$ and ε/D values for the new graphs were selected and friction factor f was calculated with both equations (4.16 and 4.17). The two friction factors were then averaged. For each $\mathbf{Re}$ and ε/D, one value each of f, $(f(\mathbf{Re})^{n+5})^{1/6}$ and $\mathbf{Ro}$ (= $\varepsilon/D/\mathbf{Re}$) were calculated. Successive values of Reynolds number $\mathbf{Re}$ and ε/D were selected in harmony so that $\mathbf{Ro}$ remained constant. Graphs were then prepared.

Figures 4.3, 4.4, and 4.5 are the graphs prepared as a result of the analysis. The graphs are similar in appearance and for each, friction factor is plotted on the vertical axis. In the originally published Moody Diagram, the friction factor axis is logarithmic, but in Figures 4.3, 4.4, and 4.5, the friction factor axis is linear. Experience shows that this allows for a more accurate reading of friction factor f than if the axis were logarithmic. In all cases, the vertical axis varies from 0 to 0.08. The horizontal axis in Figure 4.3 ranges from 10^3 to 10^8 while in Figures 4.4 and 4.5, it ranges from 10^3 to 10^9. The Roughness number **Ro** in all graphs ranges from 0 (smooth wall) to $2.5/10^6$. All graphs span the transition and turbulent flow regimes. (See pages 117–119.)

EXAMPLE 4.1. Linseed oil is to be pumped from a tank to a bottling machine. The machine can fill and cap 30 two liter bottles in one minute. Determine the optimum size for the installation. Use the following parameters:

$$C_2 = \$0.05/(\text{kW·hr}) = (\$0.05 \text{ s})/(738 \text{ ft·lbf·hr})$$
$$C_1 = \$30/\text{ft}^{2.2}$$
$$t = 7000 \text{ hr/yr}$$
$$F = 6.75$$
$$n = 1.2$$
$$a = 1/(10 \text{ yr})$$
$$b = 0.01$$
$$\eta = 75\% = 0.75$$
PVC schedule 40 pipe

Solution: We now work toward calculating the optimum diameter using Equation 4.12 and the Moody Diagram (Figure 3.6). The volume and mass flow rates are:

$$Q = 30(2) \text{ l/min} = 60 \text{ l/min} = 1 \text{ l/s} = 3.53 \times 10^{-2} \text{ ft}^3/\text{s}$$

From Appendix Table B.1, we find for linseed oil

$$\rho = 0.93(62.4) \text{ lbm/ft}^3 \qquad \mu = 69 \times 10^{-5} \text{ lbf·s/ft}^2$$

The mass flow rate becomes

$$\dot{m} = \rho Q = 0.93(62.4)(3.53 \times 10^{-2}) = 2.05 \text{ lbm/s}$$

Substituting into Equation 4.12 gives

$$D_{opt} = \left[\frac{40 f \dot{m}^3 C_2 t}{n(a + b)(1 + F)C_1 \eta \pi^2 \rho^2 g_c} \right]^{\frac{1}{n+5}} \qquad (4.12)$$

$$D_{opt} = \left[\frac{40f(2.05)^3(0.05/738)(7000)}{1.2(1/10 + 0.01)(1 + 6.75)(30)(0.75)\pi^2 (0.93(62.4))^2(32.2)} \right]^{\frac{1}{6.2}}$$

or $\quad D_{opt} = 0.143f^{0.161}$ (i)

The Reynolds number of the flow is

$$Re = \frac{\rho VD}{\mu g_c} = \frac{4\rho Q}{\pi D \mu g_c}$$

Substituting,

$$Re = \frac{4(2.05)}{\pi D(69 \times 10^{-5}\,(32.2))} = \frac{117}{D}$$

For PVC pipe we use the "smooth" curve on the Moody Diagram (Figure 3.6). We now select values of diameter, calculate a Reynolds number, determine friction factor, and find a new value for diameter. Now because the numerator in the Reynolds number equation (e.g., 117) is less than 2 100, we begin by assuming laminar flow. For laminar flow of a Newtonian fluid in a circular duct, we have

$$f = \frac{64}{Re} = \frac{64D}{117} = 0.547D$$

Substituting into Equation i above, we get

$$D_{opt} = 0.143f^{0.161} = 0.143(0.547D_{opt})^{0.161} = 0.1298D^{0.161}$$

Solving yields

$$D_{opt} = 0.0874 \text{ ft} = 1.049 \text{ in.}$$

As a check on the laminar flow assumption, we calculate

$$Re = \frac{117}{D} = \frac{117}{0.0874} = 1338$$

$$f = \frac{64}{Re} = \frac{64}{1338} = 0.0478$$

and

$$D_{opt} = 0.143f^{0.161} = 0.143(0.0478) = 0.0876 \text{ ft}$$

Examination of Appendix Table D.1 shows that the correct size to use is

$$D_{opt} = 1\text{-nominal schedule 40 pipe} \qquad (D = 0.08742 \text{ ft})$$

EXAMPLE 4.2. A commercial steel pipeline is to be installed in a return line from a pump to the condenser of an air conditioner in which the rejected heat is used to preheat water and reduce energy consumption. Water is to be conveyed at a flow rate of 3.8 l/s. Determine the optimum economic pipe size for the installation given that:

$$C_2 = \$0.04/(kW \cdot hr) = \$0.04/(1\ 000\ W \cdot hr);$$
$$C_1 = \$400/m^{2.2}$$
$$t = 4\ 000\ hr/yr$$
$$F = 7.0$$
$$n = 1.2$$
$$a = 1/(7\ yr)$$
$$b = 0.01$$
$$\eta = 75\% = 0.75$$

Solution: In the last example, we worked with Equation 4.12 directly in order to find the optimum diameter D_{opt} The same procedure will be used here. The appropriate economic diameter selection graph (Figures 4.3, 4.4, and 4.5 on pages 117–119) will also be used in order to illustrate the method and to compare the results. We begin by substituting all known parameters into Equation 4.12 to formulate the trial and error procedure. The volume flow rate is

$$Q = 3.8 \, l/s = 0.003\ 8 \, m^3/s$$

From Appendix Table B.1, we read for water

$$\rho = 1\ 000 \, kg/m^3 \qquad\qquad \mu = 0.89 \times 10^{-3} \, N \cdot s/m^2$$

The mass flow rate is calculated as

$$\dot{m} = \rho Q = 1\ 000(0.003\ 8) = 3.8 \, kg/s$$

Equation 4.12 is

$$D_{opt} = \left[\frac{40 \, f \, \dot{m}^3 \, C_2 \, t}{n(a + b)(1 + F)C_1 \eta \pi^2 \, \rho^2 g_c} \right]^{\frac{1}{n+5}} \qquad (4.12)$$

Substituting gives

$$D_{opt} = \left[\frac{40 f(3.8)^3 (0.04/1\ 000)(4\ 000)}{1.2(1/7 + 0.01)(1 + 7.0)(400)(0.75)\pi^2 (1\ 000)^2} \right]^{\frac{1}{6.2}}$$

or $\qquad D_{opt} = 0.071\ 8 f^{0.161}$ $\qquad\qquad\qquad\qquad$ (i)

The Reynolds number of the flow is

$$Re = \frac{\rho VD}{\mu g_c} = \frac{4\rho Q}{\pi D \mu g_c} = \frac{4\dot{m}}{\pi D \mu g_c}$$

$$Re = \frac{4(3.8)}{\pi D(0.89 \times 10^{-3})} = 5.44 \times 10^3/D$$

From Table 3.1 for commercial steel,

$$\varepsilon = 0.004\ 6\ cm = 0.000\ 046\ m$$

We assume diameters selected from a table of pipe sizes (such as Appendix Table D.1) or diameters selected at random. Here we choose the latter. Assume

1st trial: $D = 4$ cm; then

$$\left. \begin{array}{l} Re = 5.44 \times 10^3/0.04 = 1.36 \times 10^5 \\[2mm] \dfrac{\varepsilon}{D} = \dfrac{0.004\ 6}{4} = 0.001\ 15 \end{array} \right\} \quad f = 0.023 \ \text{(Figure 3.2)}$$

Using Equation i above,

$$D_{opt} = 0.071\ 8(0.023)^{0.161} = 0.039\ 1\ m$$

2nd trial: $D = 0.039\ 1$ m; then

$$\left. \begin{array}{l} Re = 5.44 \times 10^3/0.039\ 1 = 1.39 \times 10^5 \\[2mm] \dfrac{\varepsilon}{D} = \dfrac{0.004\ 6}{3.91} = 0.001\ 18 \end{array} \right\} \quad f \approx 0.024 \ \text{(Figure 3.2)}$$

and $\quad D_{opt} = 0.071\ 8(0.024)^{0.161}$

or $\quad D_{opt} = 0.039\ 3\ m \quad$ (close enough)

In the following portion of this example, we will use the modified figures directly. Thus we now work towards calculating $(f(Re)^{n+5})^{1/6} = (f(Re)^{6.2})^{1/6}$ using Equation 4.14. Substituting into Equation 4.14 gives

$$(f(\mathrm{Re})^{n+5})^{1/6} = \left[\frac{128}{5\pi^3 g_c^4} \frac{\dot{m}^2}{\mu^5} \left(\frac{4\dot{m}}{\pi\mu g_c} \right)^n \left(\frac{n\,(a+b)(1+F)C_1\eta\rho^2}{C_2 t} \right) \right]^{1/6}$$

$$(4.14)$$

$$(f(\mathrm{Re})^{6.2})^{1/6} = \left[\frac{128}{5\pi^3(1)^4} \frac{3.8^2}{(0.89 \times 10^{-3})^5} \left(\frac{4(3.8)}{\pi(0.89 \times 10^{-3})(1)} \right)^{1.2} \cdot \right.$$

$$\left. \left(\frac{1.2(1/7 + 0.01)(1+7)(400)(0.75)(1\,000)^2}{(0.04/1\,000)(4\,000)} \right) \right]^{1/6}$$

Solving,

$$(f(\mathrm{Re})^{6.2})^{1/6} = 1.10 \times 10^5$$

Also, the roughness number is calculated to be

$$\mathrm{Ro} = \frac{\pi\varepsilon\mu g_c}{4\dot{m}} = \frac{\pi(0.000\,046)(0.89 \times 10^{-3})(1)}{4(3.8)} = 8.46/10^9$$

With these values,

$$\left. \begin{array}{c} (f(\mathrm{Re})^{6.2})^{1/6} = 1.10 \times 10^5 \\[2mm] \mathrm{Ro} = 8.46/10^9 \end{array} \right\} \qquad f \approx 0.022 \quad \text{(Figure 4.4)}$$

Therefore,

$$(f(\mathrm{Re})^{6.2})^{1/6} = (0.022(\mathrm{Re})^{6.2})^{1/6} = 1.10 \times 10^5$$

and the Reynolds number is calculated to be

$$\mathrm{Re} = \left(\frac{(1.10 \times 10^5)^6}{0.022} \right)^{1/62} = 1.40 \times 10^5$$

From the definition of Reynolds number,

$$\mathrm{Re} = \frac{4\dot{m}}{\pi D \mu g_c}$$

we write

$$D = \frac{4\dot{m}}{\pi \mathrm{Re}\mu g_c} = \frac{4(3.8)}{\pi(1.40 \times 10^5)(0.89 \times 10^{-3})(1)}$$

or $\qquad D = 3.9 \times 10^{-2}\,\mathrm{m} = 3.9\,\mathrm{cm}$

Using Equation 4.12 for D_{opt}, we obtained 3.93 cm. Any discrepancy in these two results is due to roundoff errors and to errors in reading the graphs. Use of the modified figures has now been illustrated.

EXAMPLE 4.3. Suppose that we plan to use the result above (3.93 cm) in order to select a pipe size for the application given in the problem statement. Suppose further that we are required to use a schedule 40 pipe. Determine the pipe size that should be used.

Solution: Referring to a table of pipe sizes (Appendix Table D.1), it is apparent that 3.93 cm does not appear explicitly as a schedule 40 pipe size. The above diameter falls between the diameters that correspond to the following:

$$1\text{-nominal schedule 40} \qquad D = 2.664 \text{ cm}$$
$$1^1/_2\text{-nominal schedule 40} \qquad D = 4.09 \text{ cm}$$

In order to make a decision on which of these yields minimum cost, it is necessary to actually calculate the costs using Equation 4.10 which is

$$C_T = LC_{PT} + C_{OP}$$

$$= (a + b)(1 + F)C_1D^nL + \frac{\dot{m} \, C_2 \, t}{\eta}(H_2 - H_1)\frac{g}{g_c} + \frac{8fL \, \dot{m}^3}{\pi^2\rho^2D^5g_c} \frac{C_2 t}{\eta}$$

$$(4.10)$$

At this point $H_2 - H_1$ is unknown, so cost calculated with the above equation will have to be done assuming that $H_2 - H_1 = 0$. Furthermore, length is unspecified, so cost per unit length will be evaluated for both pipe sizes. Equation 4.10 becomes

$$\frac{C_T}{L} = (a + b)(1 + F)C_1D^n + \frac{8f \, \dot{m}^3}{\pi^2\rho^2D^5g_c} \frac{C_2 t}{\eta}$$

For 1-nominal schedule 40 pipe,

$$\frac{C_T}{L} = (1/7 + 0.01)(1 + 7)(400)(0.026 \, 64)^{1.2} +$$

$$\frac{8(0.022)(3.8)^3}{\pi^2(1 \, 000)^2(0.026 \, 64)^5(1)} \frac{(0.04/1 \, 000)(4 \, 000)}{0.75}$$

or $\dfrac{C_T}{L} = \$21.86/(\text{yr·m of pipe})$ $\left(\begin{array}{c}\text{1-nominal schedule 40}\\ D = 2.664 \text{ cm}\end{array}\right)$

For $1^1/_2$-nominal schedule 40 pipe,

$$\frac{C_T}{L} = (1/7 + 0.01)(1 + 7)(400)(0.040\ 9)^{1.2} +$$

$$\frac{8(0.022)(3.8)^3}{\pi^2(1\ 000)^2(0.040\ 9)^5(1)}\quad \frac{(0.04/1\ 000)(4\ 000)}{0.75}$$

or $\dfrac{C_T}{L} = \$12.38/(\text{yr·m of pipe})$ $\left(\begin{array}{c}1^1/_2\text{-nominal schedule 40}\\ D = 4.09 \text{ cm}\end{array}\right)$

Comparing the two results leads to the conclusion that the least cost pipe is

$1^1/_2$-nominal schedule 40	$D = 4.09$ cm

    The above calculations were made assuming that friction factor is the same for both pipe sizes ($f = 0.022$ from the last example) conveying the same flow rate. Calculations made using 1-, $1^1/_2$-, and 2-nominal schedule 40 pipe yielded the following:

nominal size	ID	Re	ε/D	f
1-nom sch 40	0.026 64 m	8.66×10^4	0.001 72	0.024
$1^1/2$-nom sch 40	0.040 9 m	1.33×10^5	0.001 12	0.022
2-nom sch 40	0.052 52 m	1.04×10^5	0.000 876	0.022

Thus friction factor remains nearly constant over the range of interest. In fact, for most calculations of this type, friction factor varies only from 5 to 30% for all cases, even for non-Newtonian fluids.
    Figure 4.6 (page 120) is a graph of the cost data in this example, provided to illustrate the behavior of the cost equations. Diameter ranging from 0.02 to 0.1 m appears on the horizontal axis. Cost per length (per year) is on the vertical axis and three curves are shown. The pipe cost curve is calculated using the data of the above example with

$$C_{PT} = (a + b)(1 + F)C_1 D^n = (1/7 + 0.01)(1 + 7)(400)D^{1.2} = 489D^{1.2}$$

The operating cost curve is calculated (assuming $H_2 - H_1 = 0$ and $f = 0.022$) as

$$\frac{C_{OP}}{L} = \frac{8f\dot{m}^3}{\pi^2 \rho^2 D^5 g_c} \frac{C_2 t}{\eta} = \frac{8(0.022)(3.8)^3}{\pi^2(1\,000)^2 D^5(1)} \frac{(0.04/1\,000)(4\,000)}{0.75}$$

or $$\frac{C_{OP}}{L} = \frac{2.088 \times 10^{-7}}{D^5}$$

The total cost curve is determined by summing C_{PT} and C_{OP}/L at each diameter. As seen in the figure, the total cost is a minimum at $D \approx 0.04$ m. The behavior of these functions is now apparent.

The above equations and derivations for determining the optimum economic diameter have two obvious shortcomings: the problem of minor losses; and the difficulty in applying the results to noncircular cross sections. These shortcomings can be easily overcome in both cases with a slightly different approach to the problem. For a given set of parameters, the optimum economic diameter is first calculated for a straight run of a constant diameter circular pipe. When diameter is known, cross sectional area is calculated and divided into the flow rate. The result is what is known as the **optimum economic velocity**. The optimum economic velocity is then used as necessary to size the dimensions of a noncircular duct, and fittings can be added to the pipeline as appropriate.

4.2 Equivalent Length of Fittings

As seen in the last chapter, minor losses can consume significant amounts of energy (in the form of a pressure loss) when the length of pipe is relatively short. Also, it is remembered that the inclusion of minor losses in the Modified Bernoulli Equation can make the iterative (or trial and error) type problems exceedingly less popular than if minor losses could be ignored. Consequently efforts have been made to represent minor losses in a different way using the concept of what is called an **equivalent length**.

Consider for the moment the Modified Bernoulli Equation

$$\frac{p_1 g_c}{\rho g} + \frac{V_1^2}{2g} + z_1 = \frac{p_2 g_c}{\rho g} + \frac{V_2^2}{2g} + z_2 + \Sigma \frac{fL}{D_h} \frac{V^2}{2g} + \Sigma K \frac{V^2}{2g} \qquad (4.18)$$

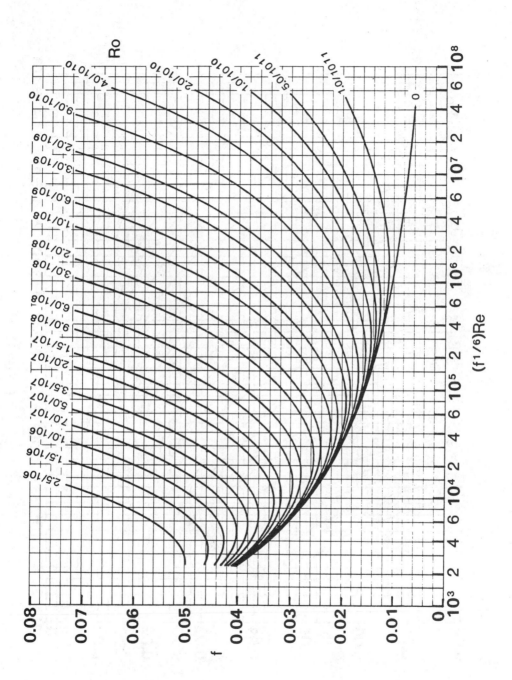

FIGURE 4.3. *Friction factor graph for n = 1.*

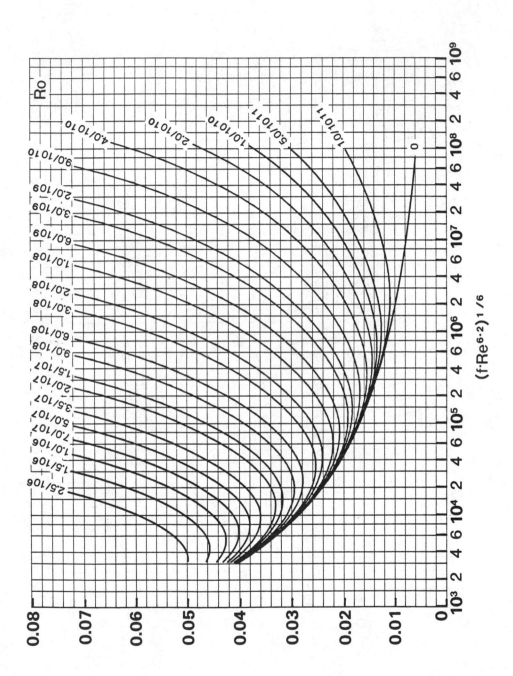

FIGURE 4.4. *Friction factor graph for n = 1.2.*

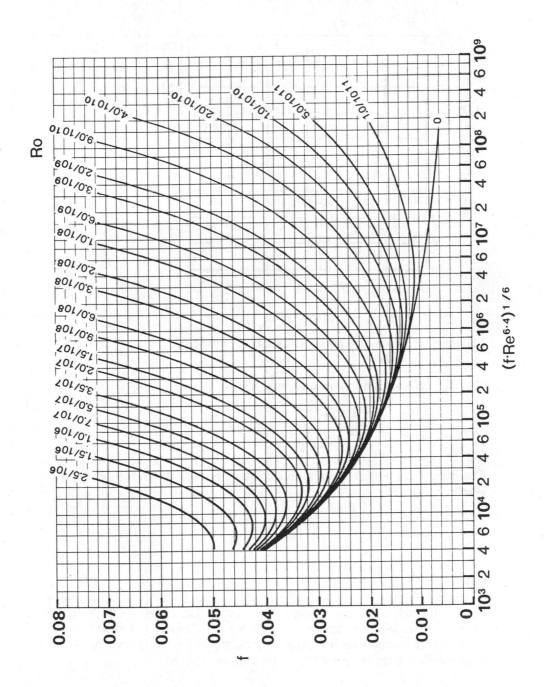

FIGURE 4.5. *Friction factor graph for n = 1.4.*

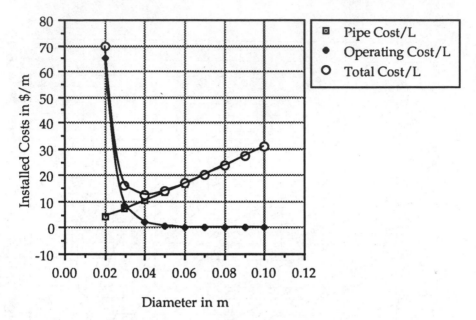

FIGURE 4.6. *Graph of the costs data of Example 4.3.*

The friction and minor loss terms are

$$\frac{fL}{D_h}\frac{V^2}{2g} + \Sigma K \frac{V^2}{2g} = \left(\frac{fL}{D_h} + \Sigma K\right)\frac{V^2}{2g}$$

The concept of equivalent length allows us to replace the minor loss term
with

$$\Sigma K = \frac{fL_{eq}}{D_h} \tag{4.19}$$

where f is the friction factor that applies to the entire pipe, D is the pipe
diameter (characteristic length), and L_{eq} is the equivalent length.
Physically, what we are doing is calculating the length of pipe (of the
original material, size, and schedule) we can "replace" the fitting with
to obtain the same pressure loss. For example, let us review the data of
Example 3.6. Data from that example are as follows:

Liquid Properties (Ethyl alcohol, Appendix Table B.1):

$$\rho = 0.787(62.4) \text{ lbm/ft}^3 \qquad\qquad \mu = 2.29 \times 10^{-5} \text{ lbf·s/ft}^2$$

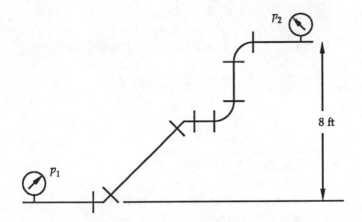

FIGURE 4.7. *The piping system of Example 3.6.*

Conduit Dimensions (12-nom sch 40, Appendix Table D.1):

$$D = 0.9948 \text{ ft} \qquad A = 0.773 \text{ ft}^2$$

Flow velocity $V = Q/A$:

$$Q = 750 \text{ gpm} = 1.67 \text{ ft}^3/\text{s} \qquad V = \frac{1.67}{0.773} = 2.16 \text{ ft/s}$$

Reynolds number $\mathbf{Re} = \rho VD/\mu g_c$:

$$\mathbf{Re} = \frac{0.787(62.4)(2.16)(0.9948)}{2.29 \times 10^{-5} \, (32.2)} = 1.43 \times 10^5$$

Relative roughness (commercial steel, Table 3.1):

$$\varepsilon = 0.00015 \text{ ft} \qquad \frac{\varepsilon}{D} = \frac{0.00015}{0.9948} = 0.00015$$

Friction factor (Figure 3.6):

$$\left.\begin{array}{l} \mathbf{Re} = 1.43 \times 10^5 \\[2ex] \dfrac{\varepsilon}{D} = 0.00015 \end{array}\right\} \quad f = 0.018$$

Modified Bernoulli Equation (4.18)

$$\frac{p_1 g_c}{\rho g} + \frac{V_1^2}{2g} + z_1 = \frac{p_2 g_c}{\rho g} + \frac{V_2^2}{2g} + z_2 + \Sigma \frac{fL}{D_h} \frac{V^2}{2g} + \Sigma K \frac{V^2}{2g} \qquad (4.18)$$

Property evaluation:

$$V_1 = V_2; \qquad z_1 = 0; \qquad z_2 = 8 \text{ ft}; \qquad L = 180 \text{ ft}$$

$$\Sigma K = 2K_{45° \, elbow} + 2K_{90° \, elbow} = 2(0.17) + 2(0.22) = 0.78$$

Equation of motion:

$$\frac{p_1 g_c}{\rho g} = \frac{p_2 g_c}{\rho g} + 8 + \left(\frac{0.018(180)}{0.9948} + 0.78 \right) \frac{(2.16)^2}{2(32.2)}$$

At this point, suppose we replace the minor loss term (0.78) and use Equation 4.19 to solve for equivalent length:

$$\Sigma K = \frac{fL_{eq}}{D_h} \tag{4.19}$$

or $\qquad 0.78 = \dfrac{0.018L_{eq}}{0.9948}$

where we use the friction factor and diameter for the pipe itself. Solving for equivalent length, we get

$$L_{eq} = 43.1 \text{ ft}$$

Thus if we replace the 2-45° elbows and the 2-90° elbows with a 43.1 ft length of 12-nominal schedule 40 commercial steel pipe (to make a straight run of length 180 + 43.1 = 223.1 ft), we would obtain the same pressure loss as would be obtained in the original configuration with the fittings. Some minor loss tables provide equivalent length data rather than K data. It would not be uncommon, for example, to find that a 90° elbow has an equivalent length of 0.14 diameters (= 0.14D) as its loss factor.

4.3 Graphical Symbols for Piping Systems

Due to the large number of fittings that are available for piping systems and the various ways of joining pipes to fittings, standards for representing piping systems have been developed. The American National Standards Institute (ANSI) provides charts (ANSI Z32.2.3) showing graphical symbols that are accepted standards in industry. Table 4.3 shows representations of some common fittings and of the usual attachment methods.

TABLE 4.3. *Graphical symbols for piping system.*
(Condensed from ANSI Z32.2.3.)

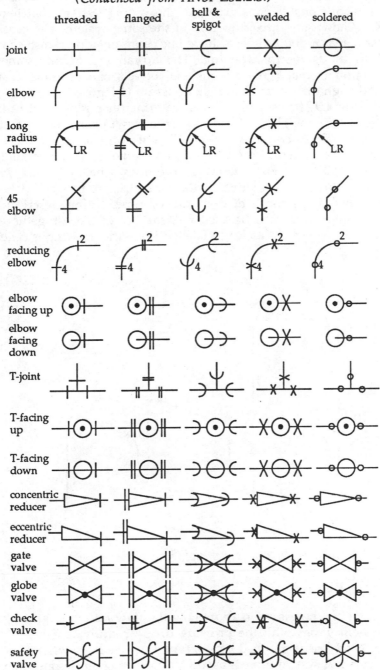

	threaded	flanged	bell & spigot	welded	soldered
joint					
elbow					
long radius elbow					
45 elbow					
reducing elbow					
elbow facing up					
elbow facing down					
T-joint					
T-facing up					
T-facing down					
concentric reducer					
eccentric reducer					
gate valve					
globe valve					
check valve					
safety valve					

Table 4.3 shows graphic symbols used in single line drawings of piping systems. There are established conventions for double line drawings also. Consider, for example, the piping system sketched in Figure 4.8. Shown are three drawings of the same system. The pipeline itself is drawn as a double line; a single line representation is also shown, as well as an isometric drawing. The advantage of an isometric representation is that three dimensional effects can be more clearly shown, although this advantage is not apparent in Figure 4.8.

Figure 4.9 shows a piping system which has pipes and fittings that are not all in one plane. Plan and profile views of the piping system are provided. These two views alone do not seem to be adequate in representing the system. Consequently, an isometric representation is provided in addition, and it definitely provides a better picture. Note the compass direction provided in the isometric. A properly labeled isometric (including length of each run of pipe, fitting descriptions, installation notes, etc.) can be used by workers to piece together the piping system. A detailed and well labeled isometric drawing is called a **spool drawing**.

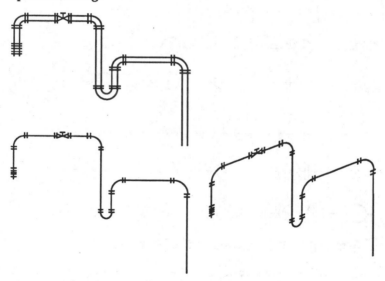

FIGURE 4.8. *Methods for drawing a piping system.*

4.4 System Behavior

It is often necessary to know how the flow rate through a given piping system varies with the pressure drop or equivalent head loss. When the head loss is graphed versus flow rate, we have what is called a **system curve**. A system curve is useful for predicting off-design behavior of the system or, in some instances, for sizing a pump.

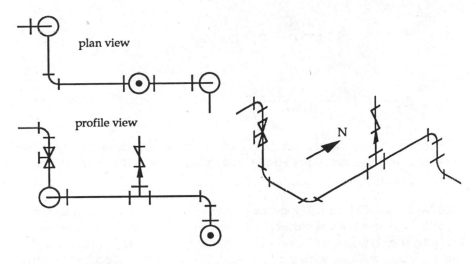

plan view

profile view

N

FIGURE 4.9. *Schematic representation of a piping system.*

To obtain the customary form for the system curve, we first refer to the Modified Bernoulli Equation (from Chapter 3):

$$\frac{p_1 g_c}{\rho g} + \frac{V_1^2}{2g} + z_1 = \frac{p_2 g_c}{\rho g} + \frac{V_2^2}{2g} + z_2 + \Sigma \frac{fL}{D_h}\frac{V^2}{2g} + \Sigma K \frac{V^2}{2g} \qquad (4.18)$$

We now define the *head* at any section to be

$$H = \frac{p g_c}{\rho g} + \frac{V^2}{2g} + z$$

where the dimension of H is L (ft or m). In terms of H, Equation 3.32 for a piping system of constant diameter becomes

$$H_1 - H_2 = \left(\Sigma \frac{fL}{D_h} + \Sigma K \right) \frac{V^2}{2g} \qquad (4.20)$$

The above equation contains velocity and our interest is in the volume flow rate. We therefore substitute for velocity from the continuity equation

$$V = \frac{Q}{A} = \frac{4Q}{\pi D^2}$$

to obtain

$$\Delta H = H_1 - H_2 = \left(\Sigma \frac{fL}{D} + \Sigma K\right)\frac{16Q^2}{2\pi^2 D^4 g}$$

or

$$\Delta H = Q^2 \left(\frac{8(\Sigma fL/D + \Sigma K)}{\pi^2 D^4 g}\right) \qquad\qquad (4.21)$$

The above equation is of a parabola (ΔH vs Q) and can be graphed for any system in which diameter is known or has been selected as a trial value.

EXAMPLE 4.4. Figure 4.10 shows a piping system that conveys water from a tank. The tank level is variable and so it is desired to have information on how the flow rate will vary through the system. It is proposed that the tank be replaced with a pump and, before such a decision is acted upon, a system curve must be drawn. Generate a system curve ΔH vs Q for the setup shown assuming the tank head can vary from 1 to 8 ft.

Solution: As indicated in the figure, the piping system is made up of 45 ft of 3-nominal schedule 40 PVC pipe. Also, the fittings are threaded. (With PVC, the fittings could be attached to the pipe with an adhesive and, in such a case, minor losses are the same as flanged fittings.) The control volume we select includes all the water in the tank and in the piping system.

Liquid Properties (Water, Appendix Table B.1):

$\rho = 62.4 \text{ lbm/ft}^3$ $\qquad\qquad\qquad\qquad \mu = 1.9 \times 10^{-5} \text{ lbf·s/ft}^2$

Pipe Dimensions (3-nominal schedule 40, Appendix Table D.1)

$ID = 0.2557 \text{ ft}$ $\qquad\qquad\qquad\qquad A = 0.05134 \text{ ft}^2$

Modified Bernoulli Equation (4.18)

$$\frac{p_1 g_c}{\rho g} + \frac{V_1^2}{2g} + z_1 = \frac{p_2 g_c}{\rho g} + \frac{V_2^2}{2g} + z_2 + \Sigma \frac{fL}{D_h}\frac{V^2}{2g} + \Sigma K \frac{V^2}{2g} \qquad (4.18)$$

or

$$H_1 = H_2 + Q^2 \left(\frac{8(\Sigma fL/D + \Sigma K)}{\pi^2 D^4 g}\right) \qquad\qquad (4.21)$$

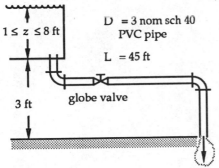

FIGURE 4.10. *The piping system of Example 4.4.*

Property Evaluation:

$$p_1 = p_2 = p_{atm} = 0; \quad V_1 = V_2 \approx 0 \text{ compared to the velocity in the pipe}$$

$$z_1 = 3 + z; \qquad 1 \le z \le 8 \text{ ft} \qquad z_2 = 0 \qquad\qquad L = 45 \text{ ft}$$

$$\Sigma K = \underset{\text{inlet}}{K_{reentrant}} + 2K_{90° \, elbow} + \underset{\text{valve}}{K_{globe}} + K_{exit}$$

$$\Sigma K = 1.0 + 2(1.4) + 10 + 1 = 14.8 \quad \text{(3-nom, threaded, regular)}$$

The head at section 1 is

$$H_1 = \frac{p_1 g_c}{\rho g} + \frac{V_1^2}{2g} + z_1 = 0 + 0 + 3 + z$$

At section 2,

$$H_2 = \frac{p_2 g_c}{\rho g} + \frac{V_2^2}{2g} + z_2 = 0$$

The change in head then becomes

$$\Delta H = H_1 - H_2 = 3 + z$$

Substituting into Equation 4.21 gives

$$3 + z = Q^2 \left(\frac{8(f(45)/(0.2557) + 14.8)}{\pi^2 (0.2557)^4 (32.2)} \right)$$

or

$$\Delta H = 3 + z = Q^2 [5.88(176f + 14.8)] \tag{i}$$

Flow velocity $V = Q/A$:

$$V = \frac{4Q}{\pi D^2}$$

Reynolds number $\mathbf{Re} = \rho VD/\mu g_c = 4\rho Q/\pi D \mu g_c$

$$\mathbf{Re} = \frac{4(62.4)Q}{\pi(0.2557)(1.9 \times 10^{-5})(32.2)} = 5.07 \times 10^5 Q \tag{ii}$$

We now select values of flow rate, then calculate Reynolds number and find the friction factor from Figure 3.6 using the "smooth" curve. Friction factor and flow rate are then substituted into Equation i above and ΔH is calculated. The procedure is repeated until the desired range of ΔH has been solved for. (It is customary in Engineering units to express flow rate in gallons per minute, gpm.) A summary of the calculations is given in Table 4.4. Note that when calculations were made, the range of Q was not known. So in the first column of Table 4.4, the assumed values of flow rate were selected in order to narrow down the range. It was sought to identify flow rates that yielded values for z that were within the required limits ($1 \le z \le 8$ ft). Graphs of the results are provided in Figure 4.11. The graph of ΔH vs Q (or z vs Q) is called a **system curve**.

TABLE 4.4. *Summary of calculations for Example 4.4.*

Q ft^3/s	$\mathbf{Re}$	f (Fig. 3.6)	ΔH, ft	z, ft	Q, gpm
0.01	5070	0.38	0.00962	<0	too low
0.1	50700	0.021	0.786	<0	too low
1	507000	0.013	70.3	67.4	too high
0.5	253000	0.0155	18.2	15.2	too high
0.3	152000	0.0165	6.66	3.66	135
0.4	203000	0.016	11.8	8.80	180
0.35	178000	0.016	9.00	6.00	157
0.25	127000	0.017	4.66	1.66	112
0.2	101000	0.018	3.02	0.02	too low
0.23	117000	0.018	3.99	0.99	103

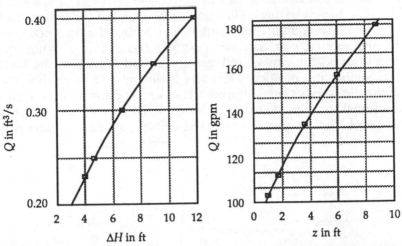

FIGURE 4.11. *System curve for the piping arrangement of Figure 4.10.*

4.5 Measurement of Flow Rate in Closed Conduits

Flow rate in a conduit can be measured by using a **rate meter**. A rate meter is a device that is inserted into a pipe (or tube), which allows for determining the volume rate of flow in the line. The meters we discuss in this section are the **turbine meter**, the **rotameter** or **variable area meter**, the **venturi meter**, and the **orifice meter**. Other meters, such as the **nozzle type**, the **elbow meter**, and the **totalizing meter**, operate on similar principles but will not be discussed here.

Turbine Meter

Figure 4.12 is a sketch of a turbine meter. It consists of a pipe or a tube with appropriate pipe fittings (not shown) at each end. Inside the tube are flow straighteners on both sides of a propeller or turbine. Flow entering the meter passes through the straighteners and causes the propeller to rotate at an angular velocity that is proportional to the flow rate. A magnetic pickup senses blade passages and transmits a signal to a readout device that totals the pulses. Turbine meters are usually made of stainless steel or brass, but special metals are available. Turbine meters are accurate typically to within ± 1%.

Rotameter or Variable Area Meter

Figure 4.13 is a schematic of a rotameter or a variable area meter. It consists of a tapered, graduated, vertically oriented, transparent tube (usually glass or plexiglas). Appropriate pipe fittings are at each end of the tube. Inside is a float that is free to move. The float can be spherical or cylindrical with axis vertical. Flow enters the meter at the bottom and raises the float to an equilibrium position. The higher the float position, the larger the annular area between the float and the tube. When the forces due to drag, buoyancy, and gravity are all balanced, the float reaches an equilibrium position. Flow rate is determined by reading the scale at the float position. Alternatively, an electronic sensor can transmit a signal to a remote readout device. Accuracy of a rotameter is usually within ± 1% on expensive units and within ± 5% on less expensive ones.

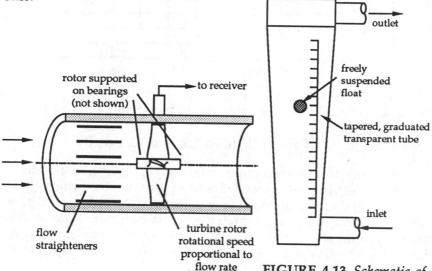

FIGURE 4.12. *Sketch of a turbine meter.*

FIGURE 4.13. *Schematic of a rotameter or variable area meter.*

Venturi Meter

Another type of rate meter is one that introduces a flow constriction, which in turn causes a change in one of the measurable properties of the flow. The measured property change is then related to the flow rate through the meter. A venturi meter is an example of this kind of meter, and is shown in two configurations in Figure 4.14.

The venturi meter consists of an upstream section, a convergent section leading to a throat, and a divergent section. The upstream section is the same diameter as the pipe line and attaches to it. At the upstream section and at the throat are static pressure taps which can be connected to two legs of a manometer. It is considered good practice to provide at least 10-diameter of approach piping upstream of the meter to ensure that the flow is fully developed and uniform at the meter entrance. The

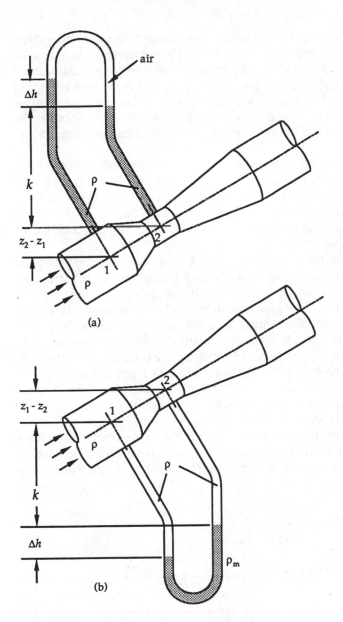

FIGURE 4.14. *A venturi meter shown in two configurations.*

size of the venturi meter is usually specified by a pipe and a throat size. For example, a 6 x 4 meter attaches to a 6-nominal pipeline and has a throat diameter corresponding to 4-nominal pipe.

When fluid flows through the meter, there is an increase in the velocity from the upstream section to the throat. There is a corresponding pressure drop also from upstream to the throat. The pressure drop increases with increasing flow rate through the meter. A graph of pressure drop versus flow rate for the meter is referred to as a **calibration curve**. Often, a calibration curve must be determined experimentally; that is, flow through the meter must be collected over a certain time interval to find volume flow rate. The corresponding pressure drop must also be measured to find flow rate Q as a function of pressure drop Δp or head loss Δh.

Standards for venturi meters (and other constriction meters) can be found in various publications. (*Fluid Meters—Their Theory and Application, 6th ed.*, published by ASME, 1971, is especially recommended.) Such publications show construction details and recommendations that allow the user to successfully select or design a meter for a specific application.

When consulting a meters handbook or another similar reference text, the information provided usually consists of meter construction details and a dimensionless graph of **discharge coefficient** versus Reynolds number. The discharge coefficient relates the actual flow rate through the meter to the theoretical flow rate predicted by the Bernoulli equation. Construction details contain dimensions which are expressed in terms of pipe diameters. For example, the upstream static pressure tap might be located one-half pipe diameter ($D_1/2$) from the edge of the convergent section. The meter length might be expressed as 10 pipe diameters ($10D_1$); and so on. Figure 4.15 shows some typical details on pressure tap locations for a venturi meter. When constructed according to the recommended dimensions, the applicable discharge coefficient versus Reynolds number relationship (graph or equation) can then be used to obtain a calibration curve for the meter, *without experimentally calibrating the meter.*

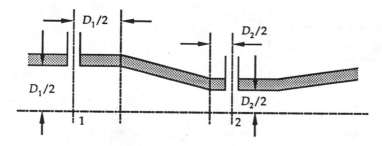

FIGURE 4.15. *Section view of a venturi meter showing tap locations.*

To investigate this point further, we derive equations for both configurations shown in Figure 4.14. For an incompressible fluid flowing through either configuration, the continuity equation is

$$Q = A_1 V_1 = A_2 V_2$$

Because the throat area A_2 is less than the upstream area A_1, continuity predicts that $V_1 < V_2$. Thus the fluid velocity must increase at the throat. For frictionless flow through the meter, the Bernoulli equation is

$$\frac{p_1 g_c}{\rho g} + \frac{V_1^2}{2g} + z_1 = \frac{p_2 g_c}{\rho g} + \frac{V_2^2}{2g} + z_2 \qquad (4.22)$$

Rearranging and substituting $V = Q/A$ gives

$$\frac{(p_1 - p_2) g_c}{\rho g} + z_1 - z_2 = \frac{Q^2}{2g}\left(\frac{1}{A_2^2} - \frac{1}{A_1^2}\right) = \frac{Q^2}{2g A_2^2}\left(1 - \frac{A_2^2}{A_1^2}\right)$$

Solving for flow rate, we get

$$Q = A_2 \sqrt{\frac{2g\{[(p_1 - p_2)g_c/\rho g] + (z_1 - z_2)\}}{(1 - A_2^2/A_1^2)}}$$

Noting that $A_2^2/A_1^2 = D_2^4/D_1^4$ and that this is the theoretical equation for the venturi meter, assuming frictionless, incompressible flow, we write:

$$Q_{th} = A_2 \sqrt{\frac{2g\{[(p_1 - p_2)g_c/\rho g] + (z_1 - z_2)\}}{(1 - D_2^4/D_1^4)}} \qquad (4.23)$$

For the configuration of Figure 4.14a, we have the following for the manometer:

$$p_1 + \frac{\rho g}{g_c}(k + \Delta h) = p_2 + \frac{\rho g}{g_c}[(z_1 - z_2) + k] + \frac{\rho_{air} g}{g_c}\Delta h$$

The density of air is small compared to the liquid density so that the term containing ρ_{air} can be neglected. Rearranging and simplifying gives

$$\frac{(p_1 - p_2) g_c}{\rho g} + z_1 - z_2 = \Delta h$$

Substituting into Equation 4.23 yields

$$Q_{th} = A_2 \sqrt{\frac{2g\Delta h}{(1 - D_2^4/D_1^4)}} \qquad \binom{\text{air over liquid}}{\text{manometer}} \qquad (4.24)$$

Thus the theoretical flow rate through the meter is related to the manometer reading such that the meter orientation is not important. The same equation results whether the meter is horizontal, inclined, or vertical.

For the two-fluid manometer of Figure 4.14b, we write

$$p_1 + \frac{\rho g}{g_c}[(z_1 - z_2) + k + \Delta h] = p_2 + \frac{\rho g}{g_c}k + \frac{\rho_m g}{g_c}\Delta h$$

Rearranging,

$$\frac{(p_1 - p_2)g_c}{\rho g} + z_1 - z_2 = \Delta h\frac{\rho_m - \rho}{\rho} = \Delta h\left(\frac{\rho_m}{\rho} - 1\right)$$

Substituting into Equation 4.23 gives

$$Q_{th} = A_2\sqrt{\frac{2g\Delta h(\rho_m/\rho - 1)}{(1 - D_2^4/D_1^4)}} \qquad \binom{\text{two-liquid}}{\text{manometer}} \qquad (4.25)$$

This equation differs from that of Equation 4.24 by the $(\rho_m/\rho - 1)$ term, which results from having a two-liquid manometer, as indicated in Figure 4.14b.

For a given meter, liquid, and manometer fluid, the variables D_1, D_2, A_2, ρ, and ρ_m are all known. Therefore a curve of theoretical flow rate Q_{th} versus head loss Δh using Equation 4.23 or 4.25 can be constructed. Consider the line labeled Q_{th} in Figure 4.16 as such a curve. Suppose next that this same meter is taken to the laboratory and actual measurements of Q_{ac} vs Δh are made (i.e., a calibration curve is determined). Typical data are plotted also in Figure 4.16 giving the line labeled as Q_{ac}.

Now for any pressure drop Δh_i, there are two corresponding flow rates: Q_{ac} and Q_{th}. The ratio of these flow rates is the venturi discharge coefficient C_v, defined as

$$C_v = \frac{Q_{ac}}{Q_{th}} \qquad (4.26)$$

The coefficient C_v can be calculated for many Δh values and will vary over the entire range. The flow rate Q_{ac} will always be less than Q_{th} because of frictional effects which are not accounted for in the Bernoulli equation. For each C_v calculated, we can also calculate a throat Reynolds number defined as

$$Re = \frac{V_2D_2}{v} = \frac{4\rho Q_{ac}}{\pi D_2\mu g_c} \qquad (4.27)$$

which is based on the actual flow rate Q_{ac} and on the throat diameter D_2. The discharge coefficient for a venturi meter with a rough cast inlet

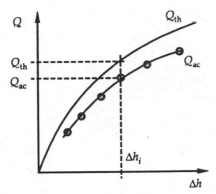

FIGURE 4.16. *Flow rate as a function of pressure drop for a venturi meter.*

varies from 0.95 to 0.984 for a Reynolds number that varies from 3×10^4 to 2×10^5. Beyond 2×10^5, the coefficient C_v is a constant and equal to 0.984. For a precisely machined meter, the discharge coefficient can be as high as 0.995.

The procedure for generating a calibration curve (Q_{ac} vs Δh) without taking data is simple but tedious:
(1) Obtain a recommended configuration for the meter from a handbook (or another appropriate source).
(2) Make physical measurements on the meter and determine liquid density as well as manometer fluid density.
(3) Construct a graph of Q_{th} vs Δh.
(4) Refer to discharge coefficient information provided with the meter.
(5) At various flow rates, calculate Δh, Re, and C_v.
(6) Use the C_v numbers and the Q_{th} to determine Q_{ac}.
(7) Graph Q_{ac} vs Δh.
The procedure for calibrating a venturi meter is illustrated in the following example.

Example 4.5. A pipeline company is responsible for pumping hexane from a manufacturer to a distributor. The company wishes to install a meter in the pipeline to monitor the flow rate. A fluid meters reference book was consulted and it has been decided to install a 6 x 4 venturi meter in the line. The discharge coefficient for the meter is a constant at 0.984 for a Reynolds number greater than 2×10^5, provided that pressure taps are located as indicated in Figure 4.15. The meter is to be installed in an upward sloping configuration that makes an angle of 20° with the horizontal, as shown in Figure 4.17. It is proposed to use a hexane over mercury manometer. Generate a calibration curve for the meter for a flow rate up to 0.15 m³/s.

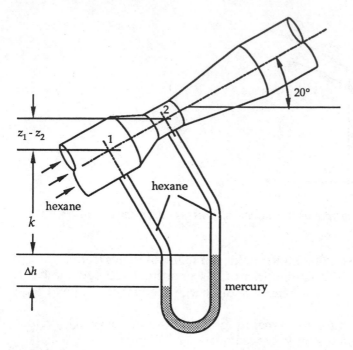

FIGURE 4.17. *A venturi meter conveying hexane.*

Solution: From Appendix Table B.1 for hexane and for mercury:

hexane $\rho = 0.657(1\ 000)\ kg/m^3$ $\mu = 0.297 \times 10^{-3}\ N \cdot s/m^2$
mercury $\rho_m = 13.6(1\ 000)\ kg/m^3$

From Appendix Table C.1, we read

6-nom sch 40 $D_1 = 15.41$ cm $A_1 = 186.50\ cm^2$
4-nom sch 40 $D_2 = 10.23$ cm $A_2 = 82.19\ cm^2$

For the manometer, we have

$$\frac{(p_1 - p_2)g_c}{\rho g} + z_1 - z_2 = \Delta h \left(\frac{\rho_m}{\rho} - 1\right) = \Delta h \left(\frac{13.6(1\ 000)}{0.657(1\ 000)} - 1\right) = 19.7\Delta h$$

Substituting into Equation 4.25 gives

$$Q_{th} = A_2 \sqrt{\frac{2g\Delta h(\rho_m/\rho - 1)}{(1 - D_2^4/D_1^4)}} \qquad \binom{\text{two liquid}}{\text{manometer}} \qquad (4.25)$$

$$Q_{th} = 82.19 \times 10^{-4} \sqrt{\frac{2(9.81)(19.7)\Delta h}{1 - \left(\frac{10.23}{15.41}\right)^4}}$$

or $Q_{th} = 0.18\sqrt{\Delta h}$

A tabulation of Δh and Q_{th} is provided in Table 4.5.
    The Reynolds number for the flowing hexane is

$$Re = \frac{V_2 D_2}{v} = \frac{4\rho Q_{ac}}{\pi D_2 \mu g_c} = \frac{4(0.657)(1\,000)C_v Q_{th}}{\pi(0.102\,3)(0.297 \times 10^{-3})}$$

or $\dfrac{Re}{C_v} = 2.75 \times 10^7\, Q_{th}$

Results of calculations made with this equation are provided also in Table 4.5.

    In this case, the discharge coefficient C_v is a constant for all flow rates in Table 4.5. If C_v was not a constant, we would need to resort to a trial and error procedure for each entry. The value of C_v (= 0.984) is listed in the table. The actual flow rate Q_{ac} (= $C_v Q_{th}$) is also listed. A graph of flow rate versus Δh is provided in Figure 4.18.

TABLE 4.5. *Results of calculations made for Example 4.5.*

Δh, (m)	Q_{th} (m³/s)	Re/C_v	C_v	$Q_{ac} = C_v Q_{th}$ (m³/s)
0	0	0	—	0
0.2	0.081	2.23×10^6	0.984	0.08
0.4	0.114	3.04×10^6	0.984	0.112
0.6	0.139	3.82×10^6	0.984	0.137
0.8	0.161	4.43×10^6	0.984	0.158

Orifice Meter

    Another type of constriction device is the orifice meter, shown schematically in Figure 4.19. The orifice meter consists of a flat plate with a hole that is inserted in a pipeline, conveniently between flanges. The hole can be drilled so that the orifice is either sharp edged or square edged. Flow through the plate follows a pattern similar to that in Figure 4.19. Downstream of the plate, the flow reaches a point of minimum area called a **vena contracta**.

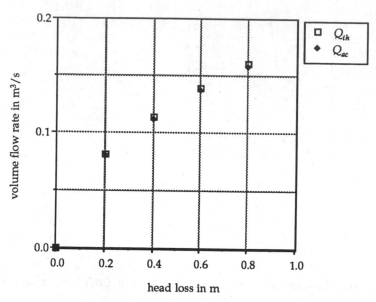

FIGURE 4.18. *Solution to Example 4.5 shown graphically.*

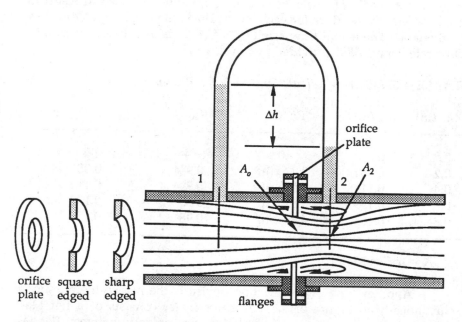

FIGURE 4.19. *Orifice plates and streamlines of flow through an orifice meter.*

Pressure taps are attached to the meter at various recommended locations, as shown in Figure 4.20: $1D$ and $\frac{1}{2}D$ locations; or $2\frac{1}{2}D$ and $8D$

locations; or flange taps where holes are drilled through the flanges; or corner taps where holes are drilled through the pipe wall leading to the very edge of the plate.

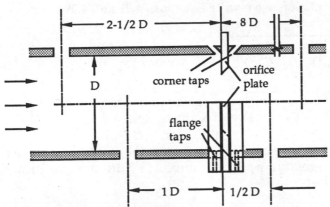

FIGURE 4.20. *Recommended locations for pressure taps for an orifice meter.*

Referring to Figure 4.19, we can identify three different areas associated with the orifice meter:

A_1 = upstream area corresponding to pipeline diameter
A_2 = flow area at the vena contracta
A_o = orifice hole area (calculated with the orifice diameter)

The area at section 2 where pressure p_2 is measured (the vena contracta), is unknown, but it can be expressed in terms of the orifice area:

$$A_2 = C_c A_o$$

where C_c is a *contraction coefficient.*

Applying Bernoulli's equation to the points where the manometer attaches (i.e., A_1 and A_2) yields the same result as for the venturi meter:

$$Q_{th} = C_c A_o \sqrt{\frac{2(p_1 - p_2)g_c}{\rho(1 - D_2^4/D_1^4)}} \qquad \left(\begin{array}{c}\text{air over liquid} \\ \text{manometer}\end{array}\right) \qquad (4.28)$$

where $C_c A_o$ has been substituted for A_2. The actual flow rate through the meter is considerably less than the theoretical flow rate. We define a discharge coefficient as

$$C = \frac{Q_{ac}}{Q_{th}}$$

and combine with Equation 4.28 to obtain

$$Q_{ac} = CC_cA_o \sqrt{\frac{2(p_1 - p_2)g_c}{\rho(1 - D_2{}^4/D_1{}^4)}} \tag{4.29}$$

To simplify this formulation, we rewrite Equations 4.28 and 4.29 to get

$$Q_{th} \approx A_o \sqrt{\frac{2(p_1 - p_2)g_c}{\rho(1 - D_2{}^4/D_1{}^4)}} \quad \left(\begin{matrix}\text{air over liquid}\\ \text{manometer}\end{matrix}\right) \tag{4.30}$$

$$Q_{ac} = C_oA_o \sqrt{\frac{2(p_1 - p_2)g_c}{\rho(1 - D_2{}^4/D_1{}^4)}} \quad \left(\begin{matrix}\text{air over liquid}\\ \text{manometer}\end{matrix}\right) \tag{4.31}$$

The area of significance in these equations is A_o, the orifice area, rather than A_2 where the pressure p_2 is measured. In addition, an orifice coefficient is defined as

$$C_o = CC_c = \frac{Q_{ac}}{Q_{th}}$$

Remember that the static pressure p_2 is not measured at the orifice area A_o, but this discrepancy and the losses encountered by the fluid are accounted for in the overall coefficient C_o. Tests on many meters have resulted in an equation (called the **Stolz equation**) for the loss coefficient:

$$C_o = 0.595\,9 + 0.031\,2\beta^{2.1} - 0.184\beta^8 + 0.002\,9\beta^{2.5}\left(\frac{10^6}{\text{Re}\,\beta}\right)^{0.75}$$

$$+ 0.09L_1\left(\frac{\beta^4}{1 - \beta^4}\right) - L_2\,(0.003\,37\beta^3) \tag{4.32}$$

where $\text{Re} = \dfrac{\rho V_o D_o}{\mu g_c} = \dfrac{4\rho Q_{ac}}{\pi D_o \mu g_c}$ $\qquad\qquad\qquad \beta = \dfrac{D_o}{D_1}$

$L_1 = 0$ $\qquad\qquad$ for corner taps
$L_1 = 1/D_1$ $\qquad\quad$ for flange taps
$L_1 = 1$ $\qquad\qquad$ for 1D & $\frac{1}{2}D$ taps

and if $L_1 \geq 0.433\,3$, the coefficient of the $\left(\dfrac{\beta^4}{1 - \beta^4}\right)$ term becomes 0.039.

$L_2 = 0$ $\qquad\qquad\qquad$ for corner taps
$L_2 = 1/D_1$ $\qquad\qquad$ for flange taps
$L_2 = 0.5 - E/D_1$ $\quad$ for 1D & $\frac{1}{2}D$ taps

E = orifice plate thickness (nominally 0.25 in. = 0.635 cm)

There are other equations, as well as graphs and tables, for the orifice coefficient C_o but Equation 4.32 is one of the simplest to use. Note that the Stolz equation is not recommended for $2\frac{1}{2}D$ and $8D$ taps.

As with the venturi meter, it is considered good practice to allow at least a 10-diameter approach section upstream of the meter.

EXAMPLE 4.6. An orifice meter is set up in a horizontal flow line with $1D$ and $\frac{1}{2}D$ taps. As water flows through the meter, pressure transducers attached to the taps read $p_1 = 17.7$ psia and $p_2 = 14.6$ psia. If the flow line is 2-nominal schedule 40, and the orifice hole diameter is 1.2 in., determine the actual flow rate through the meter.

Solution: From Appendix Table B.1 for water, we read

$$\rho = 1.94 \text{ slug/ft}^3 \qquad\qquad \mu = 1.9 \times 10^{-5} \text{ lbf·s/ft}^2$$

From Appendix Table D.1 for 2-nominal schedule 40 pipe

$$D_1 = 0.1723 \text{ ft} \qquad\qquad A_1 = 0.02330 \text{ ft}^2$$

Also, we were given $\qquad\qquad D_o = 1.2 \text{ in.} = 0.1 \text{ ft}$

The theoretical flow rate is found with

$$Q_{th} \approx A_o \sqrt{\frac{2(p_1 - p_2)g_c}{\rho(1 - D_2{}^4/D_1{}^4)}} \qquad\qquad (4.30)$$

Evaluating terms,

$$A_o = \frac{\pi D_o{}^2}{4} = \frac{\pi(0.1)^2}{4} = 0.00785 \text{ ft}^2$$

$$\frac{(p_1 - p_2)g_c}{\rho} = \frac{(17.7 - 14.6)(144)}{1.94} = 230$$

$$1 - \frac{D_o{}^4}{D_1{}^4} = 1 - \frac{(0.1)^4}{(0.1723)^4} = 0.8865$$

Substituting,

$$Q_{th} = 0.00785 \sqrt{\frac{2(230)}{0.8865}} = 0.179 \text{ ft}^3/\text{s}$$

In order to find the actual flow rate, we must first calculate the orifice coefficient C_o using the Stolz equation. The terms in that equation are evaluated as:

$$\beta = \frac{D_o}{D_1} = \frac{0.1}{0.1723} = 0.5804$$

$L_1 = 1 \geq 0.4333$ so that the coefficient of $\left(\dfrac{\beta^4}{1-\beta^4}\right)$ becomes 0.039

$E = 0.25$ in. $= 0.0208$ ft (no plate thickness given, so assume the nominal size of $\frac{1}{4}$ in.

$$L_2 = 0.5 - \frac{E}{D_1} = 0.5 - \frac{0.0208}{0.1723} = 0.3791$$

$$Re = \frac{V_o D_o}{v} = \frac{4\rho Q_{ac}}{\pi D_o \mu g_c} = \frac{4(1.94)(0.179)C_o}{\pi(0.1)(1.9 \times 10^{-5})} = 2.33 \times 10^5 C_o$$

In terms of Reynolds number, the Stolz equation becomes after substitution

$$C_o = 0.5959 + 0.0312(0.5804)^{2.1} - 0.184(0.5804)^8$$

$$+ 0.0029(0.5804)^{2.5}\left(\frac{10^6}{Re(0.5804)}\right)^{0.75}$$

$$+ 0.039\left(\frac{(0.5804)^4}{1-(0.5804)^4}\right) - 0.3791[0.00337(0.5804)^3]$$

which becomes

$$C_o = 0.5959 + 0.009954 - 0.002369 + \frac{15.65}{Re^{0.75}} + 0.004992 - 0.0002498$$

or $$C_o = 0.6082 + \frac{15.65}{Re^{0.75}}$$

The procedure for solving this problem begins by assuming a value for C_o, calculating the Reynolds number, and substituting into the Stolz equation. The calculations are repeated until convergence is achieved. Assume:

1st trial: $C_o = 0.6$; $Re = 2.33 \times 10^5 (0.6) = 1.40 \times 10^5$ and

$$C_o = 0.6082 + \frac{15.65}{(1.40 \times 10^5)^{0.75}} = 0.610$$

2nd trial: $C_o = 0.610$ $Re = 1.422 \times 10^5$ $C_o = 0.610$

The method converges rapidly. The actual flow rate then is

$$Q_{ac} = C_o Q_{th} = 0.610(0.179)$$

$$Q_{ac} = 0.109 \text{ ft}^3/s$$

Gases versus Liquids

When a compressible fluid (vapor or gas) flows through a meter, compressibility effects must be accounted for. This is done by introduction of a **compressibility factor** which can be determined analytically for some meters (venturi). For an orifice meter, on the other hand, the compressibility factor must be measured. For further information, the interested reader is referred to a fluid meters reference text.

Energy Cost

In all meter installations, there will be a permanent energy loss due to the presence of the meter, just like a minor loss associated with the presence of a fitting. The additional energy cost may influence what type of meter might be selected for a given application.

The energy lost due to the presence of a meter is found with

$$\left. \frac{d\,W}{d\,t} \right|_{lost} = \frac{\rho g Q h_m}{g_c} \qquad (4.33)$$

where h_m is the head loss due to the meter, which can be calculated with the appropriate equation found in Table 4.5.

Meter Selection Guide

Flow meter selection depends on a number of variables, so the final decision is a matter of judgment based on experience. To help in the decision-making process, consider the following when selecting a meter (information from *Flow Measurement Handbook*, by R. W. Miller, published by McGraw-Hill Book Co., 1983.):

• Fluid type—Liquid, gas, vapor, slurry, clean, dirty, corrosive.

• Limitations—Temperature, pressure, velocity.

• Installation conditions—Line size, Reynolds number, upstream approach length, vibration problems, steady or unsteady flow.

TABLE 4.6. *Loss equations for various meters. (Information from* Flow Measurement Handbook, *by R. W. Miller, published by McGraw-Hill Book Co., 1983.)*

Meter	Loss equation (liquids)
turbine meter	$h_m = 0.005\,77\,\dfrac{V^2}{g}$
venturi meter	$h_m = (0.436 - 0.86\beta + 0.59\beta^2)\Delta h$
orifice meter	$h_m = (1 - 0.24\beta - 0.52\beta^2 - 0.16\beta^3)\Delta h$

Notes:	Δh = meter reading	$\beta = \dfrac{\text{throat diameter}}{\text{upstream diameter}}$

• Performance—Required accuracy, flow rate range.

• Economics—Initial cost, operating cost, reliability, availability of parts.

• Relative initial cost—High cost (venturi); medium cost (turbine); low cost (elbow, variable area meter or rotameter, orifice meter).

EXAMPLE 4.7. The previous example dealt with an orifice meter having the following data:

$$\beta = 0.5804 \qquad p_1 = 17.7 \text{ psia} \qquad p_2 = 14.6 \text{ psia}$$
$$\rho = 1.94 \text{ slug/ft}^3 \qquad Q_{ac} = 0.109 \text{ ft}^3/\text{s}$$

If energy costs \$0.05/(kW·hr), determine the energy cost per year associated with this meter.

Solution: For the orifice meter, Table 4.6 shows the loss as

$$h_m = (1 - 0.24\beta - 0.52\beta^2 - 0.16\beta^3)\Delta h$$

The pressure drop is found with

$$\frac{(p_1 - p_2)g_c}{\rho g} = \frac{(17.7 - 14.6)(144)}{1.94(32.2)} = 7.15 \text{ ft}$$

Substituting gives

$$h_m = (1 - 0.24(0.5804) - 0.52(0.5804)^2 - 0.16(0.5804)^3)(7.15)$$

$$h_m = 0.65(7.15) = 4.65 \text{ ft of water}$$

The energy lost due to the presence of the meter is

$$\left.\frac{dW}{dt}\right|_{lost} = \frac{\rho g Q h_m}{g_c} = (1.94)(32.2)(0.109)(4.65)$$

$$\left.\frac{dW}{dt}\right|_{lost} = 31.7 \text{ ft·lbf/s} = 42.9 \text{ W}$$

The cost of energy is

$$\text{Energy cost} = (42.9 \text{ W}) \left(\frac{\$0.05}{1\,000 \text{ W·hr}}\right)(8760 \text{ hr/yr})$$

$$\text{Energy cost} = \$18.79/\text{yr}$$

4.6 Support Systems for Pipes

Unless pipes and tubes are located underground, they must be supported with a system that is practical and economical. Piping supports are usually called **pipe hangers**. The primary function of hangers is to provide support for the piping loads and movements (expansions or contractions), and to allow the building structure to safely accommodate them.

A number of companies manufacture components that can be pieced together to form a pipe support system. Figure 4.12 shows different methods of supporting pipes. When pipes are subject to expansion and contraction while in service, a hanger such as that shown in Figure 4.12a will be adequate. The pipe rests on a cylinder whose shape cradles the pipe in its center. The cylinder is attached with bar stock to an overhead support in a trapeze style configuration. In Figure 4.12b, the pipe has a thin and narrow piece of sheet metal wrapped around it which in turn is attached with a single piece of bar stock to an overhead support. Figure 4.12c shows another trapeze style installation in which clamps are

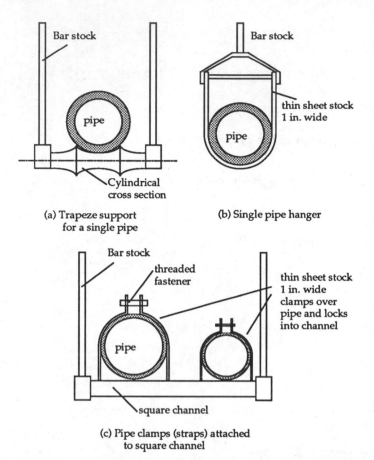

FIGURE 4.12. *Examples of pipe support hardware.*

bolted around the pipe. The bottom part of the clamps fit tightly into a U-shaped or square cross sectioned channel. Thus, in Figure 4.12c, pipes are rigidly clamped in position and are not as free to move as in Figure 4.12a. Usually, threaded bar stock is used in these hangers. Note that hangers like those in Figure 4.12a and 4.12b are made for an individual pipe while those in Figure 4.12c can be made part of a system that supports many pipes.

The question now arises as to how far apart should hangers be spaced for adequate support. For economic reasons, we wish to use larger spaces but we must consider safety of the installation. Based on many tests and calculations made for piping systems, the chart in Table 4.5 has been devised. As indicated, spaces between hangers is given as a function of nominal pipe size and type of fluid being conveyed. Remember that the hangers support the weight of the pipe *and* the fluid inside.

TABLE 4.5. *Maximum horizontal spacing between supports of pipes. (Data from Manufacturers' Standardization Society SP-69, Table 3, used with permission.)*

| Nominal Diameter | steel pipe | | | | copper tubing | | | |
| | water | | vapor | | water | | vapor | |
	ft	m	ft	m	ft	m	ft	m
$1/4$	7	2.1	8	2.4	5	1.5	5	1.5
$3/8$	7	2.1	8	2.4	5	1.5	6	1.8
$1/2$	7	2.1	8	2.4	5	1.5	6	1.8
$3/4$	7	2.1	9	2.7	5	1.5	7	2.1
1	7	2.1	9	2.7	6	1.8	8	2.4
$1^1/4$	7	2.1	9	2.7	7	2.1	9	2.7
$1^1/2$	9	2.7	12	3.7	8	2.4	10	3.0
2	10	3.0	13	4.0	8	2.4	11	3.4
$2^1/2$	11	3.4	14	4.3	9	2.7	13	4.0
3	12	3.7	15	4.6	10	3.0	14	4.3
$3^1/2$	13	4.0	16	4.9	11	3.4	15	4.6
4	14	4.3	17	5.2	12	3.7	16	4.9
5	16	4.9	19	5.8	13	4.0	18	5.5
6	17	5.2	21	6.4	14	4.3	20	6.1
8	19	5.8	24	7.3	16	4.9	23	7.0
10	22	6.1	26	7.9	18	5.5	25	7.6
12	23	7.0	30	9.1	19	5.8	28	8.5
14	25	7.6	32	9.8				
16	27	8.2	35	10.7				
18	28	8.5	37	11.3				
20	30	9.1	39	11.9				
24	32	9.8	42	12.8				
30	33	10.1	44	13.4				

Additional hangers required at concentrated loads between supports.

Plastic Pipe Fiberglass Reinforced Pipe Asbestos Cement Pipe	Follow Pipe Manufacturers' Recommendations for Spacing and Service Conditions
Glass Pipe	Follow Pipe Manufacturers' Recommendations. Use 8 ft (2.4 m) Max Spacing
Cast Iron Soil Pipe	One Hanger per Pipe Section; also at Joints; at Changes of Direction; and at Branch Connections. Use 12 ft (3.7 m) Max Spacing
Fire Protection	Follow Requirements of National Fire Protection Association

4.7 Summary

In this chapter, we have considered economics of piping systems and how costs can be minimized in the selection of a pipe size. We discussed the concept of equivalent length and examined graphical symbols used for drawing piping systems. We defined system behavior, examined flow measurement methods, and looked at commercial hardware for physically supporting a piping system.

4.8 Show and Tell

1. We have been using a nominal diameter and a schedule to specify pipe sizes. Table 4.1, however, cites what is known as an ANSI designation. To what does ANSI designation refer? Is there something similar for tubing? Give a report on these and any other alternative designations you encounter.

2. Conduct an interview with a piping engineer and report on how pipe sizing is done by a practicing engineer. Is economics important?

3. Table 4.3 refers to a "bell and spigot" pipe and gives a method for representing bell and spigot fittings. To what does bell and spigot refer? Where are such fittings used? Give a report using catalog references.

4. There exists a fitting called a "union" although it does not appear explicitly in Table 4.3. What is a union fitting and why is such a fitting necessary to have?

5. Obtain the appropriate catalog(s) and give a report on the variety of systems available for supporting pipes.

Obtain a flow meters handbook or reference text and give a brief presentation on the following meters, showing recommended construction details and the applicable loss coefficient. Include information for liquids, vapors, and gases.
6. Venturi meter.
7. Orifice meter.
8. Nozzle-type meter.
9. Elbow meter.

Give a presentation on how the following meters operate and various details about them; e.g., fluid type, materials of construction, expected energy losses, etc.
10. Nutating disk meter.
11. Target flow meter.
12. Magnetic flow meter.
13. Vortex meter.
14. Rotameter.

15. It was mentioned in the chapter that there is an ASME fluid meters handbook. Refer to various measurement textbooks and determine if other organizations publish recommendations on various flow meters. Give a brief presentation on your findings.

4.9 Problems Chapter 4

1. Using the information in Table 4.2, verify that Equation 4.10 is dimensionally consistent.

2. Beginning with Equation 4.10, derive Equation 4.12.

3. Verify that Equation 4.12 is dimensionally consistent.

4. Verify that the derivation of Equation 4.14 is correct, beginning with Equation 4.12.

5. Verify that Equation 4.14 is dimensionally correct.

6. Select 5 values on the Moody diagram (specifically, 5 Reynolds numbers and 5 ε/D's) and read the corresponding friction factors. Next substitute into Equation 4.16 and verify that it is an accurate curve fit equation.

7. Repeat Problem 6 for Equation 4.17.

8. Shown in the accompanying chart are data on the cost of PVC plastic pipe obtained from the classified section of a newspaper:

nominal diameter (schedule 40) in inches	cost per foot (clearance prices)
2	$0.141/ft
3	$0.338/ft
4	$0.390/ft
6	$0.99/ft
8	$1.99/ft

a) Construct a graph of the data on linear paper.
b) Construct a graph of the data on log-log paper.
c) Determine the parameters of the equation:

$$C_p = C_1 D^n$$

where C_p has dimensions of MU/L, C_1 has dimensions of MU/L^{n+1} evaluated numerically for 12 nominal pipe, and D has dimensions of L.

9. Repeat Problem 8 for the following data which are of PVC high pressure plastic pipe:

nominal diameter in inches	cost/meter in $/m
4	2.48
8	6.53
10	9.81
12	13.09

10. Referring to Example 4.1, suppose that electrical costs double so that $C_2 =$ $0.10/(kW·hr). Assuming that all other parameters remain the same, determine the new optimum pipe size.

11. Copper tubing (type M) is to be sized for an air conditioner that uses Freon-22 ($\rho = 1\,197 \text{ kg/m}^3$, $\mu = 0.849 \text{ lbm/ft·hr}$). The flow rate is 400 ml/s in the portion of the system that is to be selected. Determine the optimum economic diameter for the tubing when liquid is being conveyed, given the following:

$$C_2 = \$0.045/(\text{kW·hr}) \qquad t = 4500 \text{ hr/yr}$$
$$C_1 = \$42/\text{ft}^{2.2} \qquad F = 7.0$$
$$n = 1.2 \qquad a = 1/(8\,\text{yr})$$
$$b = 0.01 \qquad \eta = 62\%$$

12. Construct a graph of cost/length vs diameter, similar to that of Figure 4.6, for Problem 11.

13. Repeat Problem 11, but change the time to 2000 hr/yr.

14. Construct a graph of cost/length vs diameter, similar to that of Figure 4.6, for Problem 13.

15. Commercial steel pipe (schedule 40) conveys kerosene which ultimately is fed to a bank of diesel engines used to drive generators to produce electricity. The flow rate of kerosene is 30 gpm. Calculate the optimum economic diameter given the following conditions:

$$C_2 = \$0.01/(\text{kW·hr}) \qquad t = 90\% \text{ of } 365 \text{ days/yr}$$
$$C_1 = \$20/\text{ft}^{2.2} \qquad F = 6.5$$
$$n = 1.4 \qquad a = 1/(4\,\text{yr})$$
$$b = 0.01 \qquad \eta = 80\%$$

16. Construct a graph of cost/length vs diameter, similar to that of Figure 4.6, for Problem 15.

17. Determine the equivalent length of the globe valve in Problem 3.39. Data from that problem are as follows:

$f = 0.031$ $K = 10.0$ $L = 100$ ft $D = 0.1723$ ft

How does the equivalent length calculated for the globe valve compare to the length of the pipe itself?

18. Determine the equivalent length of each elbow in Problem 3.40. Data from that problem are as follows:

6-std	$f = 0.034$	$D = 0.1458$ m	$K = 0.31$
4-std	$f = 0.03$	$D' = 0.0908$ m	$K = 0.31$

Which size has the greater equivalent length in view of the fact that both minor loss coefficients are equal?

19. List all the fittings and attachment methods of the piping system in Figure P4.19. (The tubing material is 1-std type M copper water tube.)

20. List all the fittings and attachment methods of the piping system in Figure P4.20. (The conduit is 3-nom sch 40 wrought steel pipe.)

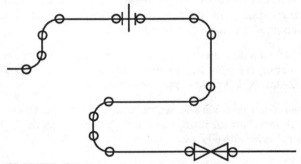

FIGURE P4.19

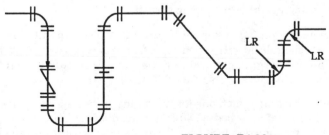

FIGURE P4.20

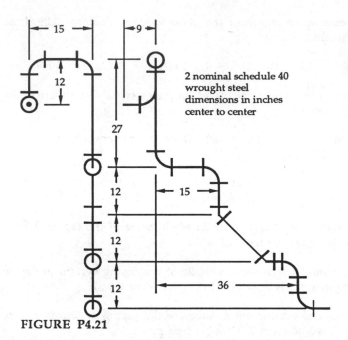

FIGURE P4.21

21. Figure P4.21 shows front and side views of a piping system. Draw an isometric view of the system and compose a list of all fittings and attachment methods. The pipe is made of 2-nominal schedule 40 wrought steel.

22. Figure P4.22 shows front and side views of a piping system. Draw an isometric view of the system and compose a list of all fittings and attachment methods. The pipe is made of 1/2-standard type M copper water tubing.

23. Figure P4.23 shows plan, front, and side views of a piping system. Draw an isometric view of the system. List all fittings and attachment methods if the pipe is made of 4-nominal schedule 80 PVC.

24. List all fittings and attachment methods for the piping system of Figure P4.24. The conduit is made up of $2^1/_2$-nominal schedule 40 galvanized pipe. Also, prepare a frontal view (as seen by someone facing east) and a profile view (as seen by someone facing north) of the piping system.

25. Water flows through the piping system of Figure P4.24. The conduit is made up of $2^1/_2$-nominal schedule 40 galvanized pipe. Construct a system curve for the water with flow rate in gpm graphed as a function of pressure drop from inlet to exit in psi. Let velocity vary from 4 to 10 ft/s.

26. List all fittings and attachment methods for the piping system of Figure P4.26 if it is made up of 3/4-nominal schedule 40 commercial steel. Also, prepare a frontal view (as seen by someone facing south) and a profile view (as seen by someone facing east) of the piping system.

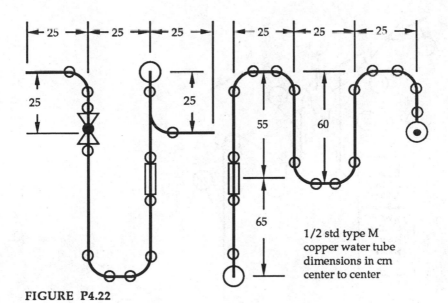

25
25
25
25
25
25
25
25
55
60
65
1/2 std type M
copper water tube
dimensions in cm
center to center

FIGURE P4.22

N
4
4 nominal schedule 80
PVC Pipe
dimensions in feet
center to center
8
angle valve
8
6
2
3
6
3
12

FIGURE P4.23

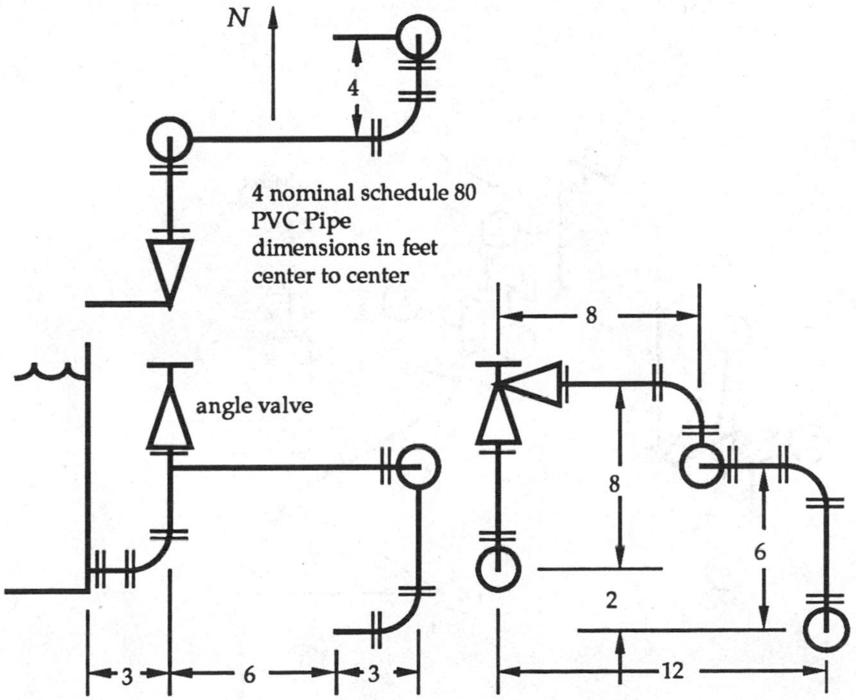

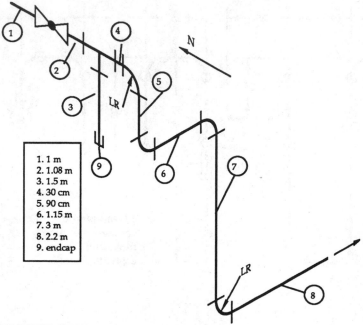

1. 1 m
2. 1.08 m
3. 1.5 m
4. 30 cm
5. 90 cm
6. 1.15 m
7. 3 m
8. 2.2 m
9. endcap

FIGURE P4.24

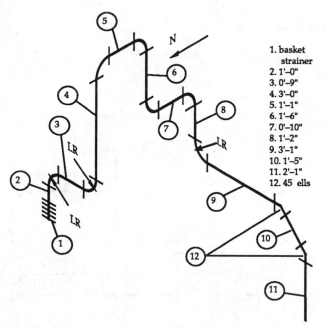

1. basket
   strainer
2. 1'–0"
3. 0'–9"
4. 3'–0"
5. 1'–1"
6. 1'–6"
7. 0'–10"
8. 1'–2"
9. 3'–1"
10. 1'–5"
11. 2'–1"
12. 45 ells

FIGURE P4.26

27. The piping system of Figure P4.26 is attached to a tank containing linseed oil. The free surface of the oil varies from 2 to 6 ft above the basket strainer. Generate a system curve of oil height above the basket strainer (6 ft $\leq h \leq 2$ ft) as a function of volume flow rate expressed in gpm. The pipe is made up of 3/4-nominal schedule 40 commercial steel. Assume that the linseed oil exits to atmospheric pressure.

28. List all the fittings and attachment methods of the piping system of Figure P4.28. The pipe material is 1-std type M drawn copper tubing. Also, draw a frontal view (as seen by someone facing west) and a profile view (as seen by someone facing south) of the piping system.

29. The piping system of Figure P4.28 conveys glycerine at a flow rate that varies from 0.3 to 1.2 1/s. Calculate the pressure drop from inlet to outlet and generate a system curve of volume flow rate in l/s vs pressure drop in kPa. The pipe material is 1-std type M drawn copper tubing.

30. Draw a system curve for the piping system of Figure P4.28. Let the flow rate be controlled by various settings of the ball valve and graph volume flow rate in liters per second as a function of the valve angle. The system is made up of 1-std type M copper tubing. The pressure drop in all cases is 400 kPa and the fluid is ethylene glycol.

31. A 12 x 10 (both schedule 20) venturi meter is placed in a horizontal flow line that conveys linseed oil. A mercury manometer is attached to the meter at locations shown in Figure 4.15. Determine the expected reading on the manometer for a volume flow rate of 1.5 m^3/s. Will a 1 m tall manometer work or should pressure gages be used instead?

32. Carbon tetrachloride is flowing in a line that contains a 10 x 8 (both schedule 40) venturi meter. The meter is inclined at an angle of 30° with the horizontal, with flow in the downhill direction. A mercury manometer attached to the meter reads 12.7 cm. For a discharge coefficient of 0.984, determine the flow rate through the meter.

33. A venturi meter (1 x $\frac{1}{2}$) is calibrated in the laboratory using water as the working fluid, and an air over water, inverted U-tube manometer. Data obtained are as follows:

Q_{ac} in ft^3/s	Δh in in. of H_2O	Q_{ac} in ft^3/s	Δh in in. of H_2O
2.00×10^{-3}	0.4	8.47×10^{-3}	2.7
4.01×10^{-3}	0.9	10.0×10^{-3}	3.8
5.57×10^{-3}	1.3	11.1×10^{-3}	4.6
6.46×10^{-3}	1.6	13.4×10^{-3}	5.8
7.80×10^{-3}	2.3		

a) Plot actual flow rate versus Δh.
b) Plot theoretical flow rate versus Δh on the same axes of part a.
c) Plot C_v (= Q_{ac}/Q_{th}) vs Re (= $4\rho Q_{ac}/\pi D_2 \mu g_c$) on semilog paper.

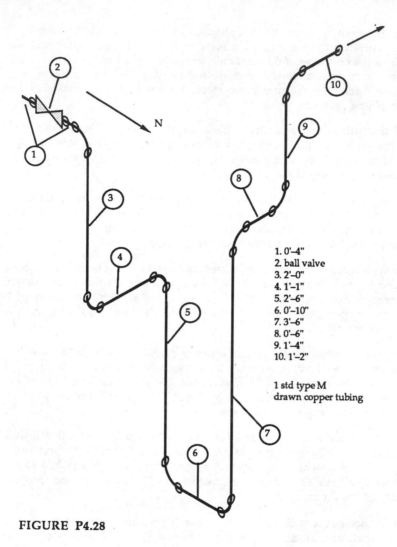

1. 0'–4"
2. ball valve
3. 2'–0"
4. 1'–1"
5. 2'–6"
6. 0'–10"
7. 3'–6"
8. 0'–6"
9. 1'–4"
10. 1'–2"

1 std type M
drawn copper tubing

FIGURE P4.28

34. An orifice meter (D_1 = 1.025 in. and D_o = 0.625 in.) is calibrated in the laboratory with water as the working fluid and with an air over water, inverted U-tube manometer. Data are as follows:

Q_{ac} in gpm	Δh in in. of H_2O	Q_{ac} in gpm	Δh in in. of H_2O
0.9	0.3	3.8	5.9
1.8	1.5	4.5	8.0
2.5	2.3	5.0	9.8
2.9	3.3	6.0	11.6
3.5	4.5		

a) Plot actual flow rate versus Δh.
b) On the same axes of part a, plot theoretical flow rate versus Δh.
c) Plot C_o (= Q_{ac}/Q_{th}) vs Re (= $4\rho Q_{ac}/\pi D_o \mu g_c$) on semilog paper.

35. A 10-nominal schedule 40 pipe contains an orifice meter with a hole diameter of 6.0 in. Heptane flows through the meter and the pressure drop measured with an air over heptane manometer is 6 ft. Determine the actual flow rate through the meter if flange taps are used.

36. Octane flows through a horizontal line containing an orifice meter with corner taps. The flow line is 6-nominal schedule 40 and the bore diameter of the orifice plate is 10.0 cm. For a flow rate of 0.03 m^3/s, determine the expected pressure drop in psi.

37. A horizontal water main is made of 12-nominal schedule 160 pipe and conveys water at 750 gpm. An orifice meter is placed in the line to measure the flow rate. The desired pressure drop for the installation is to be no more than 1.5 psi. What should the hole diameter be in the orifice plate to meet this condition? Use $1D$ and $\frac{1}{2}D$ taps.

38. Repeat Problem 37 using flange taps.

39. Repeat Problem 37 using corner taps.

40. Figure P4.40 shows a nozzle meter placed in a flow line, with pipe wall taps. One (of many) equation(s) for the discharge coefficient is

$$C_n = 0.194\ 36 + 0.152\ 884(ln\ Re) - 0.009\ 778\ 5(ln\ Re)^2 + 0.000\ 209\ 03(ln\ Re)^3$$

where Re = $4\rho Q_{ac}/\pi D_2 \mu g_c$, and D_2 = nozzle throat diameter. For the following flow parameters, determine the diameter D_2 required to meet the conditions:

       12-nominal schedule 160 pipe
       Water at 750 gpm
       Pressure drop is to be no more than 1.5 psi

41. On semilog paper, construct a graph of the orifice coefficient C_o vs Re for β = 0.2, 0.3, 0.4, 0.5, and 0.6 using the Stolz equation. Let the Reynolds number vary from 10^3 to 10^7, and assume that corner taps are used.

42. On semilog paper, construct a graph of the nozzle coefficient C_n vs Re using the following equation (same as in Problem 40):

$$C_n = 0.194\ 36 + 0.152\ 884(ln\ Re) - 0.009\ 778\ 5(ln\ Re)^2 + 0.000\ 209\ 03(ln\ Re)^3$$

Let the Reynolds number vary from 10^3 to 10^7.

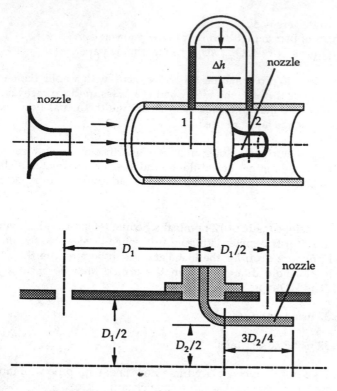

FIGURE P4.40

CHAPTER 5   Pumps and Piping Systems

Pumps are devices used to move fluid through a pipeline. There are many types of pumps designed for different applications. We will examine the types of pumps available and present guidelines useful in selecting a type of pump for a particular job. We will further discuss pump testing methods and focus exclusively on centrifugal pumps. We will show how pump test results are used to size a pump for a given piping configuration. We will cover the concept of cavitation and how it is avoided with design procedures that include calculation of net positive suction head. We conclude with a section on current design practices.

5.1 Types of Pumps

There are two general categories of pumps—dynamic and positive displacement. Dynamic pumps usually have a rotating component that imparts energy to the fluid in the form of a high velocity, high pressure, or high temperature. Positive displacement pumps have fixed volume chambers that take in and discharge the pumped fluid.

Dynamic pumps are usually classified according to the direction the fluid flows through them with respect to the axis of rotation. Fluid flows through an **axial flow pump** in a direction that is parallel to the axis of rotation of moving parts. Fluid passes through a **radial flow pump** in a direction normal to the axis of pump rotation. In a **mixed flow pump**, the fluid flow direction is neither purely axial nor purely radial but some combination of the two.

An axial flow pump (known also as a **propeller pump** or **turbine pump**) is used in low lift (short vertical pumping distance) applications. An electrically driven motor or an engine can be used to power these pumps. The motor or engine rotates a shaft onto which the impeller is attached. The rotating shaft is enclosed in a housing. The flow passage downstream of the impeller is bounded by this housing and by an outside casing. A mixed flow pump is one in which the flow at the impeller is not purely axial or purely radial. In a radial flow pump (known also as a **centrifugal pump**) flow passes through the casing where fluid enters and exits the rotating impeller in the radial direction.

159

In some pump designs, the discharge of one impeller immediately enters another. A multi-stage turbine pump operates in this way to pump water upward. The discharge from the first or lowest impeller-casing enters the second and so forth. The impeller-casings are bolted together and can consist of any number of desired stages. Also available for any of the pumps just mentioned is a number of different impeller designs including what are known as **semi open impellers** and **enclosed impellers**.

There are a number of designs of **positive displacement pumps** that have various uses. A **reciprocating pump**, for example, is made for pumping mud or cement. A reciprocating piston draws in fluid on an intake stroke and moves that fluid out on the discharge stroke. One way valves in the flow lines control the flow direction.

A **rotary gear pump** is another of the positive displacement type. It consists of two meshed gears that rotate within a housing. Fluid enters the region between the two gears and, as the gears rotate, fluid is drawn into the volumes between adjacent teeth and the housing. Fluid is discharged on the other side of the housing.

Details regarding the design of pumps are the responsibility of pump manufacturers. Our purpose here is to examine how such pumps are tested and sized for a given application.

5.2 Pump Testing Methods

In this section we examine a method for testing pumps and we will use a centrifugal pump for illustrative purposes. Figure 5.1 shows a pump and piping system. The pump contains an impeller within its housing. The impeller is attached to the shaft of the motor and the motor is mounted so that it is free to rotate, within limits. As the motor rotates and the impeller moves liquid through the pump, the motor housing tends to rotate in the opposite direction from that of the impeller. Weights are placed on the weight hanger so that at any rotational speed, the motor is kept at an equilibrium position. The amount of weight needed to balance the motor multiplied by the distance from the motor axis to the weight hanger gives the *torque* exerted by the motor.

The *rotational speed* of the motor is obtained with any number of devices available. The product of rotational speed and torque is the *input power* to the impeller from the motor.

Gages in the inlet and outlet lines about the pump give the corresponding *pressures* in gage pressure units. The gages are located at known *heights* from a reference plane. The flow meter gives a reading of the *volume flow rate* of liquid through the pump.

The valve in the outlet line is used to control the volume flow rate. As far as the pump is concerned, the resistance offered by the valve simulates a piping system with a controllable friction loss. Thus for any

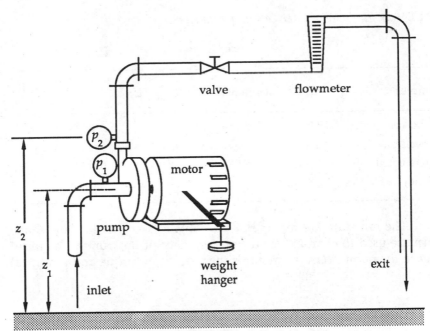

FIGURE 5.1. *Centrifugal pump testing setup.*

TABLE 5.1. *Pump testing parameters.*

Raw Data

Parameter	Symbol	Dimensions	Units	
torque	T	F·L	J = N·m	lbf·ft
rotational speed	ω	1/T	rad/s	rad/s
inlet pressure	p_1	F/L^2	kPa	psi (lbf/ft^2)
outlet pressure	p_2	F/L^2	kPa	psi (lbf/ft^2)
volume flow rate	Q	L^3/T	m^3/s	ft^3/s (gpm)

valve position in terms of % closed, the following data can be obtained: torque, rotational speed, inlet pressure, outlet pressure, volume flow rate. These parameters are summarized in Table 5.1.

The parameters used to characterize the pump are calculated with the raw data obtained from the test (listed above) and are as follows: input power to the pump, the total head difference as outlet minus inlet, the power imparted to the liquid, and the efficiency. These parameters are summarized in Table 5.2.

TABLE 5.2. *Pump characterization parameters.*

Reduced Data

Parameter	Symbol	Dimensions	Units	
input power	dW_a/dt	F·L/T	$W = J/s$	ft·lbf/s (hp)
total head diff	ΔH	L	m	ft
power to liquid	dW/dt	F·L/T	W	ft·lbf/s (hp)
efficiency	η	—	—	—

The raw data are manipulated to obtain the reduced data which in turn are used to characterize the performance of the pump. The input power to the pump from the motor is the product of torque and rotational speed:

$$-\frac{dW_a}{dt} = T\omega \qquad\qquad (5.1)$$

where the negative sign is added as a matter of convention. The total head at section 1, where the inlet pressure is measured (see Figure 5.1), is defined as

$$H_1 = \frac{p_1 g_c}{\rho g} + \frac{V_1^2}{2g} + z_1$$

where ρ is the liquid density and V_1 (= Q/A) is the velocity in the inlet line. Similarly, the total head at position 2 where the outlet pressure is measured is

$$H_2 = \frac{p_2 g_c}{\rho g} + \frac{V_2^2}{2g} + z_2$$

The *total head difference* is given by

$$\Delta H = H_2 - H_1 = \frac{p_2 g_c}{\rho g} + \frac{V_2^2}{2g} + z_2 - \left(\frac{p_1 g_c}{\rho g} + \frac{V_1^2}{2g} + z_1 \right) \qquad (5.2)$$

The dimension of the head H is L (ft or m). The power imparted to the liquid is calculated with the steady flow energy equation applied from section 1 to 2:

$$-\frac{dW}{dt} = \frac{\dot{m}g}{g_c}\left[\left(\frac{p_2 g_c}{\rho g} + \frac{V_2^2}{2g} + z_2\right) - \left(\frac{p_1 g_c}{\rho g} + \frac{V_1^2}{2g} + z_1\right)\right] \qquad (5.3)$$

In terms of total head H, we have

$$-\frac{dW}{dt} = \frac{\dot{m}g}{g_c}(H_2 - H_1) = \frac{\dot{m}g}{g_c}\Delta H \qquad (5.4)$$

The efficiency is determined with

$$\eta = \frac{dW/dt}{dW_a/dt} = \frac{\text{power imparted to liquid—Equation 5.3}}{\text{input power to impeller—Equation 5.1}} \qquad (5.5)$$

How raw data are manipulated to obtain the pump parameters is illustrated in the following example.

EXAMPLE 5.1. A pump is tested as in Figure 5.1 and, for one setting of the valve in the discharge line, the following data were read:

Torque	$= T = 0.5$ ft·lbf
Rotational speed	$= \omega = 1800$ rpm
Inlet pressure	$= p_1 = 3$ psig
Outlet pressure	$= p_2 = 20$ psig
Volume flow rate	$= Q = 6$ gpm
Height to inlet gage	$= z_1 = 2$ ft
Height to outlet gage	$= z_2 = 3$ ft
Inlet flow line	$= 2$-nominal schedule 40
Outlet flow line	$= 1^1/_2$-nominal schedule 40
Fluid	$=$ water

Calculate the pertinent pump parameters.

Solution: The power transmitted from the motor is

$$-\frac{dW_a}{dt} = T\omega \qquad (5.1)$$

$$-\frac{dW_a}{dt} = 0.5 \text{ ft·lbf}\left(1800\,\frac{\text{rev}}{\text{min}}\right)\left(2\pi\,\frac{\text{rad}}{\text{rev}}\right)\left(\frac{1\text{ min}}{60\text{ s}}\right)$$

in which it is noted that radians per second rather than revolutions per minute are the proper units to use for rotational speed. Solving,

$$-\frac{dW_a}{dt} = 94.2 \text{ ft·lbf/s} = \frac{94.2}{550} = 0.17 \text{ hp}$$

In order to calculate the change in head, we must first determine the liquid velocity in the inlet and outlet lines. For 2-nominal schedule 40 pipe, we read from Table D.1

$$ID_1 = 0.1723 \text{ ft} \qquad\qquad A = 0.02330 \text{ ft}^2$$

Also, for $1^1/_2$-nominal schedule 40,

$$ID_2 = 0.1342 \text{ ft} \qquad\qquad A = 0.01414 \text{ ft}^2$$

The volume flow rate was measured to be

$$Q = 6 \frac{\text{gal}}{\text{min}} = 13.4 \times 10^{-3} \text{ ft}^3/\text{s}$$

The inlet velocity is

$$V_1 = \frac{Q}{A_1} = \frac{13.4 \times 10^{-3}}{0.02330} = 0.573 \text{ ft/s}$$

The outlet velocity is

$$V_2 = \frac{Q}{A_2} = \frac{13.4 \times 10^{-3}}{0.01414} = 0.945 \text{ ft/s}$$

The density of water is taken here to be 1.94 slug/ft³, and so g_c is 1 in the equations. The total head at section 1 is

$$H_1 = \frac{p_1 g_c}{\rho g} + \frac{V_1^2}{2g} + z_1 = \frac{3(144)}{1.94(32.2)} + \frac{0.573^2}{2(32.2)} + 2$$

or

$$H_1 = 6.92 + 0.0051 + 2 = 8.92 \text{ ft}$$

At section 2,

$$H_2 = \frac{p_2 g_c}{\rho g} + \frac{V_2^2}{2g} + z_2 = \frac{20(144)}{1.94(32.2)} + \frac{0.945^2}{2(32.2)} + 3$$

or

$$H_2 = 46.1 + 0.0138 + 3 = 49.1 \text{ ft}$$

Gage pressures were used above. It makes no difference whether gage or absolute pressures are used, however, because our interest is in the head *difference*. The head difference would be the same for either absolute or gage pressures. The total head difference is given by

$$\Delta H = H_2 - H_1 = 49.1 - 8.92 = 40.2 \text{ ft}$$

The power imparted to the liquid (as evidenced by its change in pressure, velocity, and height) is calculated with

$$-\frac{dW}{dt} = \frac{\dot{m}\,g}{g_c}(H_2 - H_1) = \frac{\rho Q g}{g_c}(H_2 - H_1) \tag{5.4}$$

Substituting,

$$-\frac{dW}{dt} = 1.94(0.0134)(32.2)(40.2) = 33.6 \text{ ft·lbf/s}$$

The efficiency is determined with

$$\eta = \frac{dW/dt}{dW_a/dt} = \frac{33.6}{94.2}$$

or $\eta = 0.357 = 35.7\%$

The experimental technique used in obtaining data depends on the desired method of expressing performance characteristics. For example, data could be taken on only one impeller-casing-motor combination. One data point would first be taken at a certain valve setting and at a preselected rotational speed. The valve setting would then be changed and the speed control on the motor (not shown in Figure 5.1) is adjusted if necessary so that the rotational speed remains constant. The objective in all tests is to show how the pump operates over all possible variations of conditions.

Figure 5.2 illustrates a graph of data obtained from tests performed on a centrifugal pump. On the horizontal axis is the volume flow rate through the pump. On the vertical axis is the head difference ΔH defined as

$$\Delta H = H_2 - H_1$$

where

$$H = \frac{pg_c}{\rho g} + \frac{V^2}{2g} + z$$

and from Equation 5.3,

$$\Delta H = - \frac{g_c}{\dot{m}g} \frac{dW}{dt}$$

Figure 5.2 is known as a **performance map**, which in essence is a graph of power ($\propto \Delta H$) versus flow rate. The data are for a specific pump impeller-housing combination operated at four different rotational speeds. The head difference ΔH for each of the lines tends to decrease with increasing volume flow rate Q. It should be noted that the lines were drawn through discrete data points which customarily are not shown in the figure. Lines have also been drawn through data points of equal efficiency to yield **iso-efficiency** curves. The objective of the entire test is to locate the region of maximum efficiency for the pump which is now easily identified.

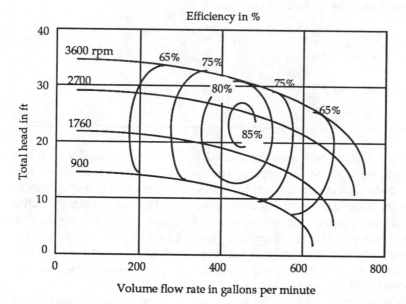

FIGURE 5.2. *Performance map of one impeller-casing-motor combination obtained at four different rotational speeds.*

Figure 5.3 represents an alternative method of obtaining data and expressing results. One pump casing-motor combination is used with four different impeller sizes. All data are obtained at only one rotational speed, however. Data for one impeller would be taken at a certain valve setting. The valve setting is then changed and motor control is adjusted if necessary so that the rotational speed is maintained constant (in this case, 1760 rpm). Again the objective is to show how the pump operates over all possible variations. The tendency for the head difference ΔH to decrease with increasing volume flow rate Q is apparent. Iso-efficiency lines have been drawn through discrete data points in order to locate the region of maximum efficiency.

A manufacturer might produce over a dozen different housings. For each housing, 4 or 5 different impellers could be used. Each combination would require testing and the production of a performance map. The reasoning is that if we were to use the pump represented in Figure 5.2 (or any pump) we would prefer to operate it in the maximum efficiency region. So with regard to all the pumps one company might produce, all that the manufacturer would need to supply to potential users is a summary or composite display of all the maximum efficiency regions.

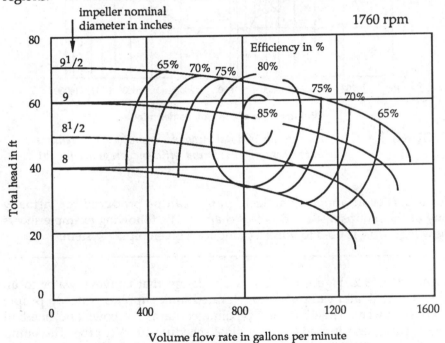

FIGURE 5.3. *Performance map of one motor-casing combination and four different impellers obtained at one rotational speed.*

Figure 5.4 is a graph of ΔH versus Q for a number of pumps showing only the maximum efficiency region for each. Within each region is a number that corresponds to the pump whose maximum efficiency region is represented. Figures like 5.4 are used to select a pump

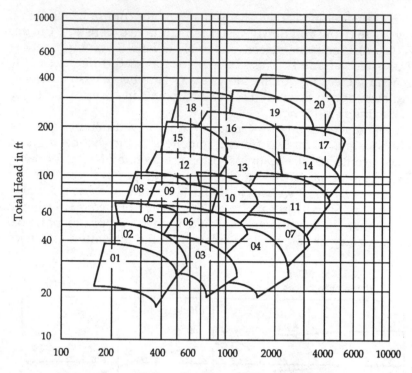

Volume flow rate in gallons per minute

FIGURE 5.4. *Composite graph of total head difference versus volume flow rate showing maximum efficiency regions for 20 pumps.*

for a particular application. Such graphs can be produced for virtually any of the pumps discussed in this chapter. The following example shows how Figure 5.4 is used to select a pump for a given piping system.

EXAMPLE 5.2. Figure 5.5 shows a pipeline that conveys water to an elevated tank at a campsite. The elevated tank supplies water to people taking showers. The 40 ft long pipeline contains 3 elbows and one ball check valve, and is made of 6-nominal schedule 40 PVC pipe. The pump must deliver 250 gpm. Use Figure 5.4 to select a pump for the system.

Solution: We will need to calculate the total head difference and enter Figure 5.4 at the given flow rate and calculated ΔH.

Liquid Properties (Appendix Table B.1, water)

$$\rho = 62.4 \text{ lbm/ft}^3 \qquad\qquad \mu = 1.9 \times 10^{-5} \text{ lbf·s/ft}^2$$

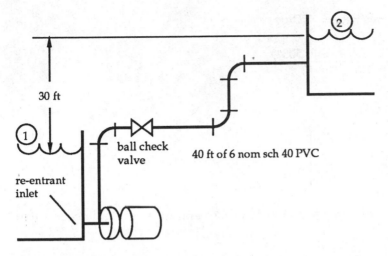

FIGURE 5.5. *The piping system of Example 5.2.*

Conduit dimensions (Appendix Table D.1, 6-nominal schedule 40)

$$ID = 0.5054 \text{ ft} \qquad\qquad A = 0.2006 \text{ ft}^2$$

For PVC, we use the ε/D = "smooth" curve on the Moody diagram. Referring to Figure 5.5, we note that the pump must lift water 30 ft and overcome minor losses and friction in a 40 ft length of pipe. The steady flow energy equation including friction for the piping system of Figure 5.5 is

$$\frac{p_1 g_c}{\rho g} + \frac{V_1^2}{2g} + z_1 = \frac{p_2 g_c}{\rho g} + \frac{V_2^2}{2g} + z_2 + \Sigma \frac{fL}{D_h}\frac{V^2}{2g} + \Sigma K \frac{V^2}{2g} + \frac{g_c}{\dot{m}g}\frac{dW}{dt}$$

$$(5.6)$$

in which we take section 1 to be the free surface in the sump tank and section 2 to be at the free surface of the elevated tank. The method involves solving the above equation for power (dW/dt) and then finding ΔH with Equation 5.4.

Property Evaluation

$$p_1 = p_2 = p_{atm} = 0 \qquad V_1 = V_2 \qquad z_1 = 0 \qquad z_2 = 30 \text{ ft}$$

$$Q = 250 \text{ gal/min} = 0.555 \text{ ft}^3/\text{s} \qquad V = \frac{Q}{A} = \frac{0.555}{0.2006} = 2.76 \text{ ft/s}$$

Reynolds number $Re = \rho VD/\mu g_c$

$$Re = \frac{62.4(2.76)(0.5054)}{1.9 \times 10^{-5} (32.2)}$$

and so

$$\left.\begin{array}{c} Re = 1.43 \times 10^5 \\[2ex] \dfrac{\varepsilon}{D} = \text{"smooth"} \end{array}\right\} \qquad f = 0.0165 \quad \text{(Figure 3.2)}$$

Minor Losses (Table 3.3, 3–90° elbows, re-entrant inlet, ball check valve, exit)

$$\Sigma K = 3K_{90°\,elbow} + \underset{inlet}{K_{re\text{-}entrant}} + \underset{valve}{K_{ball\,check}} + K_{exit}$$

$$\Sigma K = 3(0.31) + 1.0 + 70 + 1.0 = 72.9$$

Equation 5.6 reduces to

$$0 = z_2 + \left(\Sigma \frac{fL}{D_h} + \Sigma K\right)\frac{V^2}{2g} + \frac{g_c}{\dot{m}g}\frac{dW}{dt}$$

Substituting gives

$$0 = 30 + \left(\frac{0.0165(40)}{0.5054} + 72.9\right)\frac{2.76^2}{2(32.2)} + \frac{g_c}{\dot{m}g}\frac{dW}{dt}$$

Solving, we get

$$-\frac{g_c}{\dot{m}g}\frac{dW}{dt} = \Delta H = 38.8 \text{ ft}$$

Although the following step is unnecessary, we can also find the power:

$$-\frac{dW}{dt} = \rho Q\frac{g}{g_c}(38.8) = 62.4(0.555)(32.2/32.2)(38.8)$$

$$-\frac{dW}{dt} = 1344 \text{ ft·lbf/s} = 1344/550$$

$$-\frac{dW}{dt} = 2.44 \text{ hp} \approx 2.5 \text{ hp}$$

Thus the total head difference found above includes lifting the water 30 ft and overcoming friction and minor losses. At 38.8 ft and 250 gpm, Figure 5.4 shows that the pump corresponding to the region labeled "01" would be suitable for this application.

5.3 Cavitation and Net Positive Suction Head

The suction line of a centrifugal pump contains liquid at a pressure that is lower than atmospheric pressure. If this suction pressure is sufficiently low, the liquid will begin to boil at the local temperature. For example, water boils at 33°C (92°F) if its pressure is lowered to 5.1 kPa (0.75 psia). Boiling itself involves vapor bubble formation and this phenomenon when it occurs in a pump is called **cavitation**.

In a cavitating centrifugal pump, vapor bubbles usually form at the eye of the impeller and as they move radially through the impeller with the liquid, the bubbles encounter a high pressure region. It is here where they collapse and send pressure waves outward. The pressure waves have an erosive effect on the impeller and housing, known as **cavitation erosion**. As indicated in the above discussion, when cavitation occurs, the impeller is no longer moving an all-liquid fluid through the housing. As a result, the efficiency of the pump falls drastically. If the situation is not corrected, the pump may eventually fail due to metal erosion and fatigue of shaft bearings and/or seals.

Cavitation is not a problem that should be corrected after installation of the system. Instead, the inception of cavitation is predictable and the engineer should ensure that cavitation will not occur when the system is designed. Pump manufacturers perform tests on pumps and provide information useful for predicting when cavitation will occur.

Net Positive Suction Head

Consider the centrifugal pump and inlet configurations of Figure 5.6. Illustrated in both cases is a centrifugal pump moving liquid from a tank. Two configurations are shown: suction lift when the liquid level in the tank lies below the pump impeller centerline; and suction head in which the liquid level in the tank is above the pump impeller centerline.

We will now apply the Modified Bernoulli equation from section 1 (free surface of the tank liquid) to section 2 (inlet to the pump housing)

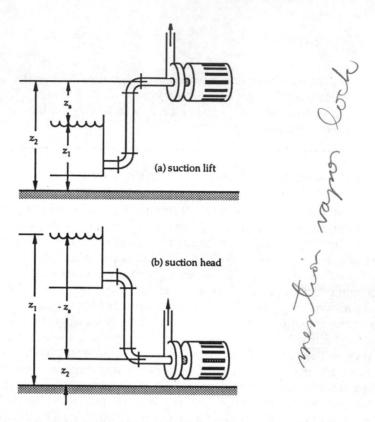

FIGURE 5.6. *Illustration of suction lift and suction head at pump inlet.*

for the case when the pump in Figure 5.6a is operating. Our objective is to determine the pressure at the pump inlet and compare it to the vapor pressure of the liquid. If the pressure at the pump inlet is less than the vapor pressure of the liquid at the local temperature, then the liquid will boil at the impeller; thus the pump will cavitate. The Modified Bernoulli Equation is

$$\frac{p_1 g_c}{\rho g} + \frac{V_1^2}{2g} + z_1 = \frac{p_2 g_c}{\rho g} + \frac{V_2^2}{2g} + z_2 + \Sigma \frac{fL}{D_h} \frac{V^2}{2g} + \Sigma K \frac{V^2}{2g} \qquad (5.7)$$

Although p_1 is atmospheric, we will not set it equal to zero. This will allow our final equation to account for an overpressure on the liquid surface in the tank. Evaluating properties yields

$$V_1 = 0 \qquad z_2 - z_1 = +z_s \qquad V_2 = V = \text{velocity in the pipe}$$

Rearranging Equation 5.7 and solving for $p_2 g_c / \rho g$, we obtain

$$\frac{p_2 g_c}{\rho g} = \frac{p_1 g_c}{\rho g} - z_s - \left(\Sigma \frac{fL}{D_h} + \Sigma K + 1 \right) \frac{V^2}{2g}$$

Next we subtract the vapor pressure from both sides of the above equation, and rearrange slightly to obtain

$$\frac{p_2 g_c}{\rho g} - \frac{p_v g_c}{\rho g} = \frac{p_1 g_c}{\rho g} - z_s - \left(\Sigma \frac{fL}{D_h} + \Sigma K + 1 \right) \frac{V^2}{2g} - \frac{p_v g_c}{\rho g} \qquad (5.8a)$$

or

$$NPSH_a = \frac{p_2 g_c}{\rho g} - \frac{p_v g_c}{\rho g} = \frac{p_1 g_c}{\rho g} - z_s - \left(\Sigma \frac{fL}{D_h} + \Sigma K + 1 \right) \frac{V^2}{2g} - \frac{p_v g_c}{\rho g}$$

$$\binom{\text{Figure 5.6a}}{\text{suction lift}}$$

The left hand side of the above equation is known as the **net positive suction head available**, $NPSH_a$. Equation 5.8a applies to the inlet configuration of Figure 5.6a. For Figure 5.6b, we would apply Equation 5.7 as before to obtain:

$$\frac{p_2 g_c}{\rho g} - \frac{p_v g_c}{\rho g} = \frac{p_1 g_c}{\rho g} + z_s - \left(\Sigma \frac{fL}{D_h} + \Sigma K + 1 \right) \frac{V^2}{2g} - \frac{p_v g_c}{\rho g} \qquad (5.8b)$$

or

$$NPSH_a = \frac{p_2 g_c}{\rho g} - \frac{p_v g_c}{\rho g} = \frac{p_1 g_c}{\rho g} + z_s - \left(\Sigma \frac{fL}{D_h} + \Sigma K + 1 \right) \frac{V^2}{2g} - \frac{p_v g_c}{\rho g}$$

$$\binom{\text{Figure 5.6b}}{\text{suction head}}$$

The dimension of the above equation is L (ft or m of liquid), and so each term can be represented as a head of liquid. Equation 5.8a in some texts is written as

$$h_{p_2} - h_{vp} = NPSH_a = h_{p_1} - h_{z_2} - h_f - h_{vp} \qquad (5.9)$$

Manufacturers perform tests on pumps and report values of **net positive suction head required**, $NPSH_r$. Cavitation is prevented when the available net positive suction head is greater than the required net positive suction head, or

$$NPSH_a > NPSH_r \qquad \text{(Cavitation prevention)} \qquad (5.10)$$

Furthermore, the net positive suction head available $NPSH_a$ should exceed the net positive suction head required $NPSH_r$ by a reasonable margin of safety (3 ft or 1 m are typical values). Data on net positive suction head for a particular pump are often shown on the pump performance map or are provided separately.

In order to make calculations on net positive suction head, it is necessary to have data on vapor pressure. Figure 5.7 is a graph of vapor pressure versus temperature for various liquids.

EXAMPLE 5.3. A certain pump delivers 900 gpm of water from a tank at a head difference ΔH of 8 ft. The net positive suction head required is 10 ft. Determine where the pump inlet should be with respect to the level of water in the tank. The water surface is exposed to atmospheric pressure. Neglect frictional effects and take the water temperature to be 90°F.

Solution: Here we apply Equation 5.8a assuming that we have a suction lift as in Figure 5.6a:

$$NPSH_a = \frac{p_1 g_c}{\rho g} - z_s - \left(\Sigma \frac{fL}{D_h} + \Sigma K + 1 \right) \frac{V^2}{2g} - \frac{p_v g_c}{\rho g} \quad \left(\begin{array}{c} \text{Figure 5.6a} \\ \text{suction lift} \end{array} \right)$$

With $NPSH_r = 10$ ft, we impose a 3 ft safety margin to get

$$NPSH_a = \frac{p_2 g_c}{\rho g} - \frac{p_v g_c}{\rho g} = 13 \text{ ft}$$

This means that the pressure of the liquid at the casing inlet at the impeller should exceed the vapor pressure of the liquid by an amount equal to or greater than 13 ft in order to prevent cavitation. Proceeding with the calculations:

Liquid Properties

$\rho = 62.4 \text{ lbm/ft}^3$ \qquad (Appendix Table B.1 for water)

$p_v = 0.55 \text{ lbf/in}^2$ \qquad (Figure 5.7 @ 90°F, water)

Property Evaluation

$p_1 = p_{atm} = 14.7 \text{ lbf/in}^2$ \qquad Frictional effects $= 0$

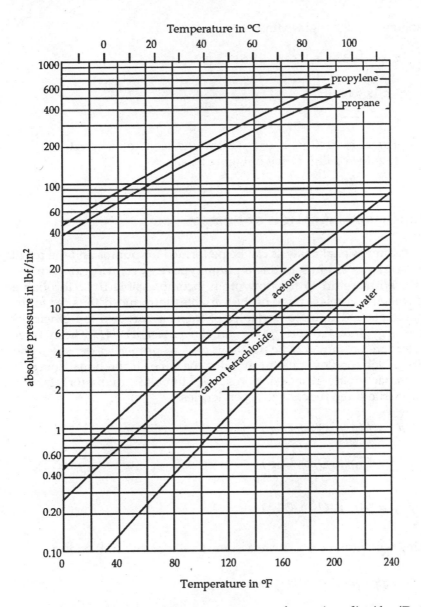

FIGURE 5.7. *Vapor pressure vs temperature for various liquids. (Data from several sources.)*

Substituting into Equation 5.8a gives

$$13 < \frac{14.7(144)(32.2)}{62.4(32.2)} - z_s - \frac{0.55(144)(32.2)}{62.4(32.2)}$$

Rearranging and solving, we find

$$- z_s > 13 - 33.9 + 1.27 = - 19.6$$

or $\quad z_s < 19.6$ ft

Thus to prevent cavitation, z_s $(= z_2 - z_1)$ would have to be equal to or less than 19.6 ft in the configuration of Figure 5.6a.

5.4 Dimensional Analysis of Pumps

A dimensional analysis can be performed for pumps and the result can be used as an aid in selecting a pump type (e.g., centrifugal vs mixed vs axial). With regard to the flow of an incompressible fluid through a pump, we wish to relate three of the variables introduced thus far to the flow parameters. The three variables of interest here are the efficiency η, the energy transfer rate $g\Delta H$, and the power dW/dt. These three parameters are presumed to be functions of fluid properties density ρ and viscosity μ, volume flow rate through the machine Q, rotational speed ω, and a characteristic dimension (usually impeller diameter) D. We therefore write three functional dependencies:

$$\eta = f_1(\rho, \mu, Q, \omega, D, g_c) \tag{5.11}$$

$$g\Delta H = f_2(\rho, \mu, Q, \omega, D, g_c) \tag{5.12}$$

$$\frac{d W}{d t} = f_3(\rho, \mu, Q, \omega, D, g_c) \tag{5.13}$$

Beginning with Equation 5.11, we assume a relationship of the form

$$\eta = a \, \rho^{\,b} \mu^{\,c} Q^{\,d} \, \omega^{\,e} D^{\,f} g_c^{\,g} \tag{5.14}$$

Next we substitute dimensions into the above equation for each parameter to obtain

$$0 = a \left(\frac{M}{L^3}\right)^b \left(\frac{F \cdot T}{L^2}\right)^c \left(\frac{L^3}{T}\right)^d \left(\frac{1}{T}\right)^e (L)^f \left(\frac{M \cdot L}{F \cdot T^2}\right)^g$$

We can now write an equation for each dimension in the equation:

M: $0 = b + g$

F: $0 = c - g$

L: $0 = -3b - 2c + 3d + f + g$

T: $0 = c - d - e - 2g$

Solving simultaneously gives

$$b = -g \qquad\qquad c = g \qquad\qquad d = -e - g \qquad\qquad f = 3e + g$$

Substituting into Equation 5.14 gives

$$\eta = a \frac{1}{\rho^8} \mu^8 \frac{1}{Q^e Q^8} \omega^e D^{3e} D^8 g_c^8$$

Grouping terms with like exponents yields

$$\eta = a \left(\frac{Dg_c\mu}{\rho Q}\right)^8 \left(\frac{\omega D^3}{Q}\right)^e \tag{5.15}$$

Although the above groups are dimensionless and the equation itself is valid, current practice calls for a slight rearrangement of the result. The groups on the right hand side can be combined to yield a more easily recognizable result:

$$\frac{\rho Q}{Dg_c\mu} \cdot \frac{\omega D^3}{Q} = \frac{\rho \omega D^2}{\mu g_c}$$

Equation 5.15 can now be rewritten in functional form as

$$\eta = f_1 \left(\frac{\rho \omega D^2}{\mu g_c}, \; \frac{Q}{\omega D^3}\right) \tag{5.16}$$

Likewise, Equations 5.12 and 5.13 become

$$\frac{g\Delta H}{\omega^2 D^2} = f_2 \left(\frac{\rho \omega D^2}{\mu g_c}, \; \frac{Q}{\omega D^3}\right) \tag{5.17}$$

$$\frac{g_c(dW/dt)}{\rho\omega^3D^5} = f_3\left(\frac{\rho\omega D^2}{\mu g_c}, \frac{Q}{\omega D^3}\right) \tag{5.18}$$

where

$$\frac{g\Delta H}{\omega^2 D^2} = \text{energy transfer coefficient}$$

$$\frac{Q}{\omega D^3} = \text{volumetric flow coefficient}$$

$$\frac{\rho\omega D^2}{\mu g_c} = \text{rotational Reynolds number}$$

$$\frac{g_c(dW/dt)}{\rho\omega^3D^5} = \text{power coefficient}$$

Experiments conducted with pumps show that the rotational Reynolds number ($\rho\omega D^2/\mu g_c$) has a smaller effect on the dependent variables than does the flow coefficient. So for incompressible flow through pumps, Equations 5.16, 5.17, and 5.18 reduce to

$$\eta = f_1\left(\frac{Q}{\omega D^3}\right) \tag{5.19}$$

$$\frac{g\Delta H}{\omega^2 D^2} = f_2\left(\frac{Q}{\omega D^3}\right) \tag{5.20}$$

$$\frac{g_c(dW/dt)}{\rho\omega^3D^5} = f_3\left(\frac{Q}{\omega D^3}\right) \tag{5.21}$$

The significance of these ratios for a centrifugal pump are apparent when performance data are being modeled. Suppose, for example, that performance data are available for a particular pump operating under certain conditions. The data can be used to predict the performance of the pump when something has been changed, such as rotational speed, impeller diameter, volume flow rate, or fluid density. The above equations are known as **similarity laws** or **affinity laws** for pumps. The method for using them is illustrated in the following example.

EXAMPLE 5.4. Actual performance data on a centrifugal pump are as follows:

$$
\begin{aligned}
\text{Rotational speed} &= \omega &&= 3500 \text{ rpm} \\
\text{Total head difference} &= \Delta H &&= 80 \text{ ft} \\
\text{Volume flow rate} &= Q &&= 50 \text{ gpm} \\
\text{Impeller diameter} &= D &&= 5^{1}/_{8} \text{ in.} \\
\text{Fluid} && &= \text{water}
\end{aligned}
$$

It is desired to change the rotational speed to 1750 rpm and the impeller diameter to $4^{5}/_{8}$ in. Determine how the new configuration will affect the pump performance with water as the working fluid.

Solution: We will use similarity to determine how the new parameters will affect the pump performance. We will refer to the original configuration of the pump with a subscript of '1' and the new configuration with a '2.' When we use similarity laws, we assume that the dimensionless ratios apply to both pumps. Equations 5.19–5.21 contain the dimensionless groups needed; the flow coefficient applied to both pumps is:

$$
\left.\frac{Q}{\omega D^{3}}\right|_{1} = \left.\frac{Q}{\omega D^{3}}\right|_{2}
$$

Substituting and noting that conversion factors would cancel if used gives

$$
\frac{50}{3500(5.125)^{3}} = \frac{Q_{2}}{1750(4.625)^{3}}
$$

Solving, we find the flow rate in the new configuration to be

$$
Q_{2} = 18.4 \text{ gpm}
$$

The head coefficient in Equation 5.20 can also be applied:

$$
\left.\frac{g\Delta H}{\omega^{2}D^{2}}\right|_{1} = \left.\frac{g\Delta H}{\omega^{2}D^{2}}\right|_{2}
$$

Substituting and noting that gravity is constant yields

$$
\frac{80}{3500^{2}(5.125)^{2}} = \frac{\Delta H_{2}}{1750^{2}(4.625)^{2}}
$$

Solving for the total head in the new configuration gives

$$
\Delta H_{2} = 16.3 \text{ ft}
$$

If necessary, the power can be found with Equation 5.4:

$$-\frac{dW}{dt} = \frac{\dot{m}g}{g_c} \Delta H = \frac{\rho Q g}{g_c} \Delta H \qquad (5.4)$$

or with the power coefficient $(g_c(dW/dt)/\rho\omega^3 D^5)$ of Equation 5.21 applied to both configurations. The efficiency remains nearly constant for small changes.

For the pump of this example, data on the original configuration were taken from an actual pump performance curve. For purposes of comparison, the actual data (from a manufacturer) on the new configuration are:

For $Q \approx 20$ gpm, $\Delta H \approx 17$ ft compared to 16.3 ft calculated above

Also from the pump performance graphs, the efficiency of the original configuration is 57% compared to 48% in the new one.

As seen in the previous example, pump affinity laws might be rewritten for two similar pumps as:

$$\frac{Q_1}{\omega_1 D_1^3} = \frac{Q_2}{\omega_2 D_2^3}$$

$$\frac{\Delta H_1}{\omega_1^2 D_1^2} = \frac{\Delta H_2}{\omega_2^2 D_2^2}$$

$$\frac{(dW/dt)_1}{\rho_1 \omega_1^3 D_1^5} = \frac{(dW/dt)_2}{\rho_2 \omega_2^3 D_2^5}$$

In industry, pump affinity laws are written as

$$\frac{Q_1}{D_1} = \frac{Q_2}{D_2} \qquad \text{or} \qquad \frac{Q_1}{\omega_1} = \frac{Q_2}{\omega_2}$$

$$\frac{\Delta H_1}{\omega_1^2} = \frac{\Delta H_2}{\omega_2^2} \qquad \text{or} \qquad \frac{\Delta H_1}{D_1^2} = \frac{\Delta H_2}{D_2^2}$$

and $\dfrac{(dW/dt)_1}{\omega_1^3} = \dfrac{(dW/dt)_2}{\omega_2^3}$ or $\dfrac{(dW/dt)_1}{D_1^3} = \dfrac{(dW/dt)_2}{D_2^3}$

5.5 Specific Speed and Pump Types

With the many types of pumps available, it is necessary to have some criteria regarding the type of pump to use for a specific application. A dimensionless group known as **specific speed** is used in the decision making process. Specific speed is found by combining head coefficient and flow coefficient in order to eliminate characteristic length D:

$$\omega_{ss} = \left(\frac{Q}{\omega D^3}\right)^{1/2} \left(\frac{\omega^2 D^2}{g \Delta H}\right)^{3/4}$$

or
$$\omega_{ss} = \frac{\omega Q^{1/2}}{(g \Delta H)^{3/4}} \qquad \text{[dimensionless]} \qquad (5.22)$$

Exponents other than 1/2 and 3/4 could be used (to eliminate D), but 1/2 and 3/4 are customarily selected for modeling pumps. Another definition for specific speed is given by

$$\omega_s = \frac{\omega Q^{1/2}}{\Delta H^{3/4}} \qquad \left[\text{rpm} = \frac{\text{rpm}(\text{gpm})^{1/2}}{\text{ft}^{3/4}}\right] \qquad (5.23)$$

in which the rotational speed ω is expressed in rpm, volume flow rate Q is in gpm, total head ΔH is in ft of liquid, and specific speed ω_s is arbitrarily assigned the unit of rpm. Equation 5.22 for specific speed ω_{ss} is dimensionless whereas Equation 5.23 for ω_s is not. Specific speed found with Equation 5.22 can be applied using any unit system but Equation 5.23 is defined exclusively in Engineering units. Moreover, results of calculations made with these equations differ by a large magnitude.

The specific speed of a pump can be calculated at any operating point. It is advantageous, however, to calculate specific speed for a pump only at its maximum efficiency. Then specific speed data can be used to select a pump with the knowledge that the pump will be operating at its maximum efficiency point. To illustrate these concepts, refer to the pump performance curve of Figure 5.2 . At maximum efficiency, we read

$$Q \approx 450 \text{ gpm} \qquad \Delta H \approx 24 \text{ ft} \qquad \omega \approx 2500 \text{ rpm} \qquad \eta \approx 85\%$$

A flow rate of 450 gpm equals 1.004 ft^3/s, and 2500 rpm equals 262 rad/s. Substituting into Equation 5.22 gives

$$\omega_{ss} = \frac{262(1.004)^{1/2}}{(32.2(24))^{3/4}} = 1.791$$

Equation 5.23, on the other hand, yields

$$\omega_s = \frac{2500(450)^{1/2}}{24^{3/4}} = 4890 \text{ rpm}$$

After manipulation of the conversion factors, we find that

$$\omega_s \approx 2736 \omega_{ss}$$

With data obtained from tests on many types of pumps (including axial, mixed, and radial or centrifugal), the data of Table 5.3 have been produced. When this table is used, a machine that operates at or near its maximum efficiency is being selected.

EXAMPLE 5.5. Determine the type of pump best suited for pumping 250 gpm (= 0.557 ft³/s) of water with a corresponding head of 6 ft. The motor to be used has a rotational speed of 360 rpm.

Solution: We calculate the specific speed using Equation 5.23:

$$\omega_s = \frac{\omega Q^{1/2}}{\Delta H^{3/4}} \qquad \left[\text{rpm} = \frac{\text{rpm}(\text{gpm})^{1/2}}{\text{ft}^{3/4}} \right] \qquad (5.23)$$

$$\omega_s = \frac{360(250)^{1/2}}{6^{3/4}} = 1480 \text{ rpm} \approx 1500 \text{ rpm}$$

We enter Table 5.3 at 1500 rpm. We could interpolate to obtain the exact value at 250 gpm but it is not necessary to do so because these values are approximate. We read an efficiency of ~ 75%. Moving to the far right column, we find that the impeller shape suggested as best for this application is that for a mixed flow pump.

We can in addition calculate the power using the energy equation which is

$$\frac{dW}{dt} = \frac{\rho g Q \Delta H}{g_c} = \frac{62.4(32.2)(0.557)(6)}{32.2}$$

or $$\frac{dW}{dt} = 209 \text{ ft·lbf/s} = 209/550 = 0.38 \text{ hp}$$

The above value is the power actually transmitted to the fluid. The power required to be transferred from the motor is

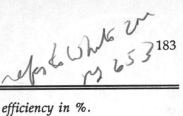
TABLE 5.3. *Pump selection chart for finding efficiency in %.*

ω_s rpm	\multicolumn{7}{c}{Flow rate in Gallons per Minute, gpm}	ω_{ss}	pump type
	100	200	500	1000	3000	10000	>10000		
500	46	54						0.18	
600	52	58						0.22	Centrifugal
700	55	62	70					0.26	or
800	58	64	72	76				0.29	Radial
900	61	66	74	77	81			0.33	flow type
1000	62	67	75	78	82			0.37	
1500	67	72	79	82	85	87	89	0.55	
2000	70	75	81	83	87	89	92	0.73	
3000			81	83	87	89	92	1.10	Mixed
4000			79	81	85	88	90	1.46	flow
5000					84	87	90	1.83	type
6000					83	85	87	2.19	
7000					82	84	86	2.56	
8000					80	84	86	2.92	
9000						83	85	3.29	
10000						82	84	3.65	Axial
15000						77	80	5.48	flow type

Equations:

$$\omega_{ss} = \frac{\omega Q^{1/2}}{(g\Delta H)^{3/4}} \qquad \text{[dimensionless]} \qquad (5.22)$$

$$\omega_s = \frac{\omega Q^{1/2}}{\Delta H^{3/4}} \qquad \left[rpm = \frac{rpm(gpm)^{1/2}}{ft^{3/4}} \right] \qquad (5.23)$$

Note on reading the table: For a specific speed of 5000 rpm and a flow rate of 3000 gpm, the maximum efficiency (any pump) that can be expected is approximately 84% and the recommended pump is a mixed flow type.

$$\frac{d W_a}{d t} = \frac{d W/dt}{\eta} = \frac{209}{0.75}$$

where $\eta = 75\%$ was obtained from Table 5.3. For the conditions given, we can expect that the maximum efficiency of any pump we select will be 75%. Solving, we get or $\dfrac{dW_a}{dt}$ = 279 ft·lbf/s = 0.51 hp   (mixed flow impeller shape)

5.6 Piping System Design Practices

Transporting a fluid from one location to another does not enhance its value. Yet a piping system can be one of the greatest cost items in an installation. For example, in a processing plant, the piping system is usually 15–20% of the total plant cost. Consequently piping should be designed to meet minimum cost requirements and still be adequate for meeting operational requirements. Fortunately, however, pipes are available in a few discreet sizes so that the economic analysis is somewhat simplified. It is the responsibility of the engineer to ensure that the optimum system be designed.

Another viewpoint that can be examined is that of the pump manufacturer. Pumps are made to last for a long time. Loss of a sale to a pump manufacturer is a permanent loss. A manufacturer that gains a sale will have income in future sales of replacement parts such as seals or bearings.

In light of the possible viewpoints associated with the design of a piping system, following now are design practices that will help an engineer in the decision making process.

As shown in the last chapter, the economic line size is the one that minimizes the total cost of a pipeline including fittings, hangers, pumps, valves, and installation. Charts were provided that should aid in determining the economic line size. Not mentioned there is one limitation that should be observed. At pressure drops greater than 25–30 psi per 1000 ft (175–200 kPa per 300 m) of pipe for liquids and 10–15 psi per 1000 ft (70–104 kPa per 300 m) of pipe for gases, excessive and objectionable vibrations in the system will result. For low pressure steam, the pressure drop should be limited to 4 psi (28 kPa) per 100 ft (30 m).

The engineer is responsible for recommending the most economic line size, unless unusual conditions exist for the system of interest. It is generally worthwhile to make a complete and thorough analysis because a bad decision during the design phase can lead to years of unnecessarily high costs which are wasteful and not recoverable. Usually when the economic diameter is calculated, the result falls between two nominal sizes. Selecting the smaller size results in a lower initial capital investment. Selecting the larger size results in a lower operating cost. From an economics viewpoint, the total yearly cost is nearly the same for either size selected. From an engineering standpoint, the larger size leads to more design flexibility. In addition, selecting the larger size allows for any late changes in the specifications for volume carrying capacity, or for errors. Lower operating costs is also an advantage as power costs are never expected to decrease. Furthermore, the smaller size pipe could have a pressure drop that is excessive leading to vibrations. Finally, the smaller pipe size is subject to corrosion and/or sediment deposition on the surface of the conduit to a greater degree than the larger size would be.

The proper design of the system contains many aspects. The economic size should be determined. The pressure drop and the power should also be calculated. If a pump is required, then it too should be selected and the net positive suction head should always be determined so that cavitation is avoided. An inadequate suction line can cause problems for an entire plant, especially if downstream conditions need liquid to be pumped at a rate that is higher than a cavitating pump can deliver.

It is necessary to determine conditions of the fluid upstream and downstream of the piping system. Specifically, pressure and/or fluid level in a container should be known. The minimum operating pressure upstream and the maximum operating pressure downstream should be calculated. When dealing with tanks the minimum liquid level in the upstream tank (usually 2 ft or 60 cm above pipe inlet) and the maximum level in the downstream tank (usually full) should be calculated. Making such calculations is vital to avoiding problems after installation. If the engineer believes that the line will someday be required to convey a flow rate that is greater than the design value, the engineer should not modify the design. Allowances for future increases in volume flow rate should not be considered unless specifically spelled out as part of the design. Determining extra capacity requirements is the responsibility of management. Notifying management that this practice may be of some value is the responsibility of the engineer.

In gravity fed systems (where gravity rather than a pump is the driving power) a line size larger than the economic diameter might be required in order to meet volume flow rate requirements. In some cases where gravity might be used, it could be more economical to install a pump and a smaller line size.

In all piping systems, it is possible that air will be trapped somewhere in the line. It is advisable to lay out pipelines with a slight grade upward in the flow direction so that air will tend not to remain in the line. Where this is not possible, a small valve should be installed at places where air (or vapor) might tend to accumulate.

Piping System Design—Suggested Procedure

1. Given the economic parameters associated with the system, determine the economic line size. If the cross section is noncircular and/or contains

fittings, perform calculations for a straight run of circular pipe. Use the calculated economic diameter to find the optimum economic velocity. Use the economic velocity to complete the details of the system design, including the placement of hangers. Table 5.4 gives results of calculations of economic or reasonable velocity ranges for many fluids.

2. Calculate the pressure drop for the system using the optimum economic line size. Check to ensure that the pressure drop is not excessive, which leads to objectionable vibrations. Prepare a system curve of ΔH vs Q.

3. In systems where the exit is lower (physically) than the inlet and where friction plus minor losses are small, the pressure at the exit might be calculated to be greater than that at the inlet for the specified flow rate. This means the fluid will flow under the action of gravity and a pump may not be needed. Further, it might be impossible to satisfy optimum velocity conditions as well. However, if a pump is to be used, determine from the appropriate chart which pump should be selected. Refer to the pump performance map if available and superimpose the system curve on it to find the exact operating point. Use NPSH data to specify the exact location of the pump.

4. If tanks are present, specify the minimum and maximum liquid heights in them.

5. Prepare a drawing for the system and a summary of specifications sheet which lists results of calculations only. Attach the calculations to the summary sheet.

While the above list is not comprehensive, it does provide a useful checklist of things that should be done during the design phase of laying out a piping system.

EXAMPLE 5.6. Figure 5.8 shows a piping system that is to convey 600 gpm of propylene glycol from a pump to a tank. Follow the suggested design procedure and make recommendations about the piping system.

Solution: We follow the design procedure in a step-by-step fashion.

1. *Economic Line Size*

Table 5.4 lists the economic velocity range for propylene glycol as

$$4.5 \leq V_{opt} \leq 9.0 \text{ ft/s}$$

TABLE 5.1. *Reasonable velocities for various fluids, calculated by using optimum economic diameter equations.*

Fluid	Economic Velocity Range ft/s	m/s
Acetone	4.9–9.8	1.5–3.0
Ethyl Alcohol	4.8–9.6	1.5–3.0
Methyl Alcohol	4.8–9.6	1.5–3.0
Propyl Alcohol	4.7–9.4	1.4–2.8
Benzene	4.6–9.2	1.4–2.8
Carbon Disulfide	4.2–8.4	1.3–2.6
Carbon Tetrachloride	3.9–7.8	1.2–2.4
Castor Oil	1.6–3.2	0.5–1.0
Chloroform	4.0–8.0	1.2–2.4
Decane	4.9–9.8	1.5–3.0
Ether	5.0–10.0	1.5–3.0
Ethylene Glycol	3.9–7.8	1.2–2.4
R-11	4.0–8.0	1.2–2.4
Glycerine	1.4–2.8	0.43–0.86
Heptane	5.1–10.2	1.5–3.0
Hexane	5.2–10.4	1.6–3.2
Kerosene	4.7–9.4	1.4–2.8
Linseed Oil	4.9–9.8	1.5–3.0
Mercury	2.1–4.2	0.64–1.3
Octane	5.0–10.0	1.5–3.0
Propane	5.6–11.2	1.7–3.4
Propylene	5.5–11.0	1.7–3.4
Propylene Glycol	4.5–9.0	1.4–2.8
Turpentine	4.6–9.2	1.4–2.8
Water	4.4–8.8	1.4–2.8

Values in the table were prepared using current and projected economic parameters to determine economic diameters. The calculated diameters were divided into flow rate to determine economic velocity. We can use any value inside the range but the velocity we use will depend on what pipe sizes are appropriate for the job. We will make initial calculations for both 4.5 and 9.0 ft/s to determine the sizes available. For a volume flow rate of 600 gpm (= 1.34 ft³/s), we calculate two flow areas:

$$A_{\substack{upper \\ limit}} = \frac{Q}{V} = \frac{1.34}{4.5} = 0.298 \text{ ft}^2$$

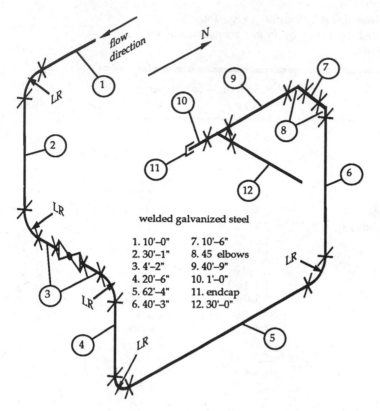

FIGURE 5.8. *The piping system of Example 5.6.*

$$A_{\substack{lower \\ limit}} = \frac{1.34}{9} = 0.150 \text{ ft}^2$$

We now refer to a table of pipe sizes. The only standard size that falls within this range of flow areas is 6-nominal schedule 40, which is the size we use for the remainder of the calculations.

Economic Line Size = 6-nom sch 40

---

2. *Pressure Drop and System Curve*

We define section 1 as the inlet of pipe length #1 in Figure 5.8 and section 2 at the exit of length #12. The Modified Bernoulli equation is

$$\frac{p_1 g_c}{\rho g} + \frac{V_1^2}{2g} + z_1 = \frac{p_2 g_c}{\rho g} + \frac{V_2^2}{2g} + z_2 + \Sigma \frac{fL}{D_h} \frac{V^2}{2g} + \Sigma K \frac{V^2}{2g} \qquad (5.7)$$

With constant diameter pipe, $V_1 = V_2$. Making height measurements from length #5, we write

$$z_1 = 20'6" + 30'1" = 50.6 \text{ ft}$$

$$z_2 = 40'3" + (10'6")\sin45° = 47.7 \text{ ft}$$

Also, $L = 10'0" + 30'1" + 4'2" + 4'2" + 20'6" + 62'4" + 40'3" + 10'6"$
$$+ 40'9" + 30'0"$$

or $L = 252.8 \text{ ft}$

From Table D.1 for 6-nominal schedule 40 pipe,

$$ID = 0.5054 \text{ ft} \qquad\qquad A = 0.2006 \text{ ft}^2$$

From Table 3.1 for a galvanized surface, $\varepsilon = 0.0005$ ft (the average value). Also,

$$V = \frac{Q}{A} = \frac{1.34}{0.2006} = 6.68 \text{ ft/s}$$

The minor loss values from Table 3.3 include:

$$\Sigma K = 5K_{\substack{long\ radius \\ 90°\ ells}} + K_{\substack{globe \\ valve}} + 2K_{45°\ ells} + K_{\substack{branch \\ T\text{-}joint}}$$

$$\Sigma K = 5(0.22) + 10 + 2(0.17) + 0.69$$

or $\Sigma K = 12.13$

Propylene glycol properties from Table B.1 are:

$$\rho = 0.968(1.94) \text{ slug/ft}^3 \qquad\qquad \mu = 88 \times 10^{-5} \text{ lbf·s/ft}^2$$

The Reynolds number is calculated to be

$$\text{Re} = \frac{\rho V D}{\mu g_c} = \frac{0.968(1.94)(6.68)(0.5054)}{88 \times 10^{-5}}$$

and so

$$\left.\begin{array}{l} Re = 7.21 \times 10^3 \\[2mm] \text{and } \varepsilon/D = \dfrac{0.0005}{0.5054} = 0.00099 \end{array}\right\} \quad f = 0.036 \quad \text{(Figure 3.2)}$$

Substituting into the Modified Bernoulli equation gives

$$\frac{p_1}{0.968(1.94)(32.2)} + 50.6 = \frac{p_2}{0.968(1.94)(32.2)} + 47.7$$

$$+ \frac{0.036(252.8)}{0.5054} \frac{(6.68)^2}{2(32.2)} + 12.13 \frac{(6.68)^2}{2(32.2)}$$

Solving,

$$p_1 - p_2 = 1090 \text{ lbf/ft}^2 = 7.6 \text{ psi at } Q = 600 \text{ gpm } (= 1.34 \text{ ft}^3/\text{s})$$

To generate a system curve of head difference ΔH versus volume flow rate Q, we use the following form of the Modified Bernoulli equation:

$$H_1 = H_2 + \frac{fL}{D} \frac{V^2}{2g} + \Sigma K \frac{V^2}{2g}$$

or

$$H_1 - H_2 = \Delta H = \frac{V^2}{2g} \left(\frac{fL}{D} + \Sigma K \right)$$

With the continuity equation, $V = Q/A$, the above equation becomes

$$\Delta H = \frac{Q^2}{2A^2 g} \left(\frac{fL}{D} + \Sigma K \right)$$

Substituting all known quantities gives

$$\Delta H = \frac{Q^2}{2(0.2006)^2(32.2)} \left(\frac{252.8f}{0.5054} + 12.13 \right)$$

or

$$\Delta H = 0.386 Q^2 (500.2f + 12.13)$$

We now select various values of flow rate, calculate Reynolds number, determine friction factor, and substitute into the above equation to find ΔH. The calculation results are summarized in the following table, and the system curve is shown in Figure 5.9.

Q, gpm	Q, ft³/s	V = Q/A, ft/s	Re	f	ΔH, ft
100	0.223	1.11	1197	0.0535	0.75
200	0.447	2.23	2405	0.0266	1.96
300	0.670	3.34	3602	0.041	5.66
400	0.893	4.45	4800	0.040	9.89
500	1.12	5.58	6018	0.037	14.8
600	1.34	6.68	7205	0.036	20.9
700	1.56	7.78	8391	0.034	27.4
800	1.79	8.92	9620	0.033	35.4
900	2.01	10.0	10790	0.032	43.9
1000	2.23	11.1	11970	0.032	54.0

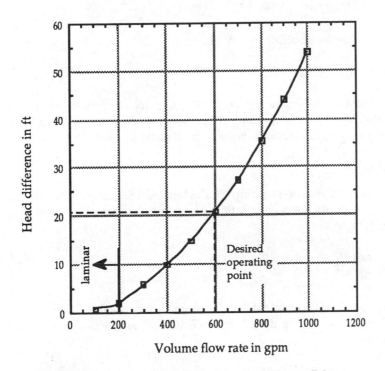

FIGURE 5.9. *System curve for the layout of Figure 5.8.*

Next we determine where the flow rate leads to a pressure drop of 25–30 psi per 1000 ft which is the lower limit of when excessive vibrations occur. We will use 30 psi per 1000 ft arbitrarily. By definition,

$$H_1 - H_2 = \Delta H = \frac{p_1 g_c}{\rho g} + \frac{V_1^2}{2g} + z_1 - \left(\frac{p_2 g_c}{\rho g} + \frac{V_2^2}{2g} + z_2\right)$$

Substituting for the data of this example, we get

$$\Delta H = \frac{p_1 - p_2}{0.968(1.94)(32.2)} + 50.6 - 47.7 = 0.0165(\Delta p) + 2.9$$

The limit we seek is 30 psi/1000 ft. The pipe length here, however, is 252.8 ft, so we set up a proportion to obtain

$$\frac{30 \text{ psi}}{1000 \text{ ft}} = \frac{\Delta p}{252.8 \text{ ft}}$$

or $\Delta p = 7.58 \text{ lbf/in}^2 = 1092 \text{ lbf/ft}^2$

The corresponding head difference ΔH is

$$\Delta H = 0.0165(1092) + 2.9 = 20.9 \text{ ft}$$

This value is just at the operating point. We specify that the flow rate of 600 gpm should not be exceeded to prevent excessive vibrations. If vibrations are excessive, the next larger pipe size would be used.

3. *Pump Selection*

We first determine the type of pump that is best suited for the job. We use Table 5.3 which requires a calculation of specific speed. Equation 5.23 defines specific speed as

$$\omega_s = \frac{\omega Q^{1/2}}{\Delta H^{3/4}}$$

The specific speed will be calculated for four different rotational speeds corresponding to those of Figure 5.2: 3600, 2700, 1760, and 900 rpm. For 3600 rpm, we have

$$\omega_{s1} = \frac{(3600)(600)^{1/2}}{(20.9)^{3/4}} = 9025 \text{ rpm}$$

Likewise,

$$\omega_{s2} = \frac{(2700)(600)^{1/2}}{(20.9)^{3/4}} = 6768 \text{ rpm}$$

$$\omega_{s3} = \frac{(1760)(600)^{1/2}}{(20.9)^{3/4}} = 4412 \text{ rpm}$$

$$\omega_{s4} = \frac{(900)(600)^{1/2}}{(20.9)^{3/4}} = 2256 \text{ rpm}$$

Trying to locate these values in Table 5.3 shows that the first 3 specific speeds (9025, 6768, and 4412 rpm) corresponding to rotational speeds of 3600, 2700, and 1760 rpm do not appear. The fourth value of specific speed (2256 rpm) coupled with the given flow rate of 600 gpm is on the chart, however. We read from Table 5.3:

$$\left.\begin{array}{c} \omega_{s4} = 2256 \text{ rpm} \\[1em] Q = 600 \text{ gpm} \end{array}\right\} \qquad \eta \approx 81\% \qquad \text{Mixed Flow Type}$$

If two or more of the specific speed values did appear, then the pump to select would depend on some other parameter such as cost or local availability or just an arbitrary decision. Continuing with the calculations, we next find the power required:

$$\frac{d\,W}{d\,t} = \frac{\rho Q g \Delta H}{g_c} = 0.968(1.94)(1.34)(32.2)(20.9)$$

or $\qquad \dfrac{d\,W}{d\,t} = 1693 \text{ ft·lbf/s} = 3.1 \text{ hp}$

The above power is that which must be transferred to the fluid. The power that is needed from the motor is

$$\frac{d\,W_a}{d\,t} = \frac{d\,W/d\,t}{\eta} = \frac{3.1}{0.81}$$

or $\qquad \dfrac{d\,W_a}{d\,t} = 3.83 \text{ hp}$

(Note: motors are not available in any desired size but, like pipe and tubing, they are available in discreet sizes. Here we might use 4 hp.)

For purposes of illustration, suppose that we have consulted a composite graph like that of Figure 5.4 and specified the appropriate mixed flow pump. We would then locate the corresponding performance map and superimpose our system curve on that performance map. This is illustrated in Figure 5.10. The point of intersection is the actual operating

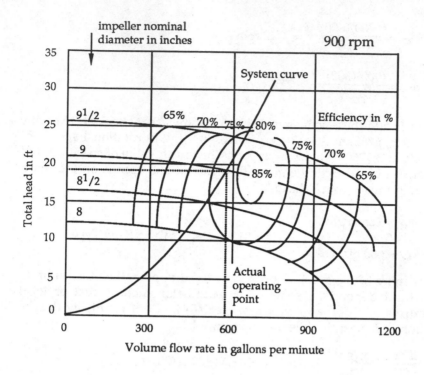

FIGURE 5.10. *System curve superimposed on performance map to find the actual operating point.*

point of the system, which is indicated in the figure with the dotted lines. If we continued in this example, we would use net positive suction head data for the pump (supplied by the manufacturer) to make further specifications regarding the pump location with respect to the liquid in the inlet.

4. Summary of Results

Economic Line Size	6-nominal schedule 40
Layout	Figure 5.8
System Curve	Figure 5.9
Specific Speed	2265 rpm at 600 gpm
Expected Pump Eff	~ 81%
Pump Type	Mixed Flow
Motor Power	≥ 3.83 hp

Pumps in Series or in Parallel

In some pump installations, it might be necessary to use two pumps in order to obtain a desired operating scheme. Suppose for example that several pumps of the same capacity are on hand and two of these pumps in the same system could accomplish what one larger pump could. The two pumps could be connected either in a series or in a parallel configuration.

When pumps are connected in series, the outlet of the first pump leads directly to the inlet of the second. The total head of the installation (both pumps) is the sum of the heads developed by each pump at the operating point. Both pumps must be sized to operate satisfactorily at the design flow rate.

Series pumping is referred to as being pressure additive. Figure 5.11 shows the effects of a series pumping installation containing two identical pumps. On the horizontal axis is volume flow rate Q while total head ΔH is on the vertical axis. The system curve is sketched on the graph along with pump curves. For only one pump operating, the intersection of the system curve and the one-pump curve is at point A— the operating point of the one-pump system. With both pumps operating, the operating point moves to B. We can project down from B (constant flow rate) to locate point B'. Projecting to the total head ΔH axis, we see that B' locates a total head that is 50% of the total head located by point B. Thus at the flow rate corresponding to B-B', the total head (both pumps) is made up of two equal contributions for two identical pumps.

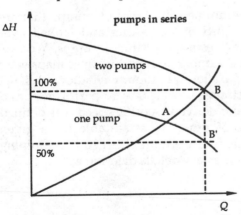

FIGURE 5.11. *A series pump installation curve showing effect on system operating point.*

When two identical pumps are set up in a parallel configuration, half the total flow (of the installation) is provided by each pump. For more than two pumps, the flow rate produced by individual pumps can represent any fraction of the total flow. Parallel pumping is described as being flow rate additive; identical pumps will always develop identical head values. Figure 5.12 shows the effects of two identical pumps in the same system, with one pump operation represented by point A, and two pumps represented at point B. We note that point B is the operating point of the installation (both pumps). If we project to the left, we locate B' on the one-pump curve. Projecting downward to the flow rate Q axis, we see that both pumps contribute equally to the total flow of the system.

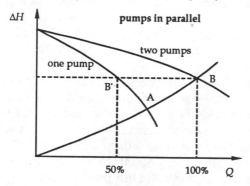

FIGURE 5.12. *A parallel pump installation curve showing effect on system operating point.*

5.7 Summary

In this chapter, we have examined the types of pumps that are commercially available. We discussed pump testing and looked over methods and results obtained from testing centrifugal pumps. We saw how such results are used to select a pump. Pump performance maps were discussed as well as series and parallel operation, cavitation, and net positive suction head. Dimensional analysis was performed and dimensionless ratios for pumps were derived. Specific speed was defined and used to determine the type of pump most suitable for a given configuration. Finally, design guidelines were provided so that a piping system could be designed and the necessary details determined.

5.8 Show and Tell

1. Obtain a catalog (or an actual model) of **axial flow pumps** and give a presentation on their operation, test methods used in evaluating their performance, and typical performance curves.

2. Obtain a catalog (or an actual model) of **mixed flow pumps** and give a presentation on their operation, test methods used in evaluating their performance, and typical performance curves.

3. Obtain a catalog (or an actual model) of **radial flow** or **centrifugal pumps** and give a presentation on their operation, test methods used in evaluating their performance, and typical performance curves.

4. Obtain a catalog (or an actual model) of **reciprocating positive displacement pumps** and give a presentation on their operation, test methods used in evaluating their performance, and typical performance curves.

5. Obtain a catalog (or an actual model) of **gear pumps** and give a presentation on their operation, test methods used in evaluating their performance, and typical performance curves.

6. Obtain a centrifugal pump catalog and locate graphs like those presented in this chapter.

7. Obtain an impeller and/or pump housing (perhaps from a manufacturer) that has been subject to cavitation erosion and prepare a presentation about it. Locate positions where the erosive effect is greatest and explain its formation.

8. What is a **vane pump**? Give a report on how a vane pump operates and cite applications where it would be useful.

5.9 Problems Chapter 5

1. A pump is tested in the laboratory and the following data were obtained:

Torque	$=$	T	$= 2$ N·m
Rotational speed	$=$	ω	$= 1760$ rpm
Inlet pressure	$=$	p_1	$= 75$ kPa
Outlet pressure	$=$	p_2	$= 210$ kPa
Volume flow rate	$=$	Q	$= 0.002$ m^3/s
Height to inlet gage	$=$	z_1	$= 0.4$ m
Height to outlet gage	$=$	z_2	$= 1$ m
Inlet flow line	$=$		$2^1/_2$-nominal schedule 40
Outlet flow line	$=$		2-nominal schedule 40
Fluid	$=$		water

Determine the pump efficiency.

2. A pump is tested in the laboratory and the following data were obtained:

Torque	$=$	T	$= 35$ ft·lbf
Rotational speed	$=$	ω	$= 3600$ rpm
Inlet pressure	$=$	p_1	$= 2.5$ psia
Outlet pressure	$=$	p_2	$= 30$ psia
Volume flow rate	$=$	Q	$= 1000$ gpm
Height to inlet gage	$=$	z_1	$= 2$ ft
Height to outlet gage	$=$	z_2	$= 5$ ft
Inlet flow line	$=$		10-nominal schedule 40
Outlet flow line	$=$		8-nominal schedule 40

       Fluid   =   octane
Determine the pump efficiency.

3. Use the pump composite curve to select a pump for the conditions of Problem 2.

4. The inlet of a centrifugal pump is 8-nominal schedule 40 while the discharge line is 6-nominal schedule 40. Inlet and outlet pressures are 4.35 psig and 36.2 psig, respectively. Volume flow rate through the pump is 40 1/s, and the pump efficiency is 0.8. Determine the pump power. Elevation differences are negligible and the fluid is water.

5. A motor operates a pump whose efficiency is 65%. Inlet and outlet lines are $1^1/_2$-nominal and 1-nominal, both schedule 40 pipe. The pressure iincrease across the pump amounts to 20 ft of ethylene glycol. The volume flow rate of ethylene glycol through the pump is 0.05 ft^3/s. The pressure gage at the outlet is 0.4 m higher than the one in the inlet. Find the pump and motor power required.

6. A pump operates at 1760 rpm. The inlet line is $1^1/_2$-nominal and the outlet is 1-nominal, both schedule 40 PVC pipe. The pressure in the inlet line is 12 kPa while the outlet line pressure is 200 kPa. The outlet pressure gage is 0.7 m higher than the inlet gage. The torque exerted is measured to be 5 N·m, and the fluid is water. If the efficiency is 0.47, determine the expected volume flow rate through the pump.

7. A pump operating at 3600 rpm has a measured input torque of 0.5 ft·lbf. The volume flow rate through the pump is 45 gpm and the outlet pressure gage is located 17 in. above the inlet pressure gage. Inlet and outlet pipes are $1^1/_2$-nominal and 1-nominal schedule 40 pipes, respectively. The fluid being pumped is water. Determine the pressure rise across the pump for an efficiency of 75%.

8. The following data were taken on a centrifugal pump operating under conditions where the efficiency is known to be 76%:

Torque	=	T	= 1 N·m
Rotational speed	=	ω	= 1800 rpm
Inlet pressure	=	p_1	= 21 kPa (absolute)
Volume flow rate	=	Q	= 0.001 5 m^3/s
Height to inlet gage	=	z_1	= 0.3 m
Height to outlet gage	=	z_2	= 1 m
Inlet flow line	=	$1^1/_2$-nominal schedule 40	
Outlet flow line	=	1-nominal schedule 40	
Fluid	=	water	

Determine the expected reading on the outlet pressure gage.

9. A centrifugal pump has an efficiency of 71% at a rotational speed of 1200 rpm. Inlet and outlet pressure gage readings are 7 and 20 psia, respectively. The volume flow rate through the pump is 20 gpm. Inlet and outlet pipes are 2- and $1^1/_2$-nominal schedule 40, respectively, galvanized steel. The elevation of the

outlet gage over that of the inlet gage is 2 ft. What is the required input torque if the fluid is carbon tetrachloride?

10. A pump was tested at 3450 rpm. The input torque was 20 ft·lbf. Inlet and outlet pressures were measured to be 7 psia and 20 psia, respectively. The inlet line is 12-nominal schedule 40 and the outlet line is 10-nominal schedule 40. The volume flow rate was 1200 gpm. The outlet pressure gage is 2 ft higher than the inlet gage. Determine the pump efficiency if the fluid is (a) water; (b) propane. Comment on how density affects efficiency.

11. Use the composite curve to select a pump for the conditions of Problem 4.

12. Problem 10 deals with one piping system but two different fluids. Does the pump composite curve of this chapter indicate that the same pump is appropriate for both fluids?

13. Derive Equation 5.8b for net positive suction head as applied to the configuration of Figure 5.6b.

14. Repeat Example 5.3 for acetone rather than water, assuming all other conditions unchanged.

15. The inlet line of a pump is 4 nominal schedule 40 PVC pipe that is 30 ft long and includes one elbow and one well–rounded inlet. The pump conveys 0.5 ft^3/s of water at 60°F from a tank. The net positive suction head required by the pump is 8 ft. Determine where the location of the water level in the tank should be with respect to the pump impeller shaft to prevent cavitation.

16. The inlet line of a centrifugal pump is made of 3-nominal schedule 40 wrought steel pipe 2 m long and includes a basket strainer. The pump delivers 8 l/s of carbon tetrachloride at 70°F. The net positive suction head required by the pump is 1 m of water. Determine where the location of the water level in the tank should be with respect to the pump impeller shaft to prevent cavitation.

17. Figure P5.17 shows a pump and piping system used to provide water to a downstream tank. The pump is to provide 2000 gpm. The pipeline is made up entirely of 10-nominal schedule 40 galvanized steel. It contains 1000 ft of pipe and one globe valve. All fittings are long radius and flanged. (a) Select a pump for the system. (b) If the rotational speed of the motor is 1170 rpm, what is the power required?

18. The pump selected in Problem 17 delivers the water at 70°F from a tank. If the $NPSH_r$ is 1 ft, determine where the pump inlet should be with respect to the liquid level in the tank. The liquid surface in the intake tank is exposed to atmospheric pressure. Neglect friction effects in the pump inlet line.

19. Figure P5.19 shows a piping system that consists of 2 line sizes, both made of type K copper tubing. A pump is to provide a flow of 0.017 m^3/s of propylene from section 1 to 2. The larger size is 6-std and the smaller size is 4-std. The 6-std line is 6 m long and the 4-std is 12 m long. Determine the pump power required. All fittings are regular and soldered.

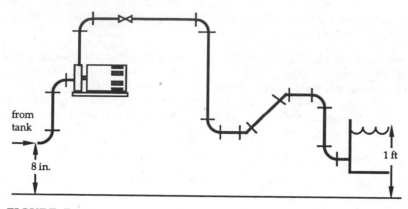

from
tank

8 in.

FIGURE P5.17

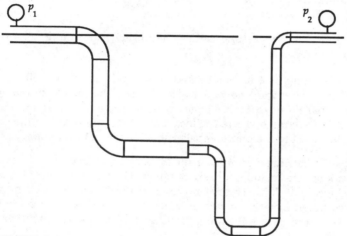

P_1 P_2

FIGURE P5.19

20. The pump selected in Problem 19 delivers the propylene from a tank which is located just upstream of section 1. If the $NPSH_r$ is - 1.1 m, determine where the pump inlet should be with respect to the liquid level in the tank. The liquid surface in the intake tank is exposed to atmospheric pressure. Neglect friction effects in the pump inlet line, and assume a liquid temperature of 80°F. Is the answer obtained for these conditions reasonable? If not, suggest a method for making the configuration work.

21. Figure P5.21 shows a tubing system used as a heat exchanger. The exchanger has fins attached (as shown) and is made of 1/2-std type M copper tubing. There are 40 feet of tubing and, in order to transfer the required amount of heat, the liquid velocity can be no less than 2 m/s. If the liquid is benzene, determine the pump power required. All fittings are soldered together.

22. Derive Equation 5.17.

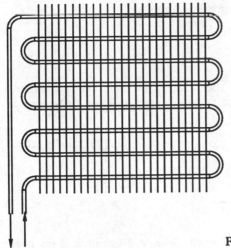

FIGURE P5.21

23. Derive Equation 5.18.

24. Verify that the energy transfer coefficient is dimensionless:

$$\frac{g \Delta H}{\omega^2 D^2} = \text{energy transfer coefficient}$$

25. Verify that the volumetric flow coefficient is dimensionless:

$$\frac{Q}{\omega D^3} = \text{volumetric flow coefficient}$$

26. Verify that the rotational Reynolds number is dimensionless:

$$\frac{\rho \omega D^2}{\mu g_c} = \text{rotational Reynolds number}$$

27. Verify that the power coefficient is dimensionless:

$$\frac{g_c (dW/dt)}{\rho \omega^3 D^5} = \text{power coefficient}$$

28. Actual performance data on a centrifugal pump are as follows:

Rotational speed = 2400 rpm
Head difference   = 70 ft
Volume flow rate = 100 gpm
Impeller diameter = 9 in
Fluid   = Water

The rotational speed of this pump is changed to 1760 rpm. What is the expected volume flow rate and head difference at the new speed?

29. Actual performance data on a pump are as follows:

Rotational speed   = 3600 rpm
Head difference   = 10 m
Volume flow rate   = 20 l/s
Impeller diameter = 20 cm

The pump is next operated at 2400 rpm with an impeller whose diameter is 18 cm. What is the expected flow rate and the expected head difference for the new configuration?

30. A pump is tested in the laboratory and the following data were obtained:

Fluid   = water
Rotational speed   = 300 rpm
Head difference   = 8 ft
Volume flow rate   = 10 gpm
Impeller diameter = 4 in.

The impeller is changed to one whose diameter is 3-1/2 in. and the pump is run at 900 rpm. Find the new volume flow rate and head difference. Calculate also the power for both configurations.

31. Tests on a pump yielded the following:

Fluid   = water
Rotational speed   = 1750 rpm
Volume flow rate   = 8 l/s
Head difference   = 15 ft
Impeller diameter = 6 in.

It is desired to run the pump with a 5 in. impeller and determine its operating characteristics when the head difference is 20 ft. Find the new volume flow rate.

32. What are the actual dimensions of specific speed in Equation 5.23? (Recall that it is *arbitrarily* given the units of rpm.)

33. Calculate the specific speed for the pump of Figure 5.3.

34. Calculate the specific speed for the pump of Figure 5.10.

35. Determine the type of pump and power required for the pump corresponding to that of Figure 5.3 when operated in its maximum efficiency region.

36. Determine the type of pump and power required for the pump represented by Figure 5.10 when operated in its maximum efficiency region.

37. Compare the "industry equations" for the pump affinity laws to those in Equations 5.19–5.21. Which equations are different? Make up a comparison chart for the equations.

38. Sketch a ΔH vs Q curve for a three pump system where the pumps operate (a) in series; and (b) in parallel. The pumps are all identical.

5.10 Group Problems

GP1. A series of tests were conducted on a centrifugal pump and it was desired to verify some iso-efficiency test results. The pump itself has a 12-nominal schedule 40 inlet line and a 10-nominal schedule 40 outlet line. Water was the liquid used, and the pressure gage in the outlet line was 2 ft higher than the inlet pressure gage. Following is a chart of data obtained from the pump, all at 75% efficiency. Fill in the blank spaces with the expected value:

torque ft·lbf	rotational speed, rpm	inlet pressure, psia	outlet pressure, psia	volume flow rate, gpm
30.0	1200	___	16	800
25.5	___	7	17	850
22.6	1800	8	___	900
16.4	2400	6	16	___
12.9	___	5	15	775
___	3600	5	15	625

GP2. A pump was tested in the laboratory to obtain data and construct a performance map. The pump had 1-nominal schedule 40 inlet and outlet lines. The inlet pressure valve was 4 in. below the outlet pressure valve. The data are as follows:

Impeller Diameter

Q gpm	4-7/8 in. Δp	Torque	4-1/2 in. Δp	Torque	4 in. Δp	Torque	3-1/2 in. Δp	Torque	3 in. Δp	Torque
0	11.4	0.000	9.5	0.000	7.5	0.000	5.6	0.000	3.7	0.000
4	11.2	0.313	9.3	0.242	7.4	0.191	5.5	0.154	3.5	0.121
8	10.6	0.425	9.1	0.335	6.9	0.262	5.2	0.214	3.2	0.152
12	10.0	0.487	8.4	0.394	6.3	0.293	4.5	0.239	2.7	0.178
16	9.1	0.554	7.6	0.452	5.4	0.337	3.5	0.256	1.4	0.136
20	7.8	0.635	6.2	0.503	4.0	0.360	1.9	0.244		
24	6.2	0.683	4.8	0.572	2.6	0.405				
28	4.5	0.743	2.8	0.460						

Construct a performance map for the data given in the above table. Plot ΔH vs Q for all impeller diameters. Show iso-efficiency lines for 30%, 35%, 40%, 43%, 45%, and 46%. The pump operates at 1750 rpm, and all pressures are in psi.

GP3. For all data in the table given in the preceding problem, calculate values of the following dimensionless groups:

$$\frac{g\Delta H}{\omega^2 D^2} = \text{energy transfer coefficient}$$

$$\frac{Q}{\omega D^3} = \text{volumetric flow coefficient}$$

$$\frac{\rho\omega D^2}{\mu g_c} = \text{rotational Reynolds number}$$

$$\frac{g_c(dW/dt)}{\rho\omega^3 D^5} = \text{power coefficient}$$

$$\eta = \text{efficiency}$$

Construct graphs of: a) energy transfer coefficient vs volumetric flow coefficient; b) energy transfer coefficient vs rotational Reynolds number; c) power coefficient vs volumetric flow coefficient; d) power coefficient vs rotational Reynolds number; e) efficiency vs volumetric flow coefficient; f) efficiency vs rotational Reynolds number. Take the fluid to be water in all cases.

GP4. For the piping system assigned, perform calculations to design the piping system, following the suggested procedure in the design practice section. Use the following economic parameters:

C_2 = \$0.04/(kW·hr) = \$ 0.04/(738 ft·lbf·hr);
C_1 = \$400/m$^{2.2}$ PVC schedule 40 pipe
t = 7880 hr/yr F = 7
n = 1.2 a = 1/(7 yr)
b = 0.01 η = 75% = 0.75

Assume that there is a pump in the system that has to be selected and that has a net positive suction head required of 8 ft. (The fluid, the volume flow rate, and the piping layout are to be assigned by the instructor.)

GP5. It is often preferable when working with noncircular cross sections and with minor losses to have information on economic velocity. The economic velocity will vary with the economic parameters and so minimum values should be selected for calculation purposes. The values that yield the minimum economic velocity are as follows:

C_2 = \$0.10/(kW·hr) = \$0.10/(738 ft·lbf·hr);
C_1 = \$300/m$^{2.2}$ n = 1.4
t = 7 880 hr/yr a = 1/(7 yr)
F = 6 b = 0.01
η = 75% = 0.75 ε = 0.000 25

For 5 fluids assigned by the instructor, calculate (a) economic diameter; and (b) economic velocity by using the continuity equation. Using the above data will provide a minimum for the economic velocity. In all cases, take the volume flow rate to be 1 l/s. Explain how to modify the results without recalculating if the volume flow rate in an actual case is different from 1 l/s.

GP6. A pipeline that is 600 ft long is to convey chloroform at a flow rate of 850 gpm. The line is to be suspended 2 ft (or so) from a ceiling. Determine: (a) the economic line size and piping material; (b) the system curve to 1000 gpm; (c) the pump power requirements; (d) the expected pump efficiency; (e) an appropriate pump to use (refer to a catalog rather than a curve from this chapter); and, (f) the location and configuration of pipe hangers. The pipeline is horizontal and minor losses can be neglected.

GP7. A pipeline conveying ether is 80 m long. The flow rate of the ether is 0.1 m^3/s. The line is horizontal and contains 3 regular elbows; all other minor losses can be neglected. The first 70 m of the line are a straight run and should be elevated a distance of 3 ft from the ground. Determine: (a) the economic line size and piping material; (b) the system curve to 0.15 m^3/s; (c) the pump power requirements; (d) the expected pump efficiency; (e) an appropriate pump to use (refer to a catalog rather than a curve from this chapter); and, (f) the placement and configuration of pipe supports for the first 70 m of the line.

GP8. A pipeline is to be designed to convey 5 gpm of glycerine a distance of 25 ft. The line contains 6 elbows and a globe valve (fully open). Determine: (a) the economic line size and piping material; (b) the system curve to 10 gpm; (c) the pump power requirements; (d) the expected pump efficiency; and, (e) an appropriate pump to use (refer to a catalog rather than a curve from this chapter).

GP9. Turpentine is stored in a 25 ft diameter tank. Inside the storage tank, the turpentine depth is 18 ft. A piping system containing a pump is to move the turpentine to a holding tank having a very large diameter. The turpentine depth in the holding tank is constant at 6 ft. The pipeline contains a basket strainer, 3 elbows, and a ball valve. At the exit, the end of the pipeline is submerged 5 ft below the free surface of the turpentine. For a volume flow rate of 50 gpm, determine: (a) the economic line size and piping material; (b) the system curve to 75 gpm; (c) the pump power requirements; (d) the expected pump efficiency; and, (e) an appropriate pump to use (refer to a catalog rather than a curve from this chapter). The pipeline is 108 ft long and is horizontal with inlet and exit at the same elevation.

GP10. A pipeline is to convey water uphill. The pipeline is 20 m long and is buried 18 in. underground. The inlet of the pipeline is to be submerged in water and should contain a basket strainer. The exit of the pipeline is submerged also in water at an elevation of 3 m above the inlet. The pipeline should convey the water at 25 gpm. The pipeline contains 8 elbows, a ball check valve, and a gate valve. Determine: (a) the economic line size and piping material; (b) the system curve to 30 gpm; (c) the pump power requirements; (d) the expected pump efficiency; and, (e) an appropriate pump to use (refer to a catalog rather than a curve from this chapter).

G11. Figure P5GP11 shows a piping system that is to convey 400 gpm of water. Follow the suggested design procedure and make recommendations about the piping system.†

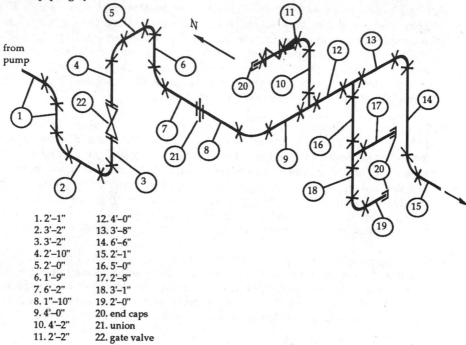

1. 2'–1"	12. 4'–0"
2. 3'–2"	13. 3'–8"
3. 3'–2"	14. 6'–6"
4. 2'–10"	15. 2'–1"
5. 2'–0"	16. 5'–0"
6. 1'–9"	17. 2'–8"
7. 6'–2"	18. 3'–1"
8. 1"–10"	19. 2'–0"
9. 4'–0"	20. end caps
10. 4'–2"	21. union
11. 2'–2"	22. gate valve

FIGURE P5GP11.

GP12. Figure P4.24 shows a piping system that is to convey 1000 gpm of octane. Follow the suggested design procedure and make recommendations about the piping system.†

GP13. Figure P4.26 shows a piping system that is to convey 600 gpm of acetone. Follow the suggested design procedure and make recommendations about the piping system.†

GP14. Figure P4.28 shows a piping system that is to convey 800 gpm of ethylene glycol. Follow the suggested design procedure and make recommendations about the piping system.†

---

† Hint: If under the action of gravity, the flow rate is greater than what is specified for optimum velocity conditions, a smaller line size might have to be used. Alternatively, the valve might have to be adjusted to give the required flow rate (not a desirable operating practice, but sometimes necessary). If such is the case, optimum velocity cannot be implemented.

CHAPTER 6   Some Heat Transfer Fundamentals

Heat is transferred from a source to a receiver in three distinct ways: conduction, convection, and radiation. Most engineering applications involve identifying one or two dominant modes and applying simplifying assumptions in order to solve the problem at hand. In this chapter, we review some of the fundamental concepts of heat transfer and define pertinent properties of substances. Solution methods for simple conduction and convection problems are illustrated. Equations are presented for various types of convection problems. Radiation heat transfer concepts are not discussed. The main purpose of the chapter, however, is to call attention to the heat transfer properties of fluids and to the solution methods in convection heat transfer. The objective is to lay a foundation for the modeling of heat exchangers which follows in chapter 7.

6.1 Conduction of Heat Through a Plane Wall

Figure 6.1 shows a planar wall in contact with a heat source on the left (a bank of heaters) and a heat sink on the right (a water jacket). Also shown imposed on the geometry of the wall are temperature T versus x axes. The heaters provide a constant heat flow per unit area per unit time, $q_x{}''$. Once this system reaches steady state, temperature within the material is measured and graphed on the axes. As shown, temperature T varies linearly with x and the slope of the temperature profile is written as dT/dx. The flow of heat per unit area (normal to the heat flow direction) per unit time is proportional to the temperature gradient

$$q_x{}'' \propto \left(-\frac{dT}{dx} \right) \tag{6.1}$$

Note that as distance x increases, temperature T decreases, so that $-dT/dx$ is actually a positive quantity. To make Equation 6.1 an equality, we introduce a proportionality constant:

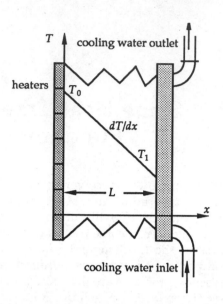

FIGURE 6.1. *Conduction of heat through a plane wall.*

$$q_x = kA \left(-\frac{dT}{dx} \right)$$
(6.2)

in which k is known as the **thermal conductivity** of the substance, with dimensions of $F \cdot L/(T \cdot L \cdot t)$ [BTU/(hr·ft·°R) or W/(m·K)]. Thermal conductivity k is a property of substances and in general is evaluated experimentally. Values of thermal conductivity for various metals and building materials are provided in Tables 6.1 and 6.2 (on pages 211 and 212). Values of thermal conductivity for various liquids and gases are found in the property tables of the appendix: B Tables for liquids and C Tables for gases.

For the planar wall of Figure 6.1, Equation 6.2 can be integrated directly to obtain

$$q_x = kA \frac{T_0 - T_1}{L}$$
(6.3)

where $(T_0 - T_1)$ is the temperature difference over the wall thickness L. Equation 6.3 can be rewritten as

$$q_x = \frac{T_0 - T_1}{L/kA} = \frac{T_0 - T_1}{R}$$
(6.4)

in which $L/kA = R$ is introduced as a resistance through which heat q_x is transferred under steady conditions due to the imposed temperature difference $(T_0 - T_1)$. The resistance R has dimensions of $t \cdot T/(F \cdot L)$ (hr·°R/BTU or K/W).

EXAMPLE 6.1. Figure 6.2 shows a cross section of a **guarded hot plate** apparatus. It is used to measure the thermal conductivity of a planar shaped material, such as plywood, insulation, sheet rock, etc. The guarded hot plate apparatus consists of a main heater and a guard heater. The guard heater completely surrounds the main heater. Two samples of the material to be tested are required. One sample is placed on each side of the heaters. Cooling water jackets are made to contact the samples. To operate the device, both heaters are activated. Heat from the main heater is made to flow in one dimension through each sample to the cooling water jackets. The guard heater supplies energy to the outer perimeter of the samples so that heat flowing from the main heater will not flow in any lateral direction. Thus a one dimensional flow of heat from the main heater to the cooling water jackets is set up.

The heaters are both heated electrically so readings of voltage and amperage on the wires to the main heater provide data from which input power to *both* samples can be calculated. Thermocouples are used to make temperature readings which are needed for several purposes. To ensure that one dimensional heat flow exists, temperature at surface 1 (Figure 6.2) of the samples must be the same at the main and the guard heater once steady state is achieved. At that time, the temperature at surface 1 of the main heater and at surfaces 2 and 3 of the cooling water jackets are recorded.

A guarded hot plate apparatus is used to measure the thermal conductivity of 3/8 in. thick plywood. The electrical input to the main heater is 110 V x 1 A. The temperature at the main heater surface is 210°F while the surface temperature of the cooling jackets is 80°F. The cross sectional area through which heat flows is 0.75 ft². (a) Determine the thermal conductivity of the plywood, and (b) calculate the value of the resistance as defined in Equation 6.4.

Solution: Half the electrical power input goes into each piece of plywood. The heat flow into each piece is given by Equation 6.4:

$$q_x = \frac{T_0 - T_1}{L/kA} = \frac{T_0 - T_1}{R} \tag{6.4}$$

in which q_x is calculated to be

$$q_x = 0.5(110)(1) = 55 \text{ W}/0.2928 = 187 \text{ BTU/hr}$$

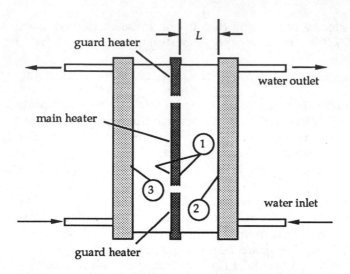

FIGURE 6.2. *Schematic of a guarded hot-plate heater for measuring thermal conductivity.*

The conversion factor was obtained from Appendix Table A.2. Rearranging Equation 6.4 to solve for thermal conductivity and substituting gives

$$k = \frac{L q_x}{A(T_0 - T_1)} = \frac{3/(8(12))(187)}{0.75(210 - 80)}$$

(a) $k = 0.060$ BTU/(hr·ft·°R)

_____

The resistance is found from the definition

$$R = \frac{L}{k A} = \frac{3/(8(12))}{0.060(0.75)}$$

(b) $R = 0.692$ hr·°R/BTU

_____

_____

TABLE 6.1. *Thermal properties of selected metals and alloys.*

Material	Specific Gravity	specific heat C_p		thermal conductivity, k		diffusivity, α	
		$\dfrac{\text{J}}{\text{kg·K}}$	$\dfrac{\text{BTU}}{\text{lbm·°R}}$	$\dfrac{\text{W}}{\text{m·K}}$	$\dfrac{\text{BTU}}{\text{hr·ft·°R}}$	$\text{m}^2/\text{s} \times 10^6$	$\text{ft}^2/\text{s} \times 10^3$
Metallic Elements at 293 K = 20°C = 528°R = 68°F							
Aluminum	2.702	896	0.214	236	163	97.5	1.05
Chromium	7.160	440	0.105	91.4	52.8	29.0	0.312
Copper	8.933	383	0.0915	399	231	116.6	1.26
Gold	19.3	129	0.0308	316	183	126.9	1.37
Iron	7.870	452	0.108	31.1	18.0	22.8	0.245
Silicon	2.330	703	0.168	153	88.4	93.4	1.01
Zinc	7.140	385	0.0920	121	69.9	44.0	0.474
Selected Alloys at 293 K = 20°C = 528°R = 68°F							
Duralumin	2.787	833	0.199	164	94.7	66.76	0.7187
Bronze	8.666	343	0.0819	26	15.0	8.59	0.0925
Brass	8.522	385	0.0920	111	64.1	34.12	0.3673
Cast Iron	7.272	420	0.100	52	30.0	17.02	0.1832
Wrought Iron	7.849	460	0.110	59	34.1	16.26	0.1750
Carbon Steel	7.801	473	0.113	43	24.8	11.72	0.1262
Chrome Steel	7.865	460	0.110	61	35.2	16.65	0.1792
Silicon Steel	7.769	460	0.110	42	24.3	11.64	0.1254
Stainless Steel	7.817	461	0.110	14.3	8.26	3.87	0.0417

Notes:

Data from several sources; see references at end of text.

Density = ρ = specific gravity x 62.4 lbm/ft^3 = specific gravity x 1 000 kg/m^3
= specific gravity x 1.94 slug/ft^3.

Diffusivity of aluminum = α = 97.5 x 10^{-6} m^2/s = 1.05 x 10^{-3} ft^2/s

TABLE 6.2. *Thermal properties of selected building materials*
at 293 K = 20°C = 528°R = 68°F.

Material	Specific Gravity	specific heat C_p		thermal conductivity, k		diffusivity, α	
		$\dfrac{J}{kg \cdot K}$	$\dfrac{BTU}{lbm \cdot °R}$	$\dfrac{W}{m \cdot K}$	$\dfrac{BTU}{hr \cdot ft \cdot °R}$	$m^2/s \times 10^5$	$ft^2/s \times 10^6$
Asbestos	0.383	816	0.195	0.113	0.0653	0.036	3.88
Asphalt	2.120			0.698	0.403		
Brick							
Common	1.800	840	0.201	0.45	0.26	0.031	3.33
Masonry	1.700	837	0.200	0.658	0.38	0.046	5.0
Silica	1.9			1.07	0.618		
Cardboard				0.25	0.14		
Concrete	0.500	837	0.200	0.128	0.074	0.049	5.3
Cork	0.120	1880	0.449	0.042	0.0243	0.03	3.15
Glass fiber	0.220			0.035	0.02		
Glass							
(window)	2.800	800	0.191	0.81	0.47	0.034	3.66
Ice at 0°C	0.913	1830	0.437	2.22	1.28	0.124	13.3
Kapok	0.025			0.035	0.02		
Plexiglas	1.180			0.195	0.113		
Plywood	0.590			0.109	0.063		
Sawdust	0.215			0.071	0.041		
Wood							
Fir, Pine,							
Spruce	0.444	2720	0.650	0.15	0.087	0.0124	1.33
Oak	0.705	2390	0.571	0.19	0.11	0.0113	1.22
Wool	0.200			0.038	0.022		

Notes:

Data from several sources; see references at end of text.

Density = ρ = specific gravity x 62.4 lbm/ft^3 = specific gravity x 1 000 kg/m^3
= specific gravity x 1.94 $slug/ft^3$.

Diffusivity of asbestos = α = 0.036 x 10^{-5} m^2/s = 3.88 x 10^{-6} ft^2/s

The concept of resistance to heat flow through a planar material can be used to apply the one dimensional heat flow equation to materials in series. Consider the composite wall of Figure 6.3. As shown, heat flows from left to right through 3 substances of different thermal conductivities in contact with one another. For the entire wall, we can write

$$q_x = \frac{T_0 - T_3}{R_{03}} \tag{6.5}$$

The heat flow through all the materials is equal and is identical to the heat flow through the wall; so in addition to Equation 6.5 we can write

$$q_x = \frac{T_0 - T_1}{R_{01}} = \frac{T_1 - T_2}{R_{12}} = \frac{T_2 - T_3}{R_{23}} = \frac{T_0 - T_3}{R_{03}}$$

We therefore conclude that the overall resistance equals the sum of the individual resistances:

$$R_{03} = R_{01} + R_{12} + R_{23} \tag{6.6a}$$

or

$$R_{03} = \frac{L_{01}}{Ak_{01}} + \frac{L_{12}}{Ak_{12}} + \frac{L_{23}}{Ak_{23}} \tag{6.6b}$$

Substituting into Equation 6.5, we find that the heat flow through the wall can therefore be written as

$$q_x = \frac{T_0 - T_3}{R_{03}} = \frac{T_0 - T_3}{\dfrac{L_{01}}{Ak_{01}} + \dfrac{L_{12}}{Ak_{12}} + \dfrac{L_{23}}{Ak_{23}}} \tag{6.7}$$

EXAMPLE 6.2. An oven wall consists of three layers of brick arranged as in Figure 6.3. The inside wall is made of silica brick, 4 in. thick, covered with masonry brick 8 in. thick, while the outside layer is of common brick 6 in. thick. During operation, the inside oven wall temperature reaches 1000°F and the outside surface temperature is 130°F. Calculate the heat transferred through the wall per square foot. Determine also the interface temperatures.

Solution: We can apply the equations of the preceding section. We will calculate the resistance offered by each layer and ultimately solve for the heat transferred.

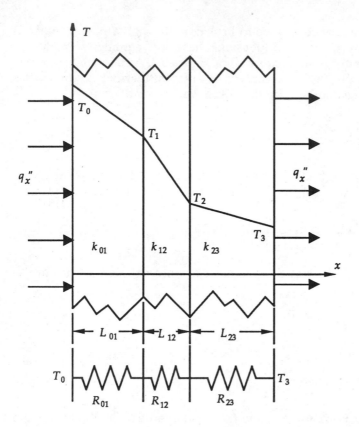

FIGURE 6.3. *Heat flow through a composite wall.*

Assumptions:
  1. The system is at steady state.
  2. The thermal properties of the materials are constant (although it is known that they vary with temperature).

From Table 6.2, we read the following values for thermal conductivity:

Silica brick k_{01} = 0.618 BTU/hr·ft·°R
Masonry brick k_{12} = 0.38 BTU/hr·ft·°R
Common brick k_{23} = 0.26 BTU/hr·ft·°R

We now calculate the resistance offered by each layer assuming a cross sectional area A of 1 ft²:

Silica brick $R_{01} = L_{01}/Ak_{01}$ = (4/12)/(1(0.618)) = 0.539 hr·°R/BTU
Masonry brick $R_{12} = L_{12}/Ak_{12}$ = (8/12)/(1(0.38)) = 1.75 hr·°R/BTU
Common brick $R_{23} = L_{23}/Ak_{23}$ = (6/12)/(1(0.26)) = 1.92 hr·°R/BTU

The total resistance for the three layers is the sum of these:

$$R_{03} = R_{01} + R_{12} + R_{23} = 0.539 + 1.75 + 1.92 \qquad (6.6a)$$

or $R_{03} = 4.21 \text{ hr} \cdot {}^{\circ}\text{R/BTU}$

The heat loss per square foot of wall cross section is

$$q_x = \frac{T_0 - T_3}{R_{03}} = \frac{1000 - 130}{4.21} \qquad (6.5)$$

or $q_x = 207 \text{ BTU/hr}$

For the first layer, we can write

$$q_x = \frac{T_0 - T_1}{R_{01}}$$

Rearranging and solving for T_1 gives

$$T_1 = T_0 - q_x R_{01} = 1000 - 207(0.539)$$

Solving,

$$T_1 = 888{}^{\circ}\text{F}$$

Similarly,

$$q_x = \frac{T_1 - T_2}{R_{12}}$$

and $T_2 = T_1 - q_x R_{12} = 888 - 207(1.75)$

Solving,

$$T_2 = 526{}^{\circ}\text{F}$$

6.2 Conduction of Heat Through a Cylindrical Wall

As seen in the last section, the cross sectional area through which heat flows in a planar geometry is constant. The model develope involved the concept of resistance to heat flow in order to make the appropriate calculations. We now extend the discussion to a cylindrical geometry.

Figure 6.4 shows a circular cylinder with inside and outside radii of R_1 and R_2, respectively. The temperature of the inside and outside surfaces are T_1 and T_2. The cylinder length is L. The figure contains a sketch of temperature versus radius (T vs r) as well as the resistance R_{12} that corresponds to the cross section. In order to determine an equation for R_{12}, we begin with Equation 6.2 written in the r direction, which is

$$q_r = kA \left(-\frac{dT}{dr} \right) \tag{6.2}$$

where dq_r represents heat flow in the radial direction, and $-dT/dr$ is the temperature gradient in the pipe wall. At any radius $R_1 < r < R_2$, the cross sectional area is given by

$$A = 2\pi rL$$

Substituting into Equation 6.2 and separating variables for integration gives

$$\int_{T_1}^{T_2} dT = \int_{R_1}^{R_2} -\frac{1}{2\pi kL} \frac{q_r \, dr}{r}$$

Integrating gives

$$T_2 - T_1 = -\frac{q_r}{2\pi kL} \ln \frac{R_2}{R_1}$$

or $$T_1 - T_2 = \frac{q_r}{2\pi kL} \ln \frac{R_2}{R_1} \tag{6.8}$$

The above equation can be rearranged in order to solve for a resistance as in the plane wall problem:

$$q_r = \frac{T_1 - T_2}{R_{12}} = \frac{T_1 - T_2}{\frac{1}{2\pi kL} \ln \frac{R_2}{R_1}}$$

We therefore conclude that in a cylindrical geometry the resistance to heat flow is given by

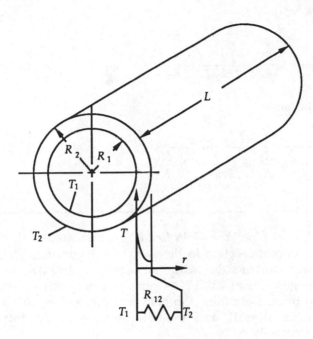

FIGURE 6.4. *Heat flow through a cylindrical wall.*

$$R_{ij} = \frac{1}{2\pi k L} \ln \frac{R_j}{R_i}$$

in which the subscripts i and j refer to surface locations.

In many practical cases, such as an insulated pipe, the geometry of interest will consist of two cylinders in series. This is illustrated in Figure 6.5. In calculating heat flow through such a geometry, we use the same approach as was done for a plane wall, namely, the concept of resistance to heat flow. From the inside surface to the outside surface, we have

$$q_r = \frac{T_1 - T_3}{R_{13}} \tag{6.9}$$

In addition, for each material,

$$q_r = \frac{T_1 - T_2}{R_{12}} = \frac{T_2 - T_3}{R_{23}}$$

Therefore,

$$R_{13} = R_{12} + R_{23}$$

or

$$R_{13} = \frac{1}{2\pi k_{12}L} \ln \frac{R_2}{R_1} + \frac{1}{2\pi k_{23}L} \ln \frac{R_3}{R_2} \tag{6.10}$$

Equation 6.9 then becomes

$$q_r = \frac{T_1 - T_3}{\frac{1}{2\pi k_{12}L} \ln \frac{R_2}{R_1} + \frac{1}{2\pi k_{23}L} \ln \frac{R_3}{R_2}} \tag{6.11}$$

EXAMPLE 6.3. A steel pipe [k = 40 W/(m·K)] is insulated with kapok insulation, similar in cross section to the sketch of Figure 6.5. The pipe carries a fluid that maintains the inside surface at 100°C. The outside surface of the insulation is at 25°C. The pipe is 4-nom sch 40 and the insulation is 6 cm thick. Determine the heat transferred per unit length through the cylindrical wall, and the temperature at the interface between the two materials.

Solution: We can apply the equations of the preceding section. We will calculate the resistance offered by each layer and ultimately solve for the heat transferred.

Assumptions:
        1. The system is at steady state.
        2. The thermal properties of the materials are constant (although it is known that they vary with temperature).

From Appendix Table D.1, we read the following dimensions of 4-nom sch 40 pipe:

OD = 11.43 cm ID = 10.23 cm

From Table 6.2, the thermal conductivity of kapok is 0.035 W/(m·K). In terms of the notation of Figure 6.5, we have for each radius

R_1 = 10.23/2 = 5.12 cm
R_2 = 11.43/2 = 5.72 cm
$R_3 = R_2 + 6$ = 11.72 cm

Also, for each material,

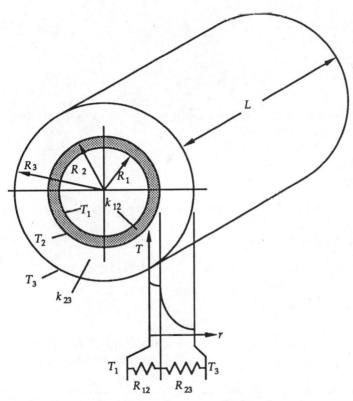

FIGURE 6.5. *Resistances for two cylinders in series.*

steel $k_{12} = 40$ W/(m·K)
kapok $k_{23} = 0.035$ W/(m·K)

Assuming a unit length, the resistances are calculated to be

steel $R_{12} = \dfrac{1}{2\pi k_{12}L}\, \ln\dfrac{R_2}{R_1} = \dfrac{1}{2\pi\,(40)(1)}\, \ln\dfrac{5.72}{5.12} = 0.000\,44$ K/W

kapok $R_{23} = \dfrac{1}{2\pi k_{23}L}\, \ln\dfrac{R_3}{R_2} = \dfrac{1}{2\pi\,(0.035)(1)}\, \ln\dfrac{11.72}{5.72} = 3.26$ K/W

As seen from these figures, the insulation offers a much greater resistance to the flow of heat than does steel. The total resistance then is

$$R_{13} = R_{12} + R_{23} = 3.26 \text{ K/W}$$

The heat transfer rate becomes

$$q_r = \frac{T_1 - T_3}{R_{13}} = \frac{100 - 25}{3.26} \qquad (6.9)$$

or $q_r = 23 \text{ W}$

In order to find the interface temperature we apply the heat flow equation to either material. For the steel,

$$q_r = \frac{T_1 - T_2}{R_{12}}$$

Rearranging and solving for the interface temperature gives

$$T_2 = T_1 - q_r R_{12} = 100 - 23(0.000\ 44)$$

Solving,

$$T_2 \approx 100°C$$

The temperature drop across the steel is virtually negligible. In many practical problems, temperature within a metal is often assumed to be a constant throughout.

6.3 Convection Heat Transfer—The General Problem

Heat transfer by convection occurs when a solid surface is in contact with a moving fluid and a temperature difference exists between the two. We identify two different ways convection heat transfer takes place: **forced convection** and **natural convection**. Forced convection occurs when the fluid motion is due to an external motive force. Natural convection (also traditionally known as **free convection**) occurs if fluid motion is induced by the transfer of heat.

The heat transferred by convection is calculated by use of a **convection coefficient** $\bar{h}$. The dimensions of the convection coefficient are $F \cdot L/(T \cdot L^2 \cdot t)$ [$W/(m^2 \cdot K)$ or $BTU/hr \cdot ft^2 \cdot °R$)]. The overbar denotes that the convection coefficient for the problem of interest is an average value, valid over the entire surface or geometry. In this text, the overbar is not used in order to simplify the notation, and because the average value of the convection coefficient h is all that will be used here.

Measurements of heat transfer rates and temperatures must be made in order to calculate the convection coefficient. It has been found that the convection coefficient is a function of temperature difference and of actual temperatures. Therefore, the convection coefficient (also called the **surface coefficient**) cannot be calculated except by trial-and-error methods. Once the convection coefficient h is known, the heat transfer rate can be found with

$$q = hA(T_s - T_\infty) \qquad\qquad (6.12)$$

It should be noted that the convection coefficient in Equation 6.12 relies on surface and free stream temperatures T_s and T_∞, respectively. Other temperature differences can be and have been devised to yield alternatives to Equation 6.12.

By rewriting Equation 6.12 in a form similar to that of Equation 6.9, we can define a resistance to convection heat transfer and treat the convection problem like the conduction problem. Equation 6.12 becomes

$$q = \frac{T_s - T_\infty}{R_{s\infty}} = \frac{T_s - T_\infty}{1/hA}$$

So the resistance to heat transfer at a convective surface is given by

$$R_{s\infty} = \frac{1}{hA} \qquad\qquad (6.13)$$

Heat Transfer Properties of Fluids

In previous chapters, the emphasis was on fluid mechanics types of problems. Fluid properties were discussed but these were isothermal properties. It is known that fluid properties do indeed vary with temperature. Although the Appendix Tables B.1 (*Properties of liquids at room temperature and pressure*) and C.1 (*Physical properties of gases at room temperature and pressure*) are sufficient for isothermal problems, it is desirable to have more extensive data available for the solution of heat transfer problems. Such data are provided in the appendix tables for several fluids. The B Tables in the appendix show properties of liquids and the C Tables show properties of gases (specific gravity, thermal conductivity, kinematic viscosity, thermal diffusivity, and Prandtl number).

6.4 Convection Heat Transfer Problems: Formulation and Solution

Exhaustive amounts of research have been performed in order to develop relations for the convection coefficient h in various geometries.

TABLE 6.3. *Some commonly encountered dimensionless groups.*

Ratio	Symbol	Name
hL/k	**Bi**	Biot number
$\mu V^2/[k_f(T_s - T_\infty)]$	**Br**	Brinkman number
$2D_f g_c/\rho V^2 D^2$	C_D	Drag coefficient
$2\Delta p D/\rho V^2 L$	f	Friction Factor
V^2/gL	**Fr**	Froude number
$g\beta(T_s - T_\infty)L^3/\nu^2$	**Gr**	Grashof number
hL/k_f	**Nu**	Nusselt number
VL/α	$\mathbf{Pe = Re \cdot Pr}$	Peclet number
$C_p\mu/k_f = \nu/\alpha$	**Pr**	Prandtl number
$2\Delta p g_c/\rho V^2$	C_p	Pressure coefficient
$g\beta(T_s - T_\infty)L^3/\nu\alpha$	$\mathbf{Ra = Gr \cdot Pr}$	Rayleigh number
$\rho VD/\mu g_c$	**Re**	Reynolds number
$h/\rho VC_p$	$St = \dfrac{\mathbf{Nu}}{\mathbf{Re \cdot Pr}}$	Stanton number
$\rho V^2 L/\sigma g_c$	**We**	Weber number

Results of such measurements are usually given in terms of dimensionless ratios. For example, the convection coefficient is traditionally represented by the **Nusselt number**, defined as

$$Nu = \frac{hL}{k_f} \tag{6.14}$$

where L is a characteristic length appropriate to the geometry of interest and k_f is the thermal conductivity of the fluid. Table 6.3 gives a list of some dimensionless groups encountered in fluid mechanics and heat transfer problems. Calculation of heat transfer rate using a convection coefficient equation is illustrated in the following example.

EXAMPLE 6.4. A 1 m tall vertical wall of a kitchen oven consists of three materials placed in series—sheet metal, insulation, and sheet metal. The sheet metal pieces are made of carbon steel and are 1 mm thick, while the (glass fiber) insulation is 4 cm thick. Inside the oven, the air temperature is 250°C and heat is convected to the wall. Determine the heat transferred through the wall if the outside surface is in contact with air at 25°C.

Solution: Figure 6.6 shows a cross section of the wall, as well as a temperature profile and the resistances to the flow of heat. Heat is convected to one surface, conducted through the wall, and convected from the outside surface. For both outside surfaces, the heat transfer process is one of natural convection. Heat is also transferred to the surroundings by radiation.

Assumptions:
> 1. The system is at steady state.
> 2. Properties of the materials are constant.
> 3. Air properties are constant and are evaluated at the appropriate temperatures.
> 4. Radiation heat transfer is neglected.
> 5. Resistance to heat flow within the sheet metal pieces is negligible so that the temperature of each is uniform throughout.

Reference to a heat transfer textbook shows that a number of equations are available to determine a natural convection coefficient for a vertical wall. Here we use the experimentally determined Churchill-Chu equation:

$$\text{Nu} = \frac{hL}{k_f} = 0.68 + \frac{0.67 \, \text{Ra}^{1/4}}{\left(1 + \left[\frac{0.492}{\text{Pr}}\right]^{9/16}\right)^{4/9}} \tag{6.15}$$

where

$$\text{Ra} = \frac{g\beta(T_s - T_\infty)L^3}{\nu\alpha} < 10^9; \qquad 0 < \text{Pr} = \frac{\nu}{\alpha} = < \infty$$

and

$$\beta = 1/T_\infty = \text{coefficient of thermal expansion}$$

For convection heat transfer on the outside of each piece of sheet metal, we will need air properties. The properties of air vary with temperature so we must first decide on the temperature to use for both cases. For the

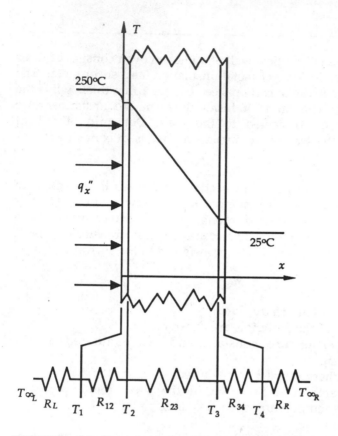

FIGURE 6.6. *Heat transfer through a vertical wall.*

left hand side (Figure 6.6), the air temperature is 250°C (+ 273 = 523 K), but near the surface of the sheet metal, the temperature is known to be somewhat lower. So for the left hand side, we *intuitively* elect to select properties at 500 K. Similarly for the right hand side, we have an air temperature of 25°C (= 298 K). We evaluate properties at 300 K. From Table C.2, we read for air at 500 K:

$$\rho = 0.705 \text{ kg/m}^3 \qquad\qquad C_p = 1\,029.5 \text{ J/(kg·K)}$$
$$k_f = 0.040\,38 \text{ W/(m·K)} \qquad \alpha = 0.556\,4 \times 10^{-4} \text{ m}^2/\text{s}$$
$$\nu = 37.90 \times 10^{-6} \text{ m}^2/\text{s} \qquad \text{Pr} = 0.68$$

For air at 300 K, we read

$$\rho = 1.177 \text{ kg/m}^3 \qquad\qquad C_p = 1\,005.7 \text{ J/(kg·K)}$$
$$k_f = 0.026\,24 \text{ W/(m·K)} \qquad \alpha = 0.221\,60 \times 10^{-4} \text{ m}^2/\text{s}$$
$$\nu = 15.68 \times 10^{-6} \text{ m}^2/\text{s} \qquad \text{Pr} = 0.708$$

Tables 6.1 and 6.2 give thermal conductivities of

sheet metal $k = 43 \text{ W}/(\text{m·K})$
glass fiber $k = 0.035 \text{ W}/(\text{m·K})$

The material thicknesses are

sheet metal $L = 0.001 \text{ m}$
glass fiber $L = 0.04 \text{ m}$

Heat is transferred through the wall, so we can write

$$q = \frac{T_{\infty L} - T_{\infty R}}{R_L + R_{12} + R_{23} + R_{34} + R_R} \tag{i}$$

Each resistance is found from

$$R_L = \frac{1}{h_L A} \qquad R_{12} = \frac{L_{12}}{k_{12} A} \qquad R_{23} = \frac{L_{23}}{k_{23} A}$$

$$R_{34} = \frac{L_{34}}{k_{34} A} \qquad R_R = \frac{1}{h_R A}$$

It is apparent that the temperature within the sheet metal pieces is uniform due to how thin they are; therefore

$$T_1 \approx T_2 \qquad T_3 \approx T_4 \qquad R_{12} = 0 \qquad R_{34} = 0$$

Each resistance contains a cross sectional area term A, and because area is unspecified, we assume an area of 1 m² and perform the calculations on a per square meter basis. The resistances are now determined as

$$R_L = \frac{1}{h_L} \qquad \text{(evaluated with Equation 6.15)} \qquad \text{(ii)}$$

$$R_{12} = 0 \qquad R_{23} = \frac{L_{23}}{k_{23} A} = \frac{0.04}{0.035(1)} = 1.143 \text{ K/W}$$

$$R_{34} = 0$$

$$R_R = \frac{1}{h_R} \qquad \text{(evaluated with Equation 6.15)} \qquad \text{(iii)}$$

To find the convection coefficients we start with the Churchill-Chu Equation:

$$\text{Nu} = \frac{hL}{k_f} = 0.68 + \frac{0.67\ \text{Ra}^{1/4}}{\left(1 + \left[\frac{0.492}{\text{Pr}}\right]^{9/16}\right)^{4/9}}$$ (6.15)

The length term in the above equation does not refer to a wall thickness but instead to a wall height. This was given as $L = 1$ m. So for the left side of the wall,

$$h_L = \frac{0.040\ 38}{1}\left\{0.68 + \frac{0.67\ \text{Ra}^{1/4}}{\left(1 + \left[\frac{0.492}{0.68}\right]^{9/16}\right)^{4/9}}\right\}$$

or $h_L = 0.027\ 46 + 0.020\ 66\ \text{Ra}^{1/4}$ (iv)

Likewise for the right side,

$$h_R = \frac{0.026\ 24}{1}\left\{0.68 + \frac{0.67\ \text{Ra}^{1/4}}{\left(1 + \left[\frac{0.492}{0.708}\right]^{9/16}\right)^{4/9}}\right\}$$

or $h_R = 0.017\ 84 + 0.013\ 49\ \text{Ra}^{1/4}$ (v)

The Rayleigh number is calculated with

$$\text{Ra} = \frac{g\beta(T_s - T_\infty)L^3}{\nu\alpha};\qquad\qquad \text{where } \beta = 1/T_\infty$$

For the left side, $\beta = 1/T_{\infty L} = 1/(250 + 273) = 0.001\ 912/\text{K}$, and for the right side, $\beta = 1/T_{\infty R} = 1/(25 + 273) = 0.003\ 356/\text{K}$. The Rayleigh numbers are now calculated as

$$\text{Ra}_L = \frac{(9.81)(0.001\ 912)(250 - T_1)}{(37.90 \times 10^{-6})(0.556\ 4 \times 10^{-4})} = 8.895 \times 10^6(250 - T_1)\quad\text{(vi)}$$

$$\text{Ra}_R = \frac{(9.81)(0.003\ 356)(T_4 - 25)}{(15.68 \times 10^{-6})(0.221\ 60 \times 10^{-4})} = 9.475 \times 10^7(T_4 - 25)\quad\text{(vii)}$$

In addition to Equation i above, we can write other equations for the heat transferred through the wall:

$$q = \frac{T_{\infty L} - T_2}{R_L + R_{12}} \qquad\qquad q = \frac{T_3 - T_{\infty R}}{R_{34} + R_R}$$

Rearranging these equations gives

$$T_2 = T_{\infty L} - q\,(R_L + R_{12}) \qquad\qquad \text{(viii)}$$

$$T_3 = T_{\infty R} + q\,(R_{34} + R_R) \qquad\qquad \text{(ix)}$$

Interface temperatures can be determined with the above equations when heat transfer rate is known.

We now formulate an iterative procedure to determine the heat transferred through the wall. The steps are as follows:

- Assume temperatures T_1 (= T_2) and T_3 (= T_4). Calculate:
- Rayleigh numbers $\mathbf{Ra}_L$ and $\mathbf{Ra}_R$ with Equations vi and vii;
- Convection coefficients h_L and h_R with Equations iv and v;
- Resistances R_L and R_R with Equations ii and iii;
- Heat transfer rate q with Equation i;
- Refined values of T_2 and T_4 with Equations viii and ix; and
- Repeat the calculations with the new interface temperatures.

The procedure is repeated until convergence within a tolerable limit is achieved. The following tables summarize the results of these calculations.

$T_1 = T_2$ (assumed)	$\mathbf{Ra}_L$ (Eq. vi)	h_L (iv)	R_L (ii)	q (i)	T_2 (viii)
225°C	2.224×10^8	2.550 W/(m²·K)	0.392 1 K/W	115.1 W	205°C
205	4.003×10^8	2.950	0.339 0	127.4	207
207	3.825×10^8	2.917	0.342 9	125.6	207
207	3.825×10^8	2.917	0.342 9	126.0	206.8
close enough					

$T_3 = T_4$ (assumed)	$\mathbf{Ra}_R$ (Eq. vii)	h_R (v)	R_R (iii)	q (i)	T_3 (ix)
35°C	9.475×10^8	2.385 W/(m²·K)	0.419 4 K/W	115.1 W	73.3°C
73.3	4.576×10^9	3.527	0.238 6	127.4	61.1
61.1	3.420×10^9	3.280	0.304 9	125.6	63.3
63.3	3.630×10^9	3.329	0.300 4	126.0	62.8
close enough					

The solution then is:

$$q = 126 \text{ W} \qquad \text{(for each } m^2 \text{ of surface)}$$

EXAMPLE 6.5. A horizontally laid 2-nom sch 40 steel pipe ($k = 25$ BTU/hr·ft·°R) is lagged with fiberglas insulation ($k = 0.02$ BTU/hr·ft·°R) that is 1 in. thick. The pipe conveys steam which maintains the inside surface temperature at 250°F. Air outside the insulation is at 80°F. Determine the heat loss through the pipe and insulation.

Solution: Figure 6.7 shows a cross section of the insulated pipe, as well as a temperature profile and the appropriate resistances to the flow of heat. Heat is transferred by conduction through the pipe wall and through the insulation. Heat is transferred by natural convection from the outside surface of the insulation to the surrounding air. Heat is also transferred to the surroundings by radiation.

Assumptions:
      1. The system is at steady state.
      2. Properties of the materials are constant.
      3. Air properties are constant and are evaluated at 80°F.
      4. Radiation heat transfer is neglected.

Reference to a heat transfer textbook shows that a number of equations are available to determine a natural convection coefficient for a horizontal cylinder. Here we use another experimentally determined equation developed by Churchill-Chu:

$$\text{Nu} = \frac{hD}{k_f} = \left\{ 0.60 + \frac{0.387 \, \text{Ra}^{1/6}}{\left(1 + \left[\dfrac{0.559}{\text{Pr}}\right]^{9/16}\right)^{8/27}} \right\}^2 \qquad (6.16)$$

where

$$10^{-5} < \text{Ra} = \frac{g\beta(T_s - T_\infty)L^3}{v\alpha} < 10^{12} ; \qquad 0 < \text{Pr} = \frac{v}{\alpha} < \infty$$

and

$$\beta = 1/T_\infty = \text{coefficient of thermal expansion}$$

From Appendix Table D.1, we read the following for 2-nom sch 40 pipe:

$$OD = 2.375 \text{ in.} \qquad ID = 0.1723 \text{ ft}$$

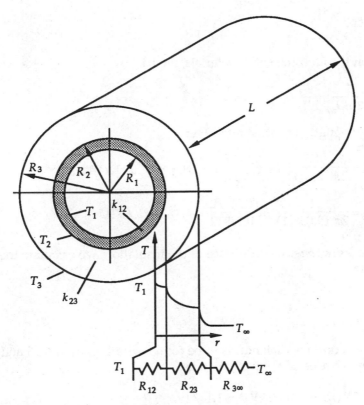

FIGURE 6.7. *Heat transfer from an insulated pipe with convection.*

In terms of the notation of this problem,

$$R_1 = 0.1723 \text{ ft}/2$$
$$R_2 = (2.375/12)/2 = 0.198 \text{ ft}/2$$
$$R_3 = ((2.375 + 2)/12)/2 = 0.365 \text{ ft}/2$$

The properties of air at $540°R = (80°F + 460)$ are obtained from Table C.2:

$$\rho = 0.0735 \text{ lbm/ft}^3 \qquad C_p = 0.240 \text{ BTU/lbm·°R}$$
$$k_f = 0.01516 \text{ BTU/hr·ft·°R} \qquad \alpha = 0.859 \text{ ft}^2/\text{hr}$$
$$v = 16.88 \times 10^{-5} \text{ ft}^2/\text{s} \qquad Pr = 0.708$$

We calculate $\beta = 1/(80 + 460) = 0.00185/°R$. The total heat transferred is

$$q = \frac{T_1 - T_\infty}{R_{12} + R_{23} + R_{3\infty}} \tag{i}$$

In addition, we write

$$q = \frac{T_1 - T_3}{R_{12} + R_{23}} \tag{ii}$$

For conduction through the solid materials, we have

$$R_{ij} = \frac{1}{2\pi k L} \, ln \frac{R_j}{R_i}$$

Assuming a unit length, we substitute to get

steel $\qquad R_{12} = \frac{1}{2\pi (25)(1)} \, ln \frac{0.198/2}{0.1723/2} = 0.000885 \text{ hr·°R/BTU}$

fiberglas $\quad R_{23} = \frac{1}{2\pi (0.02)(1)} \, ln \frac{0.365/2}{0.198/2} = 4.87 \text{ hr·°R/BTU}$

For convection from the outside surface of the insulation, we calculate the resistance using

$$R_{3\infty} = \frac{1}{h A}$$

where the convection coefficient h is to be found with Equation 6.16 and the outside surface area of the insulation is

$$A = \pi (2R_3)L = \pi (0.365)(1) = 1.147 \text{ ft}^2$$

Equation 6.16, however, depends on the outside surface temperature of the insulation (T_3) which is unknown at this point. We must therefore resort to an iterative procedure to find the convection coefficient h and ultimately the heat transferred q. We begin by substituting all known quantities into the parameters of Equation 6.16. The Rayleigh number is

$$Ra = \frac{g\beta(T_s - T_\infty)L^3}{\nu \alpha}$$

Now length L in the above equation for Rayleigh number refers to the axial length of insulated pipe. Because L was not specified, we assume a unit length (1 ft) and so our results apply on a per foot basis. Substituting gives

$$Ra = \frac{32.2(0.00185)(T_3 - 80)(1)^3}{16.88 \times 10^{-5}\,(0.859/3600)} = 1.48 \times 10^6 \, (T_3 - 80) \tag{iii}$$

The convection coefficient is found by rearranging Equation 6.16 slightly:

$$\text{Nu} = \frac{hD}{k_f} = \left\{ 0.60 + \frac{0.387 \, \text{Ra}^{1/6}}{\left(1 + \left[\frac{0.559}{\text{Pr}}\right]^{9/16}\right)^{8/27}} \right\}^2 \qquad (6.16)$$

$$h = \frac{k_f}{2R_3} \left\{ 0.60 + \frac{0.387 \, \text{Ra}^{1/6}}{\left(1 + \left[\frac{0.559}{\text{Pr}}\right]^{9/16}\right)^{8/27}} \right\}^2$$

Substituting gives

$$h = \frac{0.01516}{0.365} \left\{ 0.60 + \frac{0.387 \, \text{Ra}^{1/6}}{\left(1 + \left[\frac{0.559}{0.708}\right]^{9/16}\right)^{8/27}} \right\}^2$$

or $h = 0.0415 \, (0.6 + 0.321 \, \text{Ra}^{1/6})^2$ (iv)

The iterative procedure is as follows:

- Assume T_3; then calculate:
- Rayleigh number **Ra** from Equation iii;
- Convection coefficient h from Equation iv;
- Resistance $R_{3\infty} = 1/hA = 1/(1.147h)$;
- Total resistance $R_{1\infty} = R_{12} + R_{23} + R_{3\infty}$
  $$= 0.000885 + 4.87 + R_{3\infty} = 4.87 + R_{3\infty}$$
- Heat transferred $q = (T_1 - T_\infty)/R_{1\infty} = (250 - 80)/R_{1\infty} = 170/R_{1\infty}$
- Refined value of the surface temperature $T_3 = T_1 - q(R_{12} + R_{23})$
  $$T_3 = 250 - 4.87q \quad \text{(from Equation ii); and}$$
- Repeat the calculations until convergence is achieved.

The following table summarizes the results:

T_3	Ra (Eq iii)	h (iv)	$R_{3\infty} = 1/hA$	$R_{1\infty}$	q	T_3 (ii)
100°F	2.96×10^7	1.62	0.538	5.41	31.4	96.9
96.9	2.50×10^7	1.54	0.566	5.44	31.3	97.7
98	2.66×10^7	1.57	0.556	5.43	31.3	97.4
close enough						

The calculations show that heat transfer rate q is comparatively insensitive to large changes in temperature. For example, if the assumed value of T_3 is 200°F, then the heat transfer rate is 32.8 BTU/hr. For this example, the solution is

$q = 31.3$ BTU/hr   (for each ft of pipe)

6.5 Summary

In this chapter, we have reviewed simple one dimensional conduction heat transfer in planar and cylindrical coordinates. We defined the concept of resistance to heat transfer and derived an equation for the conduction problem. Convection heat transfer was introduced as was the pertinent fluid properties for convection problems. The concept of a resistance to heat transfer was extended to convection problems, and several example problems were solved.

6.6 Problems Chapter 6

1. An outdoor grill is made of masonry brick. Consider one wall of the grill made of a single layer of brick that is 4 in. thick. During operation, the inside surface temperature reaches 180°F, and the outside surface temperature reaches 80°F. Determine the heat transferred through the brick wall.

2. A safe is made of stainless steel with walls that are 6 cm thick. It is to be designed so that, under conditions of fire, the heat transferred through the wall is to be no greater than 100 BTU/(hr·ft^2). The temperature of the outside surface of the safe will reach 400°C. Under these conditions, what will the temperature of the inside surface be?

3. A furnace wall is to be made of two materials placed in series—common brick and masonry brick. The heat loss through the wall is to be reduced to 1 000 W/m^2. The common brick is 8 in. thick and its left surface will reach 1500°F. The masonry brick is placed next to the common brick and the right surface of the masonry brick will reach 100°F. Determine the thickness of the masonry brick required.

4. Figure P6.4 shows the one dimensional profile of two materials. The temperature of the left face is the same for both materials, the heat flow rate in both cases is identical, and the thicknesses are equal. Which material has the higher thermal conductivity? Why?

5. Referring to the composite wall of Figure 6.3, which material has the highest thermal conductivity?

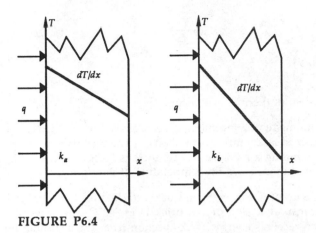

FIGURE P6.4

6. A wall of a kitchen oven is made of 3 materials placed in series—sheet metal, insulation, and plywood. The sheet metal is 0.040 in. thick. The plywood is 3/4 in. thick and the insulation thickness is to be determined. For economic reasons, the oven wall should transfer about 60 BTU/(hr·ft^2) of heat when the inside surface temperature of the sheet metal is 500°F. For safety reasons, the plywood outside surface temperature should not exceed 100°F. What is the minimum thickness of fiberglas insulation required to meet these criteria?

7. A 4-nom sch 40 steel pipe conveys steam which maintains the pipe inside surface temperature at 450°F. The pipe has a 1 in. thick layer of kapok insulation. The outside surface temperature of the kapok is 70°F. Determine the heat flow through the pipe.

8. An aluminum rod is 1 cm in diameter and is used as a handle for a grill cover. The aluminum rod temperature reaches a uniform 80°C. The rod is to be covered with a high temperature plastic such that the outside surface of the plastic is 20°C for a heat flow rate of 50 W. If the thickness of the plastic is 5 mm, determine the thermal conductivity that the plastic must have.

9. Suppose that the pipe of Example 6.5 is not insulated? Re-work the problem for no insulation and compare the heat flow rate of the insulated to the uninsulated pipe.

10. Example 6.5 was solved for a pipe that is laid horizontally. Suppose the same pipe is oriented vertically? Re-work the problem for a vertical configuration and compare the results of the two solutions. For a heated vertical cylinder losing heat to the environment, we have

$$\text{Nu} = 0.6 \left(\text{Ra}\,\frac{D}{L} \right)^{0.25} \qquad\qquad \text{Ra}\,\frac{D}{L} \ge 10^4 \qquad\qquad (6.17a)$$

$$\text{Nu} = 1.37 \left(\text{Ra}\,\frac{D}{L} \right)^{0.16} \qquad\qquad 0.05 \le \text{Ra}\,\frac{D}{L} \le 10^4 \qquad\qquad (6.17b)$$

$$\mathbf{Nu} = 0.93 \left(\mathbf{Ra}\,\frac{D}{L}\right)^{0.05} \qquad\qquad \mathbf{Ra}\,\frac{D}{L} \le 0.05 \qquad\qquad (6.17c)$$

$$\mathbf{Ra} = \frac{g\beta(T_s - T_\infty)L^3}{v\alpha}; \qquad\qquad \mathbf{Pr} = \frac{v}{\alpha}$$

and $\beta = 1/T_\infty$ = coefficient of thermal expansion.

11. Consider a vertical wall made of 2 in. thick stainless steel. The left face of the stainless steel is maintained at a temperature of 300°F. The right face convects heat away to the surrounding air whose temperature is 75°F. Determine the heat transferred through the wall and the temperature of the right face.

12. A vertical wall is made up of fiberglas (5 cm thick) attached to common brick (10 cm thick). The uninsulated side of the brick is at -40°C during winter months. The insulation receives energy by convection from the surrounding air whose temperature is 20°C. Determine the interface temperature between the two materials under these conditions.

13. A vertical pane of window glass of a house is 1/2 in. thick and 3 ft tall. On the outside, the air temperature is 95°F, while inside the air temperature is 70°F. Determine the heat flow through the glass for these conditions.

14. A new building material made of plywood with styrofoam insulation glued to it is to be tested. The plywood is 2 cm thick and the styrofoam is 7 cm thick. The test is to be performed while the sample is in a vertical configuration. On the plywood side, the sample is exposed to air at 35°C. On the styrofoam side, the sample is in contact with air at 0°C. Calculate the interface temperature between the two materials and the heat transferred through the wall.

15. A 3-nom sch 40 stainless steel pipe carries high temperature oil from a cracking tower, where crude oil is separated into components, to a packaging operation. The oil maintains the inside surface of the pipe at 180°F. The air temperature surrounding the pipe is at 60°F. Determine the heat flow through the pipe wall for the following: (a) the pipe is uninsulated, and (b) the pipe is insulated with a 1 in. thickness of kapok.

16. A horizontal pipe made of stainless steel (2-nom sch 40) carries oil (same properties as unused engine oil) whose temperature is 100°C and whose velocity is 0.1 m/s. Heat is convected from the steam to the inside surface of the stainless steel. The pipe is insulated with glass fiber that is 2 cm thick. Air surrounding the pipe is at 25°C. Determine the heat transferred from the oil through the pipe and insulation to the air. For laminar flow of fluid through a cylinder, we have

$$\mathbf{Nu} = \frac{hD}{k_f} = 1.86 \left(\frac{D\ \mathbf{Re}\ \mathbf{Pr}}{L}\right)^{1/3} \qquad\qquad \mathbf{Re} = \frac{VD}{v} < 2\,200$$

$$0.48 < \mathbf{Pr} = \frac{v}{\alpha} < 16\,700 \qquad \mu \text{ changes moderately with temperature}$$

CHAPTER 7   Analysis of Heat Exchangers I

A heat exchanger is a device used to transfer heat from one fluid to another. There are many different types of heat exchangers including **double pipe, shell and tube,** and **cross flow.** In this and in the next chapter, we will examine these three types and write equations to predict their performance. The methods presented here can be applied to the performance appraisal of any heat exchanger.

7.1 The Double Pipe Heat Exchanger

A double pipe heat exchanger consists of two concentric, different diameter tubes with fluid flowing in each, as indicated in Figures 7.1 and 7.2. If the two fluids travel in opposite directions as illustrated in Figure 7.1, the exchanger is a **counter flow** type. If the fluids travel in the same direction as shown in Figure 7.2, **parallel flow** exists. The same apparatus is used for either flow configuration. Also shown in Figures 7.1 and 7.2 are temperature versus distance graphs.

The objective in using a heat exchanger is to transfer as much heat as possible for as small a cost as necessary. In many cases involving the sizing or selection of a particular exchanger, all that will be known are the physical properties of the fluids and their inlet temperatures. If the tube sizes and areas (surface and cross sectional areas) are known, then the amount of heat transferred can be readily calculated. Conversely, if a heat transfer rate is specified, then the required surface area can be determined.

As indicated in Figures 7.1 and 7.2, a double pipe heat exchanger will bring together two fluid streams. As each fluid passes through, its temperature changes. Moreover, as temperature changes, the *local* convection coefficient between either fluid and the wall changes. Our interest here, however, is in the *overall* heat transfer coefficient and not necessarily with the instantaneous or local values.

Figure 7.3 shows a cross section of the double pipe heat exchanger and the associated temperature variation at any axial location. For purposes of discussion, we assume that heat is transferred from the fluid

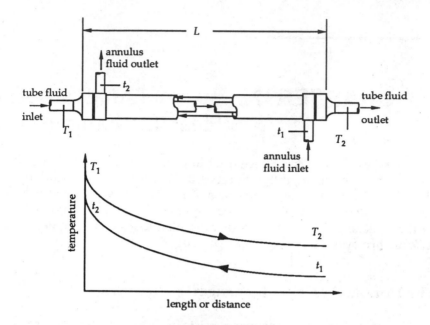

FIGURE 7.1. *A double pipe heat exchanger set up in counterflow and the corresponding temperature profile.*

within the tube to the fluid within the annulus. Also shown in Figure 7.3 are the resistances through which heat passes. The sum of the resistances is

$$\Sigma R = R_{12} + R_{23} + R_{34}$$

$$\Sigma R = \frac{1}{h_i A_i} + \frac{1}{2\pi k L} \ln \frac{OD_p}{ID_p} + \frac{1}{h_o A_o} \tag{7.1}$$

where h_i is the convection coefficient between the fluid in the tube and the tube wall, and h_o applies between the fluid in the annulus and the tube. It is remembered from the last chapter that temperature drop across a thin walled metal pipe is virtually negligible. This is true also for a tube. The implication here is that the second resistance $[(1/2\pi k L)(\ln(OD_p/ID_p))]$ may be neglected in Equation 7.1 with small error. Also, the area associated with the convection coefficient h_i is the *inside* surface area of the tube A_i. The area associated with the convection coefficient h_o is the *outside* surface area of the tube A_o. It is necessary in the analysis that h_i and h_o be referred to the same surface. The surface areas are:

$$A_i = \pi \, ID_p \, L \qquad\qquad\qquad A_o = \pi \, OD_p \, L$$

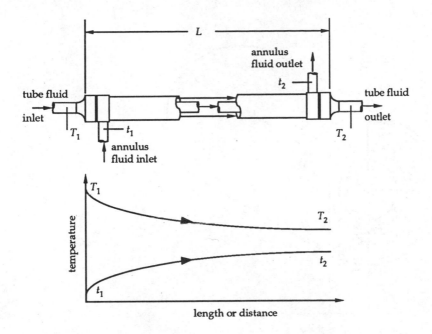

FIGURE 7.2. *A double pipe heat exchanger set up in parallel flow and the corresponding temperature profiles.*

It is standard practice to base the resistance on the outside surface area A_o. Multiplying Equation 7.1 by A_o, we get

$$A_o \Sigma R = \frac{1}{U_o} = \frac{OD_p}{h_i ID_p} + \frac{1}{h_o} \qquad (7.2)$$

in which the overall heat transfer coefficient U_o based on A_o has been introduced. The overall heat transfer coefficient has the same dimensions as h, namely [F·T/(T·L²·t)] [BTU/hr·ft²·°R or W/(m²·K)]. (Note also that $A_i U_i = A_o U_o$ and the development could proceed using U_i.)

The heat transferred within the heat exchanger equals the product of the overall heat transfer coefficient U_o, the outside surface area of the inner tube A_o, and a temperature difference. Thus

$$q = U_o A_o \Delta t \quad (= U_i A_i \Delta t) \qquad (7.3)$$

The temperature difference in the above equation is referred to as the **log mean temperature difference** defined as

$$\Delta t = \frac{\text{temp difference one end - temp diff opposite end}}{In\left(\dfrac{\text{temp diff one end}}{\text{temp diff opposite end}}\right)}$$

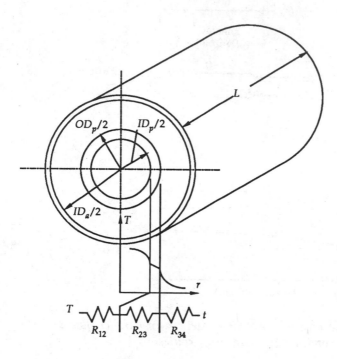

FIGURE 7.3. *Temperature profile and resistances for heat flow within a double pipe heat exchanger.*

or $\quad \Delta t = LMTD = \dfrac{(T_1 - t_2) - (T_2 - t_1)}{ln\,[(T_1 - t_2)/(T_2 - t_1)]}$ $\qquad$ (counterflow) $\qquad$ (7.4)

$\Delta t = LMTD = \dfrac{(T_1 - t_1) - (T_2 - t_2)}{ln\,[(T_1 - t_1)/(T_2 - t_2)]}$ $\qquad$ (parallel flow) $\qquad$ (7.5)

Uppercase "T" signifies the warmer fluid and a "1" subscript denotes an inlet condition. Lowercase "t" refers to the cooler fluid and a "2" subscript denotes an exit condition. Log mean temperature difference is often abbreviated as *LMTD*.

Comparison of Counterflow and Parallel Flow Configurations

At first glance, it appears that counterflow and parallel flow arrangements should yield equal heat transfer rates. To investigate this point, it is instructive to use Equations 7.4 and 7.5 to calculate the log mean temperature difference for several cases.

EXAMPLE 7.1. A fluid with a temperature of 100°C enters a double pipe heat exchanger and is cooled to 75°C by a second fluid entering at 25°C and heated to 40°C. Calculate the log mean temperature difference for counterflow and parallel flow.

Solution: Given the following temperatures,

$$T_1 = 100°C \qquad\qquad t_1 = 25°C$$
$$T_2 = 75°C \qquad\qquad t_2 = 40°C$$

we substitute into Equation 7.4 to obtain

$$LMTD = \frac{(100 - 40) - (75 - 25)}{ln\,[(100 - 40)/(75 - 25)]} = 54.8°C \qquad \text{(counterflow)}$$

Also,

$$LMTD = \frac{(100 - 25) - (75 - 40)}{ln\,[(100 - 25)/(75 - 40)]} = 52.5°C \qquad \text{(parallel flow)}$$

Equation 7.3 is used in both cases to calculate the heat transfer rate. If the overall heat transfer coefficient is constant, then the parallel flow configuration requires a greater surface area A_o than the counterflow arrangement to transfer the same energy.

EXAMPLE 7.2. A double pipe heat exchanger is to be used to exchange heat between two fluids such that their outlet temperatures are equal; specifically, a warmer fluid is cooled from 300°F to 200°F while the cooler fluid is heated from 150°F to 200°F. Calculate the log mean temperature difference for counterflow and for parallel flow.

Solution: Given the following temperatures,

$$T_1 = 300°F \qquad\qquad t_1 = 150°F$$
$$T_2 = 200°F \qquad\qquad t_2 = 200°F$$

we substitute into Equation 7.4 to get

$$LMTD = \frac{(300 - 200) - (200 - 150)}{ln\,[(300 - 200)/(200 - 150)]} = 72.5°F \qquad \text{(counterflow)}$$

Likewise,

$$LMTD = \frac{(300 - 150) - (200 - 200)}{ln\,[(300 - 150)/(200 - 200)]} = 0°F \qquad \text{(parallel flow)}$$

So, in parallel flow, the surface area of the heat exchanger would have to be infinite in order to make the outlet temperatures equal (as per Equation 7.3, $q = U_o A_o \Delta t$). This not being feasible, we conclude that there is a distinct thermal disadvantage to using parallel flow. Consequently, unless specified otherwise, all calculations made on double pipe heat exchangers will be performed using counterflow.

Referring to Figures 7.1 and 7.2, it is noted that the outlet temperatures in parallel flow can only approach each other. For counterflow, the outlet temperature of the cooler fluid can be made to exceed the outlet temperature of the warmer fluid. The counterflow apparatus has a much greater ability to transfer heat than does the parallel flow apparatus.

A double pipe heat exchanger is easily made using standard pipe or tubing fittings, or one can be purchased. It provides an inexpensive way to transfer heat between two fluids having relatively low flow rates. Double pipe heat exchangers can be assembled in any desired length and often two are used in what is known as a "hairpin" arrangement, as sketched in Figure 7.4.

7.2 Analysis of Double Pipe Heat Exchangers

Equation 7.3 relates the heat transferred in a double pipe heat exchanger to the overall heat transfer coefficient, the outside surface area of the inner tube, and to the log mean temperature difference (LMTD):

$$q = U_o A_o \Delta t \tag{7.3}$$

Equation 7.2 defines the heat transfer coefficient U_o in terms of convection coefficients h_i and h_o for the inside and outside surfaces of the inner tube. Convection coefficients are calculated from equations for Nusselt numbers. The equations to be used in the analysis were developed for flow in a circular duct. In order to find the convection coefficient for flow in an annulus, we use the same equations but modify them for the change in geometry. Modifications will involve the use of a characteristic length to replace diameter in the equations. For friction factor calculations, we defined hydraulic diameter as

$$D_h = \frac{4 \text{ flow area}}{\text{friction perimeter}} = \frac{4\pi (ID_a^2 - OD_p^2)}{4\pi (ID_a + OD_p)}$$

or $$D_h = ID_a - OD_p \tag{7.6}$$

where the diameters ID_a and OD_p are defined in Figure 7.6.

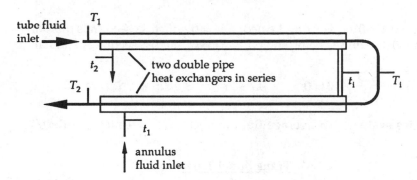

FIGURE 7.4. *Schematic of a hairpin heat exchanger.*

The above definition of hydraulic diameter applies to friction calculations. In the same way, we define an **equivalent diameter** D_e as

$$D_e = \frac{4 \text{ flow area}}{\text{heat transfer perimeter}} = \frac{4\pi (ID_a^2 - OD_p^2)}{4\pi OD_p}$$

or $$D_e = \frac{ID_a^2 - OD_p^2}{OD_p} \qquad (7.7)$$

The heat transfer perimeter is the outside surface area of the inner tube. Note carefully the difference in hydraulic and equivalent diameters. Once these diameters are calculated, they can be used in equations for Reynolds and Nusselt numbers. The equations are:

Sieder-Tate Equation for laminar flow

$$Nu = \frac{hD}{k_f} = 1.86 \left(\frac{D \, Re \, Pr}{L} \right)^{1/3} \qquad (7.8)$$

$$Re = \frac{VD}{v} < 2\,200 \qquad\qquad D = ID_p \text{ if cross section is tubular}$$
$$D = D_e \text{ if cross section is annular}$$

$$0.48 < Pr = \frac{v}{\alpha} < 16\,700; \quad \mu \text{ changes moderately with temperature}$$

Properties evaluated at average fluid temperature [= (inlet + outlet)/2]

Modified Dittus-Boelter Equation for turbulent flow

$$Nu = \frac{hD}{k_f} = 0.023(Re)^{4/5} Pr^n \qquad (7.9)$$

n = 0.4 if fluid is being heated; $D = ID_p$ if cross section is tubular
n = 0.3 if fluid is being cooled; $D = D_e$ if cross section is annular

$$\text{Re} = \frac{VD}{v} \geq 10\,000; \qquad 0.7 \leq \text{Pr} = \frac{v}{\alpha} \leq 160; \qquad L/D \geq 60$$

Properties evaluated at average fluid temperature [= (inlet + outlet)/2]

Transitional Flow

When the Reynolds number falls between 2 200 and 10 000, then interpolation can be used to find the Nusselt number. Equation 7.8 is used to find Nu at Re = 2 200. Equation 7.9 is used to find Nu at Re = 10 000. Interpolation gives a usable *estimate* of Nusselt number for transition Reynolds number.

A quantity known as the **mass velocity** is widely used in heat exchanger analyses. The mass velocity G is defined as

$$G = \frac{\dot{m}}{A} = \rho V$$

and has dimensions of $M/(L^2 \cdot T)$ [$lbm/(ft^2 \cdot s)$ or $kg/(m^2 \cdot s)$].

In Equations 7.8 and 7.9, equivalent diameter D_e would substitute for D when applying these equations to an annular duct. Note that when Reynolds number is being calculated for evaluating friction factor, D_h is used. When heat transfer effects are being modeled, D_e is used. Thus, for flow in an annulus, there will be two Reynolds numbers: one based on hydraulic diameter for finding friction factor, and one based on equivalent diameter for calculating heat transfer rates.

Outlet Temperature Equations

When a specific heat exchanger is selected, then the geometry is fixed. The only controls the operator has are flow rate and perhaps inlet temperature. In order to determine how effective the heat exchanger operates, it is necessary to calculate outlet temperature. The equations for calculating outlet temperature are presented here without detailed derivation. For the cooler fluid, we write

$$q = \dot{m}_c C_{pc} (t_2 - t_1) \tag{7.10}$$

where $\dot{m}_c$ is the mass flow rate of the cooler fluid and C_{pc} is its specific heat. Assuming counterflow, we write

$$q = U_o A_o \Delta t = U_o A_o \frac{(T_1 - t_2) - (T_2 - t_1)}{\ln \left[(T_1 - t_2)/(T_2 - t_1) \right]}$$ (7.11)

Setting these two expressions equal to one another and rearranging gives

$$\ln \frac{(T_1 - t_2)}{(T_2 - t_1)} = \frac{UA}{\dot{m}_c C_{pc}} \left(\frac{T_1 - T_2}{t_2 - t_1} - 1 \right)$$ (7.12)

For the warmer fluid,

$$q = \dot{m}_w C_{pw}(T_1 - T_2)$$ (7.13)

where $\dot{m}_w$ is the mass flow rate of the cooler fluid and C_{pw} is its specific heat. Assuming all heat lost by the warmer fluid is gained by the cooler fluid, we set Equation 7.10 equal to 7.13:

$$\dot{m}_w C_{pw} (T_1 - T_2) = \dot{m}_c C_{pc} (t_2 - t_1)$$

Rearranging and introducing a new variable R, we have

$$\frac{\dot{m}_c C_{pc}}{\dot{m}_w C_{pw}} = \frac{T_1 - T_2}{t_2 - t_1} = R$$ (7.14)

Substituting into Equation 7.12 and removing logarithms gives

$$\frac{T_1 - t_2}{T_2 - t_1} = \exp \left[\frac{U_o A_o}{\dot{m}_c C_{pc}} (R - 1) \right] = E_c$$ (7.15)

where the notation E_c has been introduced and defined. Equation 7.14 can be rearranged to give an equation for the outlet temperature of the cooler fluid:

$$t_2 = t_1 + \frac{T_1 - T_2}{R}$$ (7.16)

Substituting into Equation 7.15, we get an equation for the outlet temperature of the warmer fluid as

$$T_2 = \frac{(1 - R)T_1 + (1 - E_c)Rt_1}{1 - RE_c} \qquad \text{(counterflow)} \qquad (7.17)$$

The above equation allows us to calculate the outlet temperature of the warmer fluid knowing flow rate, fluid properties, and only the inlet temperatures. Once the outlet temperature of the warmer fluid T_2 is known, then Equation 7.16 can be used to find the outlet temperature for the cooler fluid t_2.

A similar analysis can be performed for parallel flow. We begin by defining:

$$E_p = \exp\left[\frac{U_oA_o}{\dot{m}_cC_{pc}}(R + 1)\right] \qquad (7.18)$$

Following the same line of reasoning as in the counterflow case, the outlet temperature of the warmer fluid becomes

$$T_2 = \frac{(R + E_p)T_1 + (E_p - 1)Rt_1}{(R + 1)E_p} \qquad \text{(parallel flow)} \qquad (7.19)$$

Again, once the outlet temperature of the warmer fluid T_2 is known, the outlet temperature of the cooler fluid t_2 is found with Equation 7.16.

Fouling Factors

When a heat exchanger is in service for a certain amount of time, scale and dirt will deposit on the surfaces of the tubes. These deposits reduce the rate of heat transfer between the fluids by increasing the resistance to heat flow through the inner tube wall. Figure 7.5 shows a cross section of the double pipe heat exchanger with these additional resistances.

The additional resistances on the inside and outside surfaces are identified as R_{di} and R_{do}, respectively. They affect the overall heat transfer coefficient defined earlier in Equation 7.2 as:

$$\frac{1}{U_o} = \frac{OD_p}{h_i\,ID_p} + \frac{1}{h_o} = \frac{1}{h_p} + \frac{1}{h_o} \qquad (7.2)$$

The above equation applies when the heat exchanger is new and the tubes are clean. To reflect the added resistances due to surface deposits, we define a *dirty* or *design* coefficient as

$$\frac{1}{U} = \frac{1}{U_o} + R_{di} + R_{do} \qquad (7.20a)$$

or

$$\frac{1}{U} = \frac{OD_p}{h_i \, ID_p} + \frac{1}{h_o} + R_{di} + R_{do}$$
(7.20b)

Values of the resistances for various fluids have been measured as a result of years of experience and are provided in Table 7.1. Note that the resistances are actually referred to specific areas. Because the resistance values are only best estimates, an area correction need not be used with R_{di} and R_{do} in Equation 7.20.

Pressure Drop in Pipes and Annuli

The pressure drop for flow in a tube is easily calculated with methods of previous chapters. For flow in a tube, we write

$$\Delta p_t = \frac{fL}{D_h}\frac{\rho V^2}{2g_c} = \frac{fL}{ID_p}\frac{\rho V^2}{2g_c} \qquad \text{(tube flow)} \qquad (7.21)$$

where friction factor f is obtained from Reynolds number and ε/ID_p data.

We use the same equation for finding pressure drop in an annulus except we substitute hydraulic diameter for the characteristic dimension in the Reynolds number and relative roughness expressions. In addition, we account for inlet and outlet fittings by use of a minor loss term; i.e.,

$$\Delta p_a = \left(\frac{fL}{D_h} + 1\right)\frac{\rho V^2}{2g_c} \qquad \text{(annular flow)} \qquad (7.22)$$

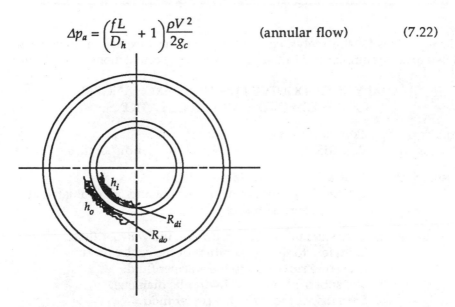

FIGURE 7.5. *Cross section of double pipe heat exchanger showing additional resistances due to fouling.*

TABLE 7.1. *Values of fouling factors for various fluids.*

Fluid	R_d (m^2·K/W)	ft^2·hr·°R/BTU
Air	0.000 4	0.002
Brine	0.000 2	0.001
Alcohol Vapor	0.000 1	0.0005
Diesel Engine Exhaust	0.002	0.01
Engine Oil	0.000 2	0.001
Organic Vapors	0.000 1	0.0005
Organic Liquids	0.000 2	0.001
Refrigerant Liquid	0.000 2	0.001
Refrigerant Vapors	0.000 4	0.002
Steam	0.000 1	0.0005
Vegetable Oil	0.000 6	0.003
Water		
City Water	0.000 2—0.000 4	0.001—0.002
Distilled	0.000 1—0.000 2	0.0005—0.001
Seawater	0.000 1—0.000 2	0.0005—0.001
Well Water	0.000 2—0.000 4	0.001—0.002

The equations for the analysis of a double pipe heat exchanger have been stated and are summarized in a suggested order of calculations procedure.

ANALYSIS OF DOUBLE PIPE HEAT EXCHANGERS
SUGGESTED ORDER OF CALCULATIONS

Problem   Complete problem statement

Discussion   Potential heat losses; other sources of difficulties

Assumptions   1. Steady state conditions exist.
                2. Fluid properties remain constant and are evaluated at a temperature of _____ .

Nomenclature   1. T refers to the temperature of the warmer fluid.
                 2. t refers to the temperature of the cooler fluid.
                 3. w subscript refers to the warmer fluid.
                 4. h subscript refers to hydraulic diameter.
                 5. c subscript refers to the cooler fluid.
                 6. a subscript refers to the annular flow area or dimension.
                 7. p subscript refers to the tubular flow area or dimension.

8. 1 subscript refers to an inlet condition.
9. 2 subscript refers to an outlet condition.
10. e subscript refers to equivalent diameter.

A. Fluid Properties

$$\dot{m}_w \;=\qquad\qquad\qquad\qquad T_1 \;=$$
$$\rho \;=\qquad\qquad\qquad\qquad\qquad C_p \;=$$
$$k_f \;=\qquad\qquad\qquad\qquad\quad \alpha \;=$$
$$\nu \;=\qquad\qquad\qquad\qquad\quad \mathrm{Pr} \;=$$

$$\dot{m}_c \;=\qquad\qquad\qquad\qquad t_1 \;=$$
$$\rho \;=\qquad\qquad\qquad\qquad\qquad C_p \;=$$
$$k_f \;=\qquad\qquad\qquad\qquad\quad \alpha \;=$$
$$\nu \;=\qquad\qquad\qquad\qquad\quad \mathrm{Pr} \;=$$

B. Tubing Sizes

$$ID_a \;=$$
$$ID_p \;=\qquad\qquad\qquad\qquad OD_p =$$

C. Flow Areas $A_p = \pi ID_p^2/4 =$
$$A_a = \pi(ID_a^2 - OD_p^2)/4 =$$

D. Fluid Velocities [Route the fluid with the higher flow rate through the flow cross section with the greater area.]

$$V_p = \dot{m}/\rho A = \qquad\qquad\qquad G_p = \dot{m}/A =$$
$$V_a = \dot{m}/\rho A = \qquad\qquad\qquad G_a = \dot{m}/A =$$

E. Annulus Equivalent Diameters

Friction $D_h = ID_a - OD_p =$

Ht Trans $D_e = (ID_a^2 - OD_p^2)/OD_p =$

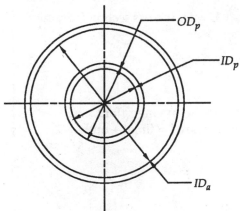

FIGURE 7.6. *Definition sketch of diameters associated with an annulus.*

F. Reynolds Numbers

$$Re_p = V_p ID_p / v =$$

$$Re_a = V_a D_e / v =$$

G. Nusselt Numbers

Modified Sieder-Tate Equation for laminar flow:

$$Nu = \frac{hD}{k_f} = 1.86 \left(\frac{D \ Re \ Pr}{L} \right)^{1/3}$$

$$Re = \frac{VD}{v} < 2\,200 \qquad D = ID_p \text{ if cross section is tubular}$$
$$D = D_e \text{ if cross section is annular}$$

$$0.48 < Pr = \frac{v}{\alpha} < 16\,700$$

μ changes moderately with temperature
Properties evaluated at the average fluid
    temperature [= (inlet + outlet)/2]

Modified Dittus-Boelter Equation for turbulent flow:

$$Nu = \frac{hD}{k_f} = 0.023 (Re)^{4/5} \ Pr^n$$

n = 0.4 if fluid is being heated
n = 0.3 if fluid is being cooled

$$Re = \frac{VD}{v} \geq 10\,000 \qquad D = ID_p \text{ if cross section is tubular}$$
$$D = D_e \text{ if cross section is annular}$$

$$0.7 \leq Pr = \frac{v}{\alpha} \geq 160$$

$$L/D \geq 60$$

Properties evaluated at the average fluid
    temperature [= (inlet + outlet)/2]

$$Nu_p =$$

$$Nu_a =$$

H. Convection Coefficients

$$h_i = Nu_p \, k_f/ID_p = \qquad\qquad h_p = h_i \, ID_p/OD_p =$$

$$h_a = Nu_a \, k_f/D_e =$$

I. Exchanger Coefficient

$$\frac{1}{U_o} = \frac{1}{h_p} + \frac{1}{h_a} \qquad\qquad U_o =$$

J. Outlet Temperature Calculations (Exchanger length $L =$)

$$R = \frac{\dot{m}_c C_{pc}}{\dot{m}_w C_{pw}} = \qquad\qquad A_o = \pi OD_p L =$$

Counterflow $\quad E_{counter} = \exp[U_o A_o (R - 1)/\dot{m}_c C_{pc}] =$

$$T_2 = \frac{T_1(R - 1) - Rt_1(1 - E_{counter})}{RE_{counter} - 1}$$

$$t_2 = t_1 + \frac{T_1 - T_2}{R}$$

Parallel Flow $\quad E_{para} = \exp[U_o A_o (R + 1)/\dot{m}_c C_{pc}] =$

$$T_2 = \frac{(R + E_{para})T_1 + Rt_1(E_{para} - 1)}{(R + 1)E_{para}}$$

$$t_2 = t_1 + \frac{T_1 - T_2}{R}$$

$$T_2 =$$

$$t_2 =$$

K. Log Mean Temperature Difference

Counterflow $\quad LMTD = \dfrac{(T_1 - t_2) - (T_2 - t_1)}{\ln\,[(T_1 - t_2)\,/(T_2 - t_1)]} =$

Parallel Flow $\quad LMTD = \dfrac{(T_1 - t_1) - (T_2 - t_2)}{\ln\,[(T_1 - t_1)\,/(T_2 - t_2)]} =$

L. Heat Balance

$$q_w = \dot{m}_w C_{pw}(T_1 - T_2) =$$

$$q_c = \dot{m}_c C_{pc}(t_2 - t_1) =$$

$$q = U_o A_o LMTD =$$ (clean)

M. Fouling Factors and Design Coefficient

$$R_{di} = \qquad\qquad R_{do} =$$

$$\frac{1}{U} = \frac{1}{U_o} + R_{di} + R_{do} \qquad\qquad U =$$

N. Heat Transfer Area and Tube Length (unless already known)

$$A_o = \frac{q}{U\,(LMTD)} =$$

$$L = \frac{A_o}{\pi\,(OD_p)} =$$

O. Friction Factors

$$Re_p = V_p ID_p/\nu =$$

$$\frac{\varepsilon}{ID_p} =$$

$$\left.\right\} \quad f_p =$$

$$Re_a = V_a D_h/\nu =$$

$$\frac{\varepsilon}{D_h} =$$

$$\left.\right\} \quad f_a =$$

<u>Laminar Flow Equations</u>

Laminar flow in a tube $f_p = \dfrac{64}{Re_p}$ $\qquad Re_p = \dfrac{V_p ID_p}{\nu} \le 2\,200$

Laminar flow in an annulus $\kappa = \dfrac{OD_p}{ID_a}$ $\qquad Re_a = \dfrac{V_a D_h}{\nu} \ge 10\,000$

$$\frac{1}{f_a} = \frac{Re_a}{64}\left[\frac{1 + \kappa^2}{(1 - \kappa)^2} + \frac{1 + \kappa}{(1 - \kappa)\ln(\kappa)}\right]$$

<u>Turbulent Flow Equations</u> $D = ID_p$ if cross section is tubular

$D = D_h$ if cross section is annular

Chen Equation

$$\frac{1}{\sqrt{f}} = -2.0\log\left\{\frac{\varepsilon}{3.7065D} - \frac{5.0452}{Re}\log\left[\frac{1}{2.8257}\left(\frac{\varepsilon}{D}\right)^{1.1098} + \frac{5.8506}{Re^{0.8981}}\right]\right\}$$

Churchill Equation

$$f = 8\left[\left(\frac{8}{Re}\right)^{12} + \frac{1}{(B + C)^{1.5}}\right]^{1/12}$$

$$\text{where } B = \left[2.457\ln\frac{1}{(7/Re)^{0.9} + (0.27\varepsilon/D)}\right]^{16}$$

$$C = \left(\frac{37\,530}{Re}\right)^{16}$$

P. Pressure Drop Calculations

$$\Delta p_p = \frac{f_p L}{ID_p}\frac{\rho_p V_p^2}{2g_c} =$$

$$\Delta p_a = \left(\frac{f_a L}{D_h} + 1\right)\frac{\rho_a V_a^2}{2g_c} =$$

Q. Summary of Information Requested in Problem Statement

EXAMPLE 7.3. Water at a temperature of 195°F and a mass flow rate of 5000 lbm/hr is to be used to heat ethylene glycol. The ethylene glycol is available at 85°F with a mass flow rate of 12,000 lbm/hr. A double pipe heat exchanger consisting of a $1^1/_4$-standard type M copper tubing inside of 2-standard type M copper tubing is to be used. The exchanger is 6 ft long. Determine the outlet temperature of both fluids using counterflow and again using parallel flow.

Discussion   Water loses energy only to the ethylene glycol and, as
                  heat is transferred, fluid properties change with
                  temperature changes. Outlet temperatures are unknown,
                  so in order to evaluate properties, we use either the inlet
                  temperatures or the average of both inlet temperatures.
                  The fluid with the higher flow rate should be placed in
                  the passage (annular or tubular) having the greater cross
                  sectional area. For this configuration, pressure losses are
                  minimized.

Assumptions   1. Steady state conditions exist.
                  2. Fluid properties remain constant and are evaluated at
                  140°F [= (195 + 85)/2].

Nomenclature   1. T refers to the temperature of the warmer fluid.
                  2. t refers to the temperature of the cooler fluid.
                  3. w subscript refers to the warmer fluid.
                  4. h subscript refers to hydraulic diameter.
                  5. c subscript refers to the cooler fluid.
                  6. a subscript refers to the annular flow area or dimension.
                  7. p subscript refers to the tubular flow area or dimension.
                  8. 1 subscript refers to an inlet condition.
                  9. 2 subscript refers to an outlet condition.
                  10. e subscript refers to equivalent diameter.

A. Fluid Properties

Water	$\dot{m}_w$	= 5000 lbm/hr	T_1	= 195°F
140°F	ρ	= 0.985(62.4) lbm/ft³	C_p	= 0.9994 BTU/lbm·°R
	k_f	= 0.376 BTU/hr·ft·°R	α	= 6.02 x 10⁻³ ft²/hr
	ν	= 0.514 x 10⁻⁵ ft²/s	Pr	= 3.02

Ethylene	$\dot{m}_c$	= 12,000 lbm/hr	t_1	= 85°F
Glycol	ρ	= 1.087(62.4) lbm/ft³	C_p	= 0.612 BTU/lbm·°R
140°F	k_f	= 0.150 BTU/hr·ft·°R	α	= 3.61 x 10⁻³ ft²/hr
	ν	= 5.11 x 10⁻⁵ ft²/s	Pr	= 51

B. Tubing Sizes

2-std M ID_a = 0.1674 ft
1¼-std M ID_p = 0.1076 ft OD_p = 1.375/12 = 0.1146 ft

C. Flow Areas

$A_p = \pi ID_p^2/4 = 0.00909$ ft²
$A_a = \pi(ID_a^2 - OD_p^2)/4 = 0.0117$ ft²

D. Fluid Velocities [Because $A_a > A_p$, we route the ethylene glycol (the fluid with the higher flow rate) through the annulus.]

Water $\quad V_p = \dot{m}_w / \rho A_p = 2.48$ ft/s $\qquad G_p = \dot{m}_w / A_p = 152$ lbm/(ft·s)

Ethylene $\quad V_a = \dot{m}_c / \rho A_a = 4.20$ ft/s $\qquad G_a = \dot{m}_c / A_a = 285$ lbm/(ft·s)
Glycol

E. Annulus Equivalent Diameters

Friction $\qquad D_h = ID_a - OD_p = 0.0528$ ft

Ht Transfer $\qquad D_e = (ID_a{}^2 - OD_p{}^2)/OD_p = 0.1299$ ft

F. Reynolds Numbers

Water $\qquad Re_p = V_p ID_p / \nu = 5.2 \times 10^4$

Ethylene $\qquad Re_a = V_a D_e / \nu = 1.07 \times 10^4$
Glycol

G. Nusselt Numbers

Water $\qquad Nu_p = 0.023 (Re_p)^{4/5}\, Pr^{0.3} = 190$

Ethylene $\qquad Nu_a = 0.023 (Re_a)^{4/5}\, Pr^{0.4} = 185$
Glycol

H. Convection Coefficients

Water $\quad h_i = Nu_p\, k_f / ID_p = 664 \qquad\qquad h_p = h_i\, ID_p / OD_p = 623$

Ethylene $\quad h_a = Nu_a\, k_f / D_e = 214$ BTU/hr·ft^2·°R
Glycol

I. Exchanger Coefficient

$$\frac{1}{U_o} = \frac{1}{h_p} + \frac{1}{h_a} \qquad\qquad U_o = 159 \text{ BTU/hr·ft}^2\cdot°R$$

J. Outlet Temperature Calculations (Exchanger length $L = 6$ ft)

$$R = \frac{\dot{m}_c C_{pc}}{\dot{m}_w C_{pw}} = 1.47 \qquad\qquad A_o = \pi OD_p L = 2.16 \text{ ft}^2$$

Counterflow $\quad E_{counter} = \exp[U_o A_o (R - 1)/\dot{m}_c C_{pc}] = 1.02$

$$T_2 = \frac{T_1(R - 1) - Rt_1(1 - E_{counter})}{RE_{counter} - 1} = 189°F$$

$$t_2 = t_1 + \frac{T_1 - T_2}{R} = 89.1°F$$

$$Parallel \ Flow \quad E_{para} = \exp[U_o A_o (R + 1)/\dot{m}_c C_{pc}] = 1.12$$

$$T_2 = \frac{(R + E_{para})T_1 + Rt_1(E_{para} - 1)}{(R + 1)E_{para}} = 188°F$$

$$t_2 = t_1 + \frac{T_1 - T_2}{R} = 89.8°F$$

K. Summary of Requested Information

Water	$T_2 = 189°F$	Counterflow
Ethylene		
Glycol	$t_2 = 89.1°F$	Counterflow
Water	$T_2 = 188°F$	Parallel Flow
Ethylene		
Glycol	$t_2 = 89.8°F$	Parallel Flow

L. Heat Balance (as a check on the results)

Counterflow

Water $q_w = \dot{m}_w C_{pw}(T_1 - T_2) = 3.00 \times 10^4$ BTU/hr

Ethylene $q_c = \dot{m}_c C_{pc}(t_2 - t_1) = 3.01 \times 10^4$ BTU/hr (discrepancy due to
Glycol   roundoff error)

Parallel Flow

Water $q_w = \dot{m}_w C_{pw}(T_1 - T_2) = 3.50 \times 10^4$ BTU/hr

Ethylene $q_c = \dot{m}_c C_{pc}(t_2 - t_1) = 3.53 \times 10^4$ BTU/hr (discrepancy due to
Glycol   roundoff error)

The results show that little difference exists between parallel flow and counterflow for this example. This is not always the case, however. Counterflow is usually the preferred flow configuration.

7.3 Shell and Tube Heat Exchangers

A double pipe heat exchanger consists of two concentric tubes with a heat transfer area equal to the outer surface area of the inner tube. The flow and heat transfer rates in these exchangers are moderate because the apparatus is comparatively small. For large flow rates and in applications where great heat transfer rates are required, larger cross sectional and surface areas are necessary. These conditions are addressed through the use of what is known as a **shell and tube heat exchanger**.

Figure 7.7 is a sketch of a shell and tube heat exchanger. It consists of a **shell** which in essence is a cylinder of diameter that ranges from less than 12 in. (30 cm) to over 39 in. (100 cm). The shell can be made as long as necessary to deliver the required heat transfer rate. The shell is used to house a number of tubes (up to 1200 in a 39 in. inside diameter shell) which are pressed into what are called **tube sheets**. The tube sheets hold the tubes in position, and the tube-to-tube-sheet connection must be made leakproof. Attached to the ends of the shell are **channels** and within the shell are **baffles** to control the flow of the fluid that passes through the shell and around the tubes.

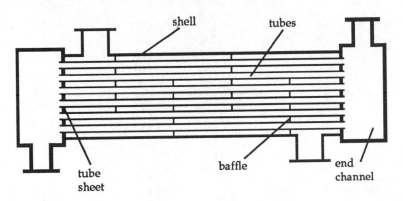

FIGURE 7.7. *Sketch of a shell and tube heat exchanger.*

Shells

The shell of a shell and tube exchanger is usually made of wrought iron or steel pipe but special metals can be used when corrosion might be a problem. Operating pressures greatly influence the wall thickness required. Appendix Table D.1 gives pipe dimensions which apply to shells. In some applications, the inside surface of the shell is machined.

Tubes

The tubes used in a shell and tube heat exchanger are specified differently from water tubing. Specifications of **heat exchanger tubes** known also as **condenser tubes** follow a standard called the **Birmingham**

TABLE 7.2. *Physical dimensions of condenser tubes in terms of BWG.* *(From* Process Heat Transfer *by D. Q. Kern, McGraw-Hill Book Co., 1950, pg. 843, with permission from the publisher.)*

Tube *OD* in inches	cm	BWG	Tube *ID* in inches	cm
3/4	1.91	10	0.482	1.22
		11	0.510	1.29
		12	0.532	1.35
		13	0.560	1.42
		14	0.584	1.48
		15	0.606	1.54
		16	0.620	1.57
		17	0.634	1.61
		18	0.652	1.66
1	2.54	8	0.670	1.70
		9	0.704	1.79
		10	0.732	1.86
		11	0.760	1.93
		12	0.782	1.99
		13	0.810	2.06
		14	0.834	2.12
		15	0.856	2.17
		16	0.870	2.21
		17	0.884	2.25
		18	0.902	2.29

Wire Gage, abbreviated BWG. A 1 in. condenser tube will have an outside diameter of 1 in. Table 7.2 gives dimensions of condenser tubes. Heat exchanger tubes are available in a variety of metals.

The tubes are held in position within the shell by holes drilled in the tube sheets. Tube holes cannot be drilled too close to one another because the tube sheet becomes structurally weakened, although it is desirable to use as many tubes as possible. The distance between adjacent tube centers, called the **tube pitch**, has been standardized. Tubes are laid out so that adjacent tube centers form square or triangular patterns.

Figure 7.8 shows tubes laid out on a square pitch pattern while Figure 7.9 shows a triangular configuration. Common square layouts are made up of 3/4 in. *OD* tubes on a 1 in. square pitch, and 1 in. *OD* on a 1-1/4 in. square pitch. Common triangular layouts are 3/4 in. *OD* on a 15/16 in. triangular pitch, 3/4 in. *OD* on a 1 in. triangular pitch, and 1 in. *OD* on a 1-1/4 in. triangular pitch.

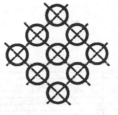

a) square pitch   b) square pitch rotated

FIGURE 7.8. *Square pitch layout.*

FIGURE 7.9. *Triangular pitch layout.*

Table 7.3 gives what are called **tube counts**. For a given shell diameter, the tube count is the maximum number of tubes that can be placed within the shell and not significantly weaken the tube sheet.

Flow Configuration

Figure 7.10 shows the exchanger of Figure 7.7 with fluid flow lines. The two fluid streams transfer heat within the exchanger. One of the fluids, referred to as the **tube fluid**, enters an end channel and is routed through all the tubes. The tube fluid then exits through the other end channel. The second fluid, referred to as the **shell fluid**, enters at the shell fluid inlet and is routed around the exterior of the tubes by the baffles. The exchanger can be set up so that the back-and-forth motion of the shell fluid can be side-to-side or up-and-down. The shell fluid is made to make multiple passes over the tubes by the presence of what are known as **segmental baffles**.

Baffles

Baffles are placed within the shell of the heat exchanger in order to direct the flow of the shell fluid, and to support the tubes. The distance between adjacent baffles is usually a constant and called the **baffle pitch** or **baffle spacing**. It is usually never greater than the shell diameter or less than 1/5th the shell diameter. Baffles are held securely by baffle spacers (not shown in Figure 7.10). Figures 7.11, 7.12, and 7.13 (located on page 261) show 3 types of baffles: the segmental baffle, the disc and doughnut baffle, and the orifice baffle. Our analysis will be for the segmental baffle only.

Figure 7.10 is of a counterflow exchanger. The shell fluid passes through only once, and the tube fluid passes through only once. The exchanger would traditionally be referred to as a **1-1 shell and tube heat exchanger**. Figure 7.14 (page 262) shows a similar exchanger with modified end channels. A partition has been placed in the left end channel.

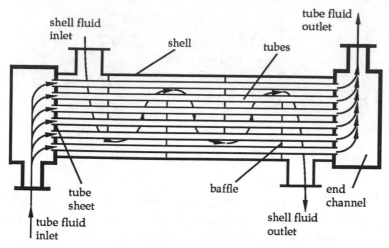

FIGURE 7.10. *The shell and tube heat exchanger of Figure 7.7 with fluid flow lines.*

TABLE 7.3. *Maximum number of tubes (tube counts) for shell and tube equipment.*

3/4 in. *OD* on a 1 in. square pitch						3/4 in. *OD* on a $1^1/_4$ in. square pitch					
Shell *ID* inches	1-P	2-P	4-P	6-P	8-P	Shell *ID* inches	1-P	2-P	4-P	6-P	8-P
8	32	26	20	20		8	21	16	14		
10	52	52	40	36		10	32	32	26	24	
12	81	76	68	68	60	12	48	45	40	38	36
$13^1/_4$	97	90	82	76	70	$13^1/_4$	61	56	52	48	44
$15^1/_4$	137	124	116	108	108	$15^1/_4$	81	76	68	68	64
$17^1/_4$	177	166	158	150	142	$17^1/_4$	112	112	96	90	82
$19^1/_4$	224	220	204	192	188	$19^1/_4$	138	132	128	122	116
$21^1/_4$	277	270	246	240	234	$21^1/_4$	177	166	158	152	148
$23^1/_4$	341	324	308	302	292	$23^1/_4$	213	208	192	184	184
25	413	394	470	356	346	25	260	252	238	226	222
27	481	460	432	420	408	27	300	288	278	268	260
29	553	526	480	468	456	29	341	326	300	294	286
31	657	640	600	580	560	31	406	398	380	368	358
33	749	718	688	676	648	33	465	460	432	420	414
35	845	824	780	766	748	35	522	518	488	484	472
37	934	914	886	866	838	37	596	574	562	544	532
39	1049	1024	982	968	948	39	665	644	624	612	600

TABLE 7.3. *Maximum number of tubes (tube counts) for shell and tube equipment, continued.*

3/4 in. OD on a 15/16 in. triangular pitch						3/4 in. OD on a 1 in. triangular pitch					
Shell ID inches	1-P	2-P	4-P	6-P	8-P	Shell ID inches	1-P	2-P	4-P	6-P	8-P
8	36	32	26	24	18	8	37	30	24	24	
10	62	56	47	42	36	10	61	52	40	36	
12	109	98	86	82	78	12	92	82	76	74	70
13 1/4	127	114	96	90	86	13 1/4	109	106	86	82	74
15 1/4	170	160	140	136	128	15 1/4	151	138	122	118	110
17 1/4	239	224	194	188	178	17 1/4	203	196	178	172	166
19 1/4	301	282	252	244	234	19 1/4	262	250	226	216	210
21 1/4	361	342	314	306	290	21 1/4	316	302	278	272	260
23 1/4	442	420	386	378	364	23 1/4	384	376	352	342	328
25	532	506	468	446	434	25	470	452	422	394	382
27	637	602	550	536	524	27	559	534	488	474	464
29	721	692	640	620	594	29	630	604	556	538	508
31	847	822	766	722	720	31	745	728	678	666	640
33	974	938	878	852	826	33	856	830	774	760	732
35	1102	1068	1004	988	958	35	970	938	882	864	848
37	1240	1200	1144	1104	1072	37	1074	1044	1012	986	870
39	1377	1330	1258	1248	1212	39	1206	1176	1128	1100	1078

The shell fluid passes through only once. The tube fluid enters one of the end channels and is routed into only half the tubes. At the other end channel, the tube fluid turns and is routed through the other half of the tubes. The fluid then exits through the same end channel it entered, and so the tube fluid has passed through the exchanger twice. This exchanger would then be called a **1-2 shell and tube heat exchanger**. Modifications to the end channels control the number of times the tube fluid passes through the exchanger. Accordingly there are 1-4, 1-6, and 1-8 exchangers. An odd number of tube passes is seldom used.

It is likely that, in some applications, the tubes might like to expand more than the shell would. Consequently, it is necessary to use modified forms of the shell and tube heat exchanger to accommodate thermal expansion effects. Figures 7.15 and 7.16 (located on pages 262-263) illustrate only two of the many alternative designs that have been used successfully to solve this problem.

TABLE 7.3. *Maximum number of tubes (tube counts) for shell and tube equipment, continued.*

1 in. *OD* on a $1^1/_4$ in. triangular pitch

Shell *ID* inches	1-P	2-P	4-P	6-P	8-P
8	21	16	16	14	
10	32	32	26	24	
12	55	52	48	46	44
$13^1/_4$	68	66	58	54	50
$15^1/_4$	91	86	80	74	72
$17^1/_4$	131	118	106	104	94
$19^1/_4$	163	152	140	136	128
$21^1/_4$	199	188	170	164	160
$23^1/_4$	241	232	212	212	202
25	294	282	256	252	242
27	349	334	302	296	286
29	397	376	338	334	316
31	472	454	430	424	400
33	538	522	486	470	454
35	608	592	562	546	532
37	674	664	632	614	598
39	766	736	700	688	672

7.4 Analysis of Shell and Tube Heat Exchangers

Figures 7.17 and 7.18 (located on pages 263-264) show the temperature variation of two fluids as they flow through a 1-2 shell and tube heat exchanger. The amount of heat exchanged is given by

$$q = U_o A_o \Delta t = \dot{m}_w C_{pw}(T_1 - T_2) = \dot{m}_c C_{pc}(t_2 - t_1) \qquad (7.23)$$

where U_o is the overall heat transfer coefficient, A_o is the outside surface area of all the tubes, and Δt is the temperature difference that applies to the 1-2 shell and tube heat exchanger. If the flow through the exchanger is entirely counterflow or parallel flow, then Δt would be the log mean temperature difference for counterflow or parallel flow, respectively. The 1-2 shell and tube heat exchanger is a combination of both flows, however, as indicated in Figures 7.17 and 7.18. The established method of

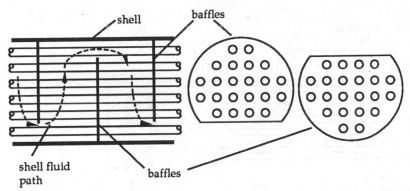

FIGURE 7.11. *Sketch of segmental cut baffles and location in shell.*

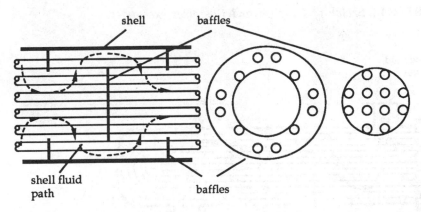

FIGURE 7.12. *Sketch of doughnut and disc baffles and location in shell.*

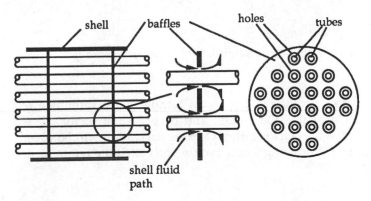

FIGURE 7.13. *Sketch of orifice baffle and location in shell.*

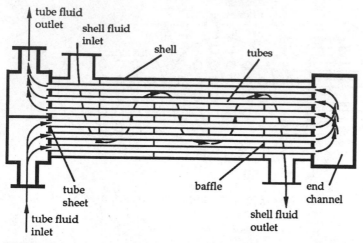

FIGURE 7.14. Sketch of 1-2 shell and tube heat exchanger.

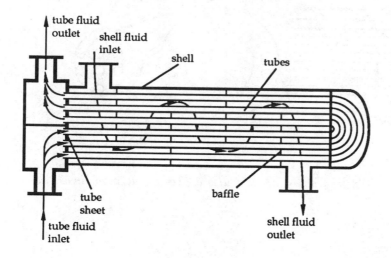

FIGURE 7.15. Sketch of 1-2, U-bend shell and tube heat exchanger.

analysis involves the use of the log mean temperature difference for counterflow, as a "best possible" case, and a correction factor, F. An equation for the correction factor rather than its derivation will be given.

We begin by introducing a new parameter, called the **temperature factor**, defined as

$$S = \frac{t_2 - t_1}{T_1 - t_1}$$

(7.24)

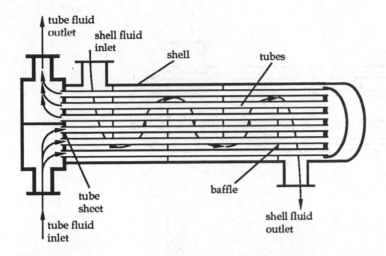

FIGURE 7.16. *Sketch of 1-2, pull through, floating head shell and tube heat exchanger.*

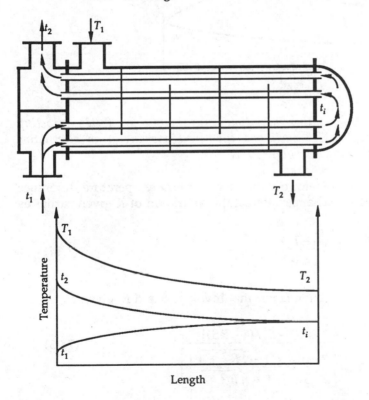

FIGURE 7.17. *Temperature variation with length for fluids traveling through a 1-2 shell and tube heat exchanger.*

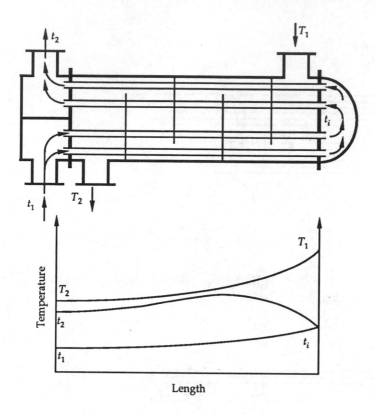

FIGURE 7.18. *Temperature variation with length for fluids traveling through a 1-2 shell and tube heat exchanger; an alternative configuration to Figure 7.17.*

Notice that the denominator of S is the maximum temperature difference associated with the exchanger. Recall the definition of R given earlier as

$$R = \frac{\dot{m}_c C_{pc}}{\dot{m}_w C_{pw}} = \frac{(T_1 - T_2)}{(t_2 - t_1)} \tag{7.14}$$

Next we state the correction factor that involves S and R, given as

$$F = \frac{\sqrt{R^2 + 1}\, ln\,[(1 - S)/(1 - RS)]}{(R - 1)ln\left[\dfrac{2 - S(R + 1 - \sqrt{R^2 + 1}\,)}{2 - S(R + 1 + \sqrt{R^2 + 1}\,)}\right]} \tag{7.25}$$

The temperature difference of Equation 7.23 now becomes

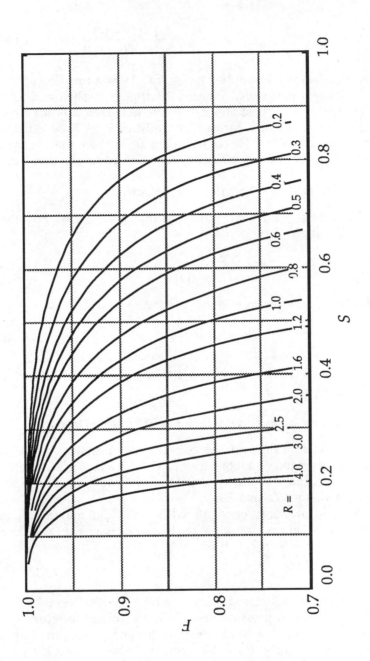

FIGURE 7.19. *Correction factor graph for a shell and tube heat exchanger with 1 shell pass and 2 or more tube passes.*

$$\Delta t = F \, LMTD_{counterflow} = F \, \frac{(T_1 - t_2) - (T_2 - t_1)}{\ln\left[(T_1 - t_2)/(T_2 - t_1)\right]} \tag{7.26}$$

The correction factor F is graphed in Figure 7.19 as a function of R with S as an independent parameter. Figure 7.19 applies to shell and tube heat exchangers with 1 shell pass and 2 or more tube passes. From a practical viewpoint, the correction factor F is indicative of how efficient the exchanger is thermally. If F is less than 0.75, then the exchanger is operating at a costly and low efficiency. Thus for good practice, $F \geq 0.75$.

The Overall Heat Transfer Coefficient

The equation for heat transfer within a 1-2 shell and tube heat exchanger is written as

$$q = U_o A_o F \, \frac{(T_1 - t_2) - (T_2 - t_1)}{\ln\left[(T_1 - t_2)/(T_2 - t_1)\right]} \tag{7.27}$$

The overall heat transfer coefficient is found with

$$\frac{1}{U_o} = \frac{OD_t}{h_i \, ID_t} + \frac{1}{h_o} = \frac{1}{h_t} + \frac{1}{h_o} \tag{7.28}$$

where OD_t and ID_t are the outside and inside diameters, respectively, of the tubes used.

Tube Side Convection Coefficient and Pressure Drop

In order to calculate the overall heat transfer coefficient in Equation 7.28, we need equations for h_i and h_o. These surface coefficients are found with the equations written earlier for double pipe heat exchangers, Equations 7.8 and 7.9.

The pressure drop encountered by the fluid making N_p passes through the exchanger is a multiple of the kinetic energy of the flow:

$$\Delta p_{tubes} = N_p \frac{fL}{ID_t} \frac{\rho V^2}{2g_c} \tag{7.29}$$

where f is the friction factor and L is the tube length. The tube fluid, in addition, experiences a pressure loss when it is forced to return through the exchanger. There is a sudden expansion and a sudden contraction encountered by the tube fluid. The return pressure loss for the fluid making N_p passes is treated like a minor loss and is given by

$$\Delta p_{return} = 4N_p \frac{\rho V^2}{2g_c} \tag{7.30}$$

where the 4 is found from empirical measurements. The tube fluid therefore experiences a pressure loss given as

$$\Delta p_t = \Delta p_{tubes} + \Delta p_{return} = N_p \left(\frac{fL}{ID_t} + 4 \right) \frac{\rho V^2}{2g_c} \qquad (7.31)$$

Shell Side Convection Coefficient and Pressure Drop

The velocity of the fluid passing through the shell varies continuously because the flow area is not constant. The shell fluid travels around tubes and baffles, and although it is not constant, we desire to identify a single representative velocity. In order to do this, we must use a single characteristic length or dimension for the geometry of the shell.

Figure 7.20 shows a cross section of a square pitch layout. The tube pitch P_T and the clearance between adjacent tubes C are both defined. We now develop an equation for the equivalent length, as was done for the double pipe heat exchanger:

$$D_e = \frac{4 \cdot \text{area}}{\text{heat transfer perimeter}}$$

The area in the above equation is that of a square minus the area of 4 quarter circles; i.e., the shaded area of Figure 7.20. The heat transfer perimeter is that of 4 quarter circles. Thus, for the square pitch,

$$D_e = \frac{4(P_T^2 - \pi OD_t^2/4)}{\pi OD_t}$$

or $\qquad D_e = \frac{4P_T^2}{\pi OD_t} - OD_t \qquad$ (square pitch) $\qquad (7.32)$

Figure 7.21 shows a cross section of a triangular pitch layout. By substituting into the expression for equivalent diameter, we obtain

$$D_e = \frac{3.44 P_T^2}{\pi OD_t} - OD_t \qquad \text{(triangular pitch)} \qquad (7.33)$$

We next define a characteristic flow area for the shell geometry. It is remembered that the shell area is not constant, but defining an area is useful in finding the film coefficient. The characteristic area A_s of the shell is defined as

$$A_s = \frac{D_s CB}{P_T} \qquad (7.34)$$

where D_s is the inside diameter of the shell, C is the clearance between adjacent tubes, B is the baffle spacing, and P_T is the tube pitch. The shell

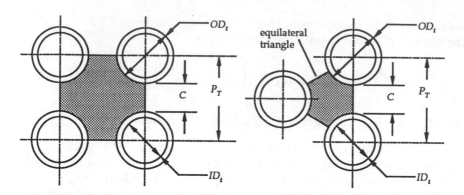

FIGURE 7.20. *Square pitch layout.*   **FIGURE 7.21.** *Triangular pitch layout.*

fluid velocity is found with

$$V_s = \frac{\dot{m}}{\rho A_s} \qquad (7.35)$$

In addition, the mass velocity of the shell fluid is given by

$$G = \frac{\dot{m}}{A_s} = \rho V_s \qquad (7.36)$$

The Nusselt number for the shell fluid is given by an equation that is based on experimental results obtained on a number of heat exchanger tests as:

$$Nu = \frac{h_o D_e}{k_f} = 0.36 \, Re^{0.55} \, Pr^{1/3} \qquad (7.37)$$

valid for $2 \times 10^3 \le Re_s = V_s D_e/v \le 1 \times 10^6$ $Pr = v/\alpha > 0$

μ changes moderately with temperature

Properties evaluated at the average fluid temperature

[= (inlet + outlet)/2]

The shell fluid experiences a pressure drop as it passes through the exchanger, over tubes, and around baffles. If the shell fluid nozzles (inlet and outlet ports) are on the same side of the exchanger, then the shell fluid makes an even number of tube bundle crossings. If the shell nozzles are on opposite sides, then the shell fluid makes an odd number of bundle crossings. The number of bundle crossings influences the pressure drop. Based on experiment, the pressure drop experienced by the shell fluid is

$$\Delta p_s = f(N_b + 1)\frac{D_s}{D_e}\frac{\rho V_s^2}{2g_c}$$ (7.38)

where N_b is the number of baffles and $N_b + 1$ is the number of times that the shell fluid crosses the tube bundle. The friction factor in the above equation is given by

$$f = \exp(0.576 - 0.19 \, ln \, Re_s)$$ (7.39)

valid for $400 \leq Re_s = V_s D_e/v \leq 1 \times 10^6$

Equations 7.38 and 7.39 are formulated to include inlet and exit losses experienced by the shell fluid.

Outlet Temperature Calculation

In a number of applications where a shell and tube heat exchanger can be used, the inlet temperatures and flow rates would be known and the outlet temperatures must be calculated. An equation has been derived to predict outlet temperatures in terms of the previously defined quantities R and S:

$$\frac{UA_o}{\dot{m}_c C_{pc}} = \frac{1}{\sqrt{R^2 + 1}} \, ln \left[\frac{2 - S(R + 1 - \sqrt{R^2 + 1})}{2 - S(R + 1 + \sqrt{R^2 + 1})}\right]$$ (7.40)

where

$$R = \frac{\dot{m}_c C_{pc}}{\dot{m}_w C_{pw}} = \frac{T_1 - T_2}{t_2 - t_1}$$ (7.15)

and $$S = \frac{t_2 - t_1}{T_1 - t_1}$$ (7.24)

Figure 7.22 is a graph of Equation 7.40. As shown, the ratio $UA_o/\dot{m}_c C_{pc}$ is plotted on the horizontal axis, which varies from 0.1 to 8. The temperature difference ratio S is graphed on the vertical axis, which varies from 0 to 1. Lines corresponding to constant values of R appear on the graph and they vary from 0.2 to 4.

For a given exchanger, the quantities R and $UA_o/\dot{m}_c C_{pc}$ can be calculated. Figure 7.22 is then used to find S. The outlet temperature of the cooler fluid is then determined with Equation 7.24 rearranged to give

$$t_2 = (T_1 - t_1)S + t_1$$

The outlet temperature of the warmer fluid is finally calculated with Equation 7.15 rearranged to yield

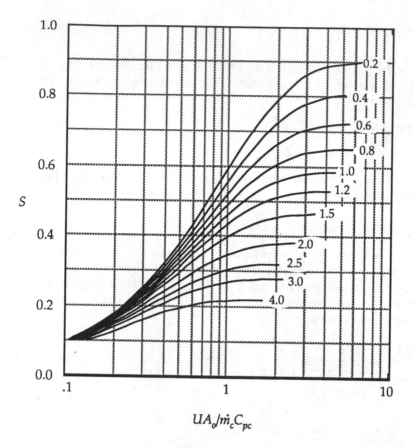

FIGURE 7.22. *Graph of Equation 7.40 for determining outlet temperature in a 1-2 shell and tube heat exchanger.*

$$T_2 = T_1 - R(t_2 - t_1)$$

Like double pipe heat exchangers, shell and tube heat exchangers are subject to mineral deposits on the tube and shell surfaces. The effect is that heat must be transferred through additional resistances. We define a "dirty" or "design" overall heat transfer coefficient as

$$\frac{1}{U} = \frac{1}{U_o} + R_{di} + R_{do}$$

The design coefficient U is used when determining the area required to transfer heat. The above equations have been organized into a suggested order which now follows.

SUGGESTED ORDER OF CALCULATIONS
FOR A SHELL AND TUBE HEAT EXCHANGER

Problem   Complete problem statement

Discussion   Potential heat losses; other sources of difficulties

Assumptions   1. Steady state conditions exist.
                   2. Fluid properties remain constant and are evaluated at
                      a temperature of:

Nomenclature   1. T refers to the temperature of the warmer fluid.
                   2. t refers to the temperature of the cooler fluid.
                   3. w subscript refers to the warmer fluid.
                   4. h subscript refers to hydraulic diameter.
                   5. c subscript refers to the cooler fluid.
                   6. s subscript refers to the shell flow area or dimension.
                   7. t subscript refers to the tubular flow area or dimension.
                   8. 1 subscript refers to an inlet condition.
                   9. 2 subscript refers to an outlet condition.
                   10. e subscript refers to equivalent diameter.

A. Fluid Properties

$$\dot{m}_w \;=\qquad\qquad\qquad\qquad T_1 \;=$$
$$\rho \;=\qquad\qquad\qquad\qquad\quad C_p \;=$$
$$k_f \;=\qquad\qquad\qquad\qquad\quad \alpha \;=$$
$$v \;=\qquad\qquad\qquad\qquad\quad \text{Pr} \;=$$

$$\dot{m}_c \;=\qquad\qquad\qquad\qquad t_1 \;=$$
$$\rho \;=\qquad\qquad\qquad\qquad\quad C_p \;=$$
$$k_f \;=\qquad\qquad\qquad\qquad\quad \alpha \;=$$
$$v \;=\qquad\qquad\qquad\qquad\quad \text{Pr} \;=$$

B. Tubing Sizes

$$ID_t \;=\qquad\qquad\qquad\qquad OD_t \;=$$
$$N_t \;= \text{no. of tubes} \;=$$
$$N_p \;= \text{no. of passes} \;=$$

C. Shell Data

$$D_s \;= \text{shell inside diameter} \qquad =$$
$$B \;= \text{baffle spacing} \qquad\qquad\quad =$$
$$N_b \;= \text{number of baffles} \qquad\quad =$$
$$P_T \;= \text{tube pitch} \qquad\qquad\qquad =$$
$$C \;= \left(\begin{array}{c}\text{clearance between}\\ \text{adjacent tubes}\end{array}\right) \;= P_T - OD_t \;=$$

D. Flow Areas $A_t = N_t \pi \, (ID_t^2)/4N_p =$
$$A_s = D_s CB/P_T \;=$$

E. Fluid Velocities [Route the fluid with the higher flow rate through the flow cross section with the greater area.]

$$V_t = \dot{m}/\rho A = \qquad\qquad\qquad G_t = \dot{m}/A \;=$$

$$V_s = \dot{m}/\rho A = \qquad\qquad\qquad G_s = \dot{m}/A \;=$$

F. Shell Equivalent Diameter

$\left(\begin{matrix}\text{square}\\ \text{pitch}\end{matrix}\right)$ $\quad D_e = \dfrac{4P_T{}^2 - \pi \, OD_t^2}{\pi \, OD_t} =$

$\left(\begin{matrix}\text{triangular}\\ \text{pitch}\end{matrix}\right)$ $\quad D_e = \dfrac{3.44 P_T{}^2 - \pi \, OD_t^2}{\pi \, OD_t} =$

G. Reynolds Numbers
$$Re_t = V_t \, ID_t/\nu =$$
$$Re_s = V_s \, D_e/\nu =$$

H. Nusselt Numbers

Tube Side

Modified Sieder-Tate Equation for laminar flow:

$$Nu_t = \frac{h_i \, ID_t}{k_f} = 1.86 \left(\frac{ID_t Re_t Pr}{L}\right)^{1/3}$$

$Re_t < 2\,200 \qquad\qquad 0.48 < Pr = \nu/\alpha < 16\,700$

Modified Dittus-Boelter Equation for turbulent flow:

$$Nu_t = \frac{h_i \, ID_t}{k_f} = 0.023 \, Re_t^{4/5} \, Pr^n$$

n = 0.4 if fluid is being heated
n = 0.3 if fluid is being cooled

$Re_t > 10\,000; \qquad 0.7 < Pr = \nu/\alpha < 160; \qquad L/D > 60$

Conditions for both equations:
μ changes moderately with temperature
Properties evaluated at the average fluid
temperature [= (inlet + outlet)/2]

<u>Shell Side</u>

$$\text{Nu}_s = \frac{h_o D_e}{k_f} = 0.36 \, \text{Re}_s^{0.55} \, \text{Pr}^{1/3}$$

$$2 \times 10^3 < \text{Re}_s = V_s D_e / v < 1 \times 10^6 \quad \text{Pr} = v/\alpha > 0$$

μ changes moderately with temperature
Properties evaluated at the average fluid
temperature [= (inlet + outlet)/2]

$$\text{Nu}_t =$$

$$\text{Nu}_s =$$

I. Convection Coefficients

$$h_i = \text{Nu}_t \, k_f / ID_t = \qquad\qquad h_t = h_i ID_t / OD_t =$$

$$h_o = \text{Nu}_s \, k_f / D_e =$$

J. Exchanger Coefficient

$$\frac{1}{U_o} = \frac{1}{h_t} + \frac{1}{h_o} \qquad\qquad U_o =$$

K. Outlet Temperatures Calculations (Exchanger length L =)

$$R = \frac{\dot{m}_c C_{pc}}{\dot{m}_w C_{pw}} = \qquad\qquad A_o = N_t \pi OD_t \, L =$$

$$\frac{U_o A_o}{\dot{m}_c C_{pc}} = \qquad\qquad S = \qquad\qquad \text{(Figure 7.22)}$$

$$t_2 = S(T_1 - t_1) + t_1 =$$

$$T_2 = T_1 - R(t_2 - t_1) =$$

L. Log Mean Temperature Difference

$$Counterflow \qquad LMTD = \frac{(T_1 - t_2) - (T_2 - t_1)}{ln\,[(T_1 - t_2)/(T_2 - t_1)]} =$$

M. Heat Balance for Fluids

$$q_w = \dot{m}_w C_{pw} (T_1 - T_2) =$$

$$q_c = \dot{m}_c C_{pc} (t_2 - t_1) =$$

N. Overall Heat Balance for the Exchanger

$$F =$$ (Figure 7.19)

$$q = U_o A_o F \; LMTD =$$

O. Fouling Factors and Design Coefficient

$$R_{di} =$$ $$R_{do} =$$

$$\frac{1}{U} = \frac{1}{U_o} + R_{di} + R_{do} =$$ $$U =$$

P. Area Required to Transfer Heat

$$A_o = \frac{q}{UF \; LMTD} =$$

$$L = \frac{A_o}{N_t \pi \, OD_t} =$$

Q. Friction Factors

<u>Tube Side</u>

Laminar flow in a tube:

$$f_t = \frac{64}{Re_t}$$ $Re_t < 2\,200$ (step G above)

Turbulent flow in a tube:

Chen Equation

$$\frac{1}{\sqrt{f_t}} = -2.0 \log\left\{\frac{\varepsilon}{3.7065D} - \frac{5.0452}{Re} \log\left[\frac{1}{2.8257}\left(\frac{\varepsilon}{D}\right)^{1.1098} + \frac{5.8506}{Re^{\,0.8981}}\right]\right\}$$

Churchill Equation

$$f_t = 8\left[\left(\frac{8}{Re}\right)^{12} + \frac{1}{(B + C)^{1.5}}\right]^{1/12}$$

$$\text{where } B = \left[2.457 \, ln\frac{1}{(7/Re)^{0.9} + (0.27\varepsilon/D)}\right]^{16}$$

and $\quad C = \left(\dfrac{37\,530}{Re}\right)^{16}$

<u>Shell Side</u>

$f_s = \exp(0.576 - 0.19\,\ln Re_s)$ $\qquad\qquad Re_s$ from step G

$Re_t =$

$\left.\dfrac{\varepsilon}{ID_t} = \right\}$ $\qquad f_t =$

$Re_s =$ $\qquad\qquad\qquad f_s =$

R. Pressure Drop Calculations

$$\Delta p_t = \frac{\rho V_t^2}{2g_c}\left(\frac{f_t L}{ID_t} + 4\right)N_p =$$

$$\Delta p_s = \frac{\rho V_s^2}{2g_c}\frac{D_s}{D_e}f_s(N_b + 1) =$$

S. Summary of Information Requested in Problem Statement

EXAMPLE 7.4. In a facility where electricity is generated, condensed (distilled) water is to be cooled by means of a shell and tube heat exchanger. Distilled water enters the exchanger at 110°F at a flow rate of 170,000 lbm/hr. Heat will be transferred to raw water (from a nearby lake) which is available at 65°F and 150,000 lbm/hr. Preliminary calculations indicate that it may be appropriate to use a heat exchanger that has a 17-1/4 in. inside diameter shell, and 3/4 in. OD, 18 BWG tubes that are 16 ft long. The tubes are laid out on a 1 in. triangular pitch, and the tube fluid will make two passes. The exchanger contains baffles that are spaced 1 ft apart. Analyze the proposed heat exchanger to determine its suitability.

Discussion   The flow rates are greater here than with double pipe heat exchangers.

Assumptions   1. Steady state conditions exist.
2. Raw and distilled water properties can be obtained

from the same property table.
3. Raw water properties are evaluated at 68°F.
4. Distilled water properties evaluated at 104°F.
5. All heat lost by the distilled water is transferred to the raw water.

Nomenclature 1. T refers to the temperature of the warmer fluid.
2. t refers to the temperature of the cooler fluid.
3. w subscript refers to the warmer fluid.
4. h subscript refers to hydraulic diameter.
5. c subscript refers to the cooler fluid.
6. s subscript refers to the shell flow area or dimension.
7. t subscript refers to the tubular flow area or dimension.
8. 1 subscript refers to an inlet condition.
9. 2 subscript refers to an outlet condition.
10. e subscript refers to equivalent diameter.

A. Fluid Properties

Dist Water at 104°F		
$\dot{m}_w$ = 170,000 lbm/hr	T_1 = 110°F
ρ = 0.994(62.4) lbm/ft³	C_p = 0.998 BTU/lbm·°R
k_f = 0.363 BTU/hr·ft·°R	α = 5.86 x 10⁻³ ft²/hr
ν = 0.708 x 10⁻⁵ ft²/s	Pr = 4.34

Dist Water $\dot{m}_w$ = 170,000 lbm/hr T_1 = 110°F
at 104°F ρ = 0.994(62.4) lbm/ft³ C_p = 0.998 BTU/lbm·°R
           k_f = 0.363 BTU/hr·ft·°R α = 5.86 x 10⁻³ ft²/hr
           ν = 0.708 x 10⁻⁵ ft²/s   Pr = 4.34

Raw Water $\dot{m}_c$ = 150,000 lbm/hr t_1 = 65°F
at 68°F ρ = 62.4 lbm/ft³ C_p = 0.9988 BTU/lbm·°R
           k_f = 0.345 BTU/hr·ft·°R α = 5.54 x 10⁻³ ft²/hr
           ν = 1.083 x 10⁻⁵ ft²/s   Pr = 7.02

B. Tubing Sizes

ID_t = 0.0543 ft OD_t = 0.06255 ft
N_t = no. of tubes = 196
N_p = no. of passes = 2

C. Shell Data

D_s = shell inside diameter   = 17.25 in. = 1.438 ft
B = baffle spacing   = 1 ft
N_b = number of baffles   = 15
P_T = tube pitch   = 1 in. = 0.08333 ft
C = $\begin{pmatrix}\text{clearance between}\\\text{adjacent tubes}\end{pmatrix}$ = $P_T - OD_t$ = 0.02078 ft

D. Flow Areas

$A_t = N_t \pi (ID_t{}^2)/4N_p = 0.2269$ ft²
$A_s = D_s CB/P_T = 0.3586$ ft² $A_s > A_t$

E. Fluid Velocities [Route the fluid with the higher flow rate through the flow cross section with the greater area.]

Raw Water $V_t = \dot{m}/\rho A = 2.943$ ft/s $\qquad G_t = \dot{m}/A = 183.6$ lbm/ft²·s

Dist Water $V_s = \dot{m}/\rho A = 2.123$ ft/s $\qquad G_s = \dot{m}/A = 131.7$ lbm/ft²·s

F. Shell Equivalent Diameter

$\begin{pmatrix} \text{triangular} \\ \text{pitch} \end{pmatrix}$ $D_e = \dfrac{3.44 P_T{}^2 - \pi\, OD_t{}^2}{\pi\, OD_t} = 0.05901$ ft

G. Reynolds Numbers

Raw Water $\quad Re_t = V_t\, ID_t/v = 1.476 \times 10^4$

Dist Water $\quad Re_s = V_s\, D_e/v = 1.769 \times 10^4$

H. Nusselt Numbers

Raw Water $\quad Nu_t = 108.5$

Dist Water $\quad Nu_s = 127.4$

I. Convection Coefficients

Raw Water $\quad h_i = Nu_t\, k_f/ID_t = 689.2 \quad h_t = h_i\, ID_t/OD_t = 598.4$

Dist Water $\quad h_o = Nu_s\, k_f/D_e = 783.7$ BTU/hr·ft²·°R

J. Exchanger Coefficient

$$\frac{1}{U_o} = \frac{1}{h_t} + \frac{1}{h_o} \qquad U_o = 339.3 \text{ BTU/hr·ft²·°R}$$

K. Outlet Temperatures Calculations (Exchanger length L = 16 ft)

$$R = \frac{\dot{m}_c C_{pc}}{\dot{m}_w C_{pw}} = 0.8831 \qquad A_o = N_t\, \pi\, OD_t\, L = 616.2 \text{ ft}^2$$

$$\frac{U_o A_o}{\dot{m}_c C_{pc}} = 1.396 \qquad\qquad S = 0.55 \qquad\qquad \text{(Figure 7.22)}$$

Raw Water $\quad t_2 = S(T_1 - t_1) + t_1 = 89.3°F$

Dist Water $\quad T_2 = T_1 - R(t_2 - t_1) = 88.5°F$

L. Log Mean Temperature Difference

$$\textit{Counterflow} \qquad LMTD = \frac{(T_1 - t_2) - (T_2 - t_1)}{\ln{[(T_1 - t_2)/(T_2 - t_1)]}} = 22.1°F$$

M. Heat Balance for Fluids

Dist Water $\quad q_w = \dot{m}_w C_{pw} (T_1 - T_2) = 3.65 \times 10^6$ BTU/hr

Raw Water $\quad q_c = \dot{m}_c C_{pc} (t_2 - t_1) = 3.64 \times 10^6$ BTU/hr

N. Overall Heat Balance for the Exchanger

$$F = 0.785 \qquad\qquad\qquad \text{(Figure 7.19)}$$

$$q = U_o A_o F \, LMTD = 3.62 \times 10^6 \qquad \text{(Roundoff error)}$$

O. Fouling Factors and Design Coefficient

$$R_{di} = 0.00075 \text{ hr·ft}^2\text{·°R/BTU} \qquad R_{do} = 0.00075$$

$$\frac{1}{U} = \frac{1}{U_o} + R_{di} + R_{do} \qquad\qquad U = 224.9 \text{ BTU/hr·ft}^2\text{·°R}$$

P. Area Required to Transfer Heat—NA

Q. Friction Factors

$$Re_t = 1.476 \times 10^4$$

$$\left.\frac{\varepsilon}{ID_t} = \text{smooth} \right\} \quad f_t = 0.027$$

$$Re_s = 1.769 \times 10^4 \qquad\qquad f_s = 0.277$$

R. Pressure Drop Calculations

$$\Delta p_t = \frac{\rho V_t^2}{2g_c}\left(\frac{f_t L}{ID_t} + 4\right) N_p = 200 \text{ psf} = 1.4 \text{ psi}$$

$$\Delta p_s = \frac{\rho V_s^2}{2g_c}\frac{D_s}{D_e} f_s (N_b + 1) = 470 \text{ psf} = 3.26 \text{ psi}$$

S. Summary of Information Requested in Problem Statement

Exchanger is suitable;
Both pressure drops are less than 10 psi; and,
F is greater than 0.75.

7.5 Increased Heat Recovery in Shell and Tube Heat Exchangers

In analyzing heat exchangers, the terms **approach** and **cross** are frequently used. These terms refer to the relationship between the outlet temperatures of both fluids. The **approach** is defined as the difference in the outlet temperatures, warmer minus cooler $T_2 - t_2$, and has significance if $T_2 > t_2$. We say that as the fluid flows through the exchanger, the fluid temperatures "approach" each other. On the other hand, if $t_2 > T_2$, then $t_2 - T_2$ is called the **temperature cross**; i.e., as the fluid flows through the heat exchanger, the fluid temperatures "crossed" each other.

Calculations performed on shell and tube equipment show that the correction factor F decreases with decreasing approach. Thus the closer the outlet temperatures are to each other, the smaller the correction factor F will be. This is an important point when considering that F should be equal to or greater than 0.75 for an efficient operation. Generally, F will equal 0.75 for a temperature cross in the range of 0 to 10°F. The following example illustrates the effect that approach and cross have on the correction factor.

EXAMPLE 7.5. Determine the correction factor F for the following cases:

a) $T_1 = 270°C$ $t_1 = 20°C$
         $T_2 = 170°C$ $t_2 = 120°C$
         (approach is 170 - 120 = 50°C)

b) $T_1 = 220°C$ $t_1 = 20°C$
         $T_2 = 120°C$ $t_2 = 120°C$
         (approach is 120 - 120 = 0°C)

c) $T_1 = 200°C$ $t_1 = 20°C$
         $T_2 = 100°C$ $t_2 = 120°C$
         (cross is 120 - 100 = 20°C)

Solution: Note that in all cases, the cooler fluid temperatures are maintained the same while the warmer fluid temperatures are modified. Furthermore, the differences for each fluid (i.e., $T_1 - T_2$ and $t_1 - t_2$) is 100°C, again for all cases. In order to find the correction factor, we first calculate R and S. For the given temperatures, we have

a) 50°C temperature approach

$$R = \frac{T_1 - T_2}{t_2 - t_1} = \frac{100}{100} = 1.0$$

$$S = \frac{t_2 - t_1}{T_1 - t_1} = \frac{100}{250} = 0.4$$

$\left.\begin{array}{l}\\ \\ \\ \\ \end{array}\right\}$ $F = 0.925$ (Figure 7.19)

b) 0°C temperature approach

$$R = 1.0$$

$$S = 0.5$$

$\left.\begin{array}{l}\\ \\ \end{array}\right\}$ $F = 0.8$

c) 20°C temperature cross

$$R = 1.0$$

$$S = 0.556$$

$\left.\begin{array}{l}\\ \\ \end{array}\right\}$ $F = 0.64$

The factor R is constant for the above combinations but the temperature factor S is not. The difference in inlet temperatures $T_1 - t_1$ was changed in the calculations so that effects of temperature approach on F could be investigated. We see that, as S increased, F decreased; but primarily, as the approach decreased, the correction factor decreased. We conclude that the efficiency of a 1-2 shell and tube heat exchanger increases with increasing inlet temperature difference or increasing temperature approach.

When a 1-2 shell and tube heat exchanger has a correction factor F that is less than 0.75, then it is operating in a less than desirable application. It is possible to improve the performance by connecting two of the exchangers in series, as illustrated schematically in Figure 7.23. This configuration shows that the shell fluid passes through the *combined* exchangers twice while the tube fluid passes through them four times. This configuration is called a 2-4 shell and tube heat exchanger. The analysis of such a combination of exchangers is identical to that for the 1-2 exchanger except that the correction factor F is different. Figure 7.24 shows an alternative configuration for a 2-4 shell and tube heat exchanger using a single shell. The presence of a longitudinal baffle causes the shell fluid to pass through twice. Modifications to the end channels cause the tube fluid to pass through 4 times. Figure 7.25 is a graph of correction factor F for the 2-4 exchanger. It is similar to that of Figure 7.19. When a 1-2 shell and tube exchanger has a correction factor F that is too low, then a 2-4 exchanger will usually be acceptable.

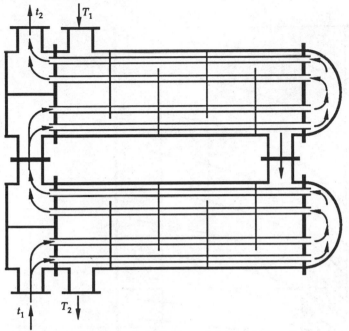

FIGURE 7.23. *Two 1-2 shell and tube heat exchangers connected in series to form a 2-4 heat exchanger.*

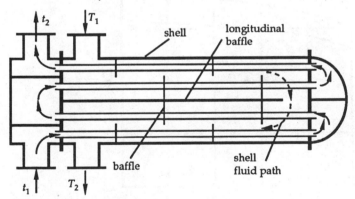

FIGURE 7.24. *A 2-4 shell and tube heat exchanger using a single shell.*

EXAMPLE 7.6. Repeat the calculations of the last example to find the correction factor F for a 2-4 shell and tube heat exchanger. The data from that example are as follows:

a) $T_1 = 270°C$ $t_1 = 20°C$
          $T_2 = 170°C$ $t_2 = 120°C$
             (approach is 170 - 120 = 50°C)

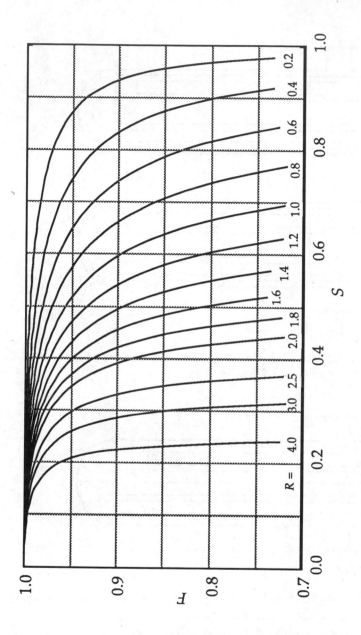

FIGURE 7.25. *Correction factor chart for a 2-4 shell and tube heat exchanger.*

b) $T_1 = 220°C$ $t_1 = 20°C$
        $T_2 = 120°C$ $t_2 = 120°C$
        (approach is 120 - 120 = 0°C)

c) $T_1 = 200°C$ $t_1 = 20°C$
        $T_2 = 100°C$ $t_2 = 120°C$
        (cross is 120 - 100 = 20°C)

Solution: The results for each case are:

a) 50°C temperature approach

$$R = \frac{T_1 - T_2}{t_2 - t_1} = \frac{100}{100} = 1.0$$

$$S = \frac{t_2 - t_1}{T_1 - t_1} = \frac{100}{250} = 0.4$$

$F = 0.98$ (Figure 7.25)

b) 0°C temperature approach
$$R = 1.0$$

$$S = 0.5$$

$F = 0.955$

c) 20°C temperature cross
$$R = 1.0$$

$$S = 0.556$$

$F = 0.925$

As shown, the correction factor is greater for the 2-4 exchanger than for a 1-2 exchanger for the same temperatures.

7.6 Show and Tell

1. Obtain two different sizes of copper tubing, the appropriate fittings, and show how a double pipe heat exchanger would be constructed.

2. Obtain a double pipe heat exchanger and illustrate how it operates.

3. Obtain a shell and tube exchanger, and illustrate its operation.

4. Obtain a catalog of shell and tube heat exchangers and discuss the various designs that have been implemented.

7.7 Problems Chapter 7

1. In many problems, the resistance of the tube (or pipe) wall is considered negligible when calculating the overall heat transfer coefficient. For the following data, determine whether this is a good assumption:

   $h_i = 1\,700\ \text{W}/(\text{m}^2 \cdot \text{K})$ $h_o = 1\,800\ \text{W}/(\text{m}^2 \cdot \text{K})$
   2-nominal schedule 40 stainless steel pipe

2. Repeat Problem 1 for the following data:

   $h_i = 175\ \text{BTU}/\text{hr} \cdot \text{ft}^2 \cdot {}^\circ\text{R}$ $h_o = 200\ \text{BTU}/\text{hr} \cdot \text{ft}^2 \cdot {}^\circ\text{R}$
   3-std type K copper tubing

3. Repeat Problem 1 for the following case:

   $h_i = 8\,500\ \text{W}/(\text{m}^2 \cdot \text{K})$ $h_o = 500\ \text{W}/(\text{m}^2 \cdot \text{K})$
   2-std type K copper tubing

4. Verify the derivation of Equation 7.17.

5. Derive Equation 7.19.

6. A hot fluid at 75°C is cooled in a double pipe heat exchanger to 25°C. A cool fluid is warmed in the exchanger from 15°C to 20°C. Calculate log mean temperature difference for (a) counterflow and (b) parallel flow.

7. A cool fluid enters a double pipe heat exchanger and is heated from 65°F to 150°F. A warmer fluid is cooled in the heat exchanger from 200°F to 180°F. Calculate LMTD for (a) counterflow and for (b) parallel flow.

8. A fluid passes through a double pipe heat exchanger and changes phase, but does not change temperature. A warm fluid at a temperature of 220°F enters a double pipe heat exchanger and leaves still at a temperature of 220°F. A cooler fluid enters the exchanger at 100°F and is heated to 150°F. Determine the LMTD for (a) counterflow and for (b) parallel flow.

9. Repeat Problem 7 for a phase change fluid temperature of 5°C and a warmer fluid temperature change of 45°C (inlet) to 10°C (outlet).

10. Calculate the LMTD for (a) counterflow and for (b) parallel flow for the following fluid temperatures:

Warmer fluid	Cooler fluid
$T_1 = 100°C$	$t_1 = 25°C$
$T_2 = 75°C$	$t_2 = 40°C$

11. Water is used to cool ethylene glycol in a 12 ft long double pipe heat exchanger made of 4-std and 2-std (both type M) copper tubing. The water inlet temperature is 60°F and the ethylene glycol inlet temperature is 200°F. The flow rate of the ethylene glycol is 20 lbm/s while that for the water is 30 lbm/s. Calculate the expected outlet temperature of the ethylene glycol and determine the pressure drop expected for both streams. Assume counterflow and place the ethylene glycol in the inner tube.

12. Problem 11 was solved with water in the annulus and ethylene glycol in the pipe. Repeat the calculations placing water in the pipe and ethylene glycol in the annulus.

13. Reverse the direction of either fluid in Problem 11 and repeat the calculations for parallel flow, with ethylene glycol in the annulus.

14. Four 2 m long double pipe heat exchangers made of 6-std type M and 4-std type M copper tubing are connected together in series to form one 8 m long exchanger. It is used to cool (unused) engine oil. The exchanger takes in water at 10°C at a flow rate of 4 l/s, and oil at a temperature of 140°C with a flow rate of 5.5 l/s. Determine the expected outlet temperature of the oil and the pressure drop encountered by both streams. Assume counterflow.

15. Ammonia is used in liquid form for a process and is needed at a temperature of 15°C. It is available at 0°C. Liquid carbon dioxide is used rather than water as a heat source (using water may lead to trouble). Liquid carbon dioxide is available at 15°C and a mass flow rate of 1.25 kg/s. A 6 m long double pipe heat exchanger, made of $2^1/_2 \times 1^1/_4$ type K copper tubing is available for this service. For an ammonia flow rate of 1.2 kg/s, analyze the heat exchanger completely.

16. In an air separation plant, air is cooled and its components are separated from the mixture. Cooled oxygen at a temperature of 20°C must be heated to a temperature of 30°C for accurate metering. The oxygen flow rate at 30°C is 0.01 kg/s. Air (available at 35°C and 0.015 kg/s) is used as the heating medium. A number of 3 x 2 schedule 40 double pipe heat exchangers that are 2 m long and made of galvanized steel are available. Determine how many are required and analyze the setup completely.

17. A 4 x 3 double pipe heat exchanger, 12 ft long and made of type M copper tubing, is used to cool air having an inlet temperature of 120°F and a flow rate of 0.03 kg/s. Methyl chloride is the cooling medium, and it is available at 10°F and a flow rate of 0.025 kg/s. Analyze the system completely.

18. Raw water is used to cool the distilled water of a small electrical generating facility by using a double pipe heat exchanger. The raw water is taken from a nearby stream and is available at a temperature that ranges from 32°F to 55°F over the course of one year. The distilled water is to be cooled from 210°F to as cold as possible. The raw water flow rate is 8500 lbm/hr while the distilled water flow rate is 8000 lbm/hr. The exchanger is 18 ft long, and is made of 2-std

and $1^1/_4$-std, both type M copper tubing. Predict outlet temperatures over the course of one year; i.e., for raw water inlet temperatures that range 32°F to 55°F.

19. The use of glycerin as a coolant is to be determined. Water at a temperature of 150°F is to be cooled in a 4 x 3-std type M copper tubing heat exchanger that is 10 ft long. The water flow rate is 10 lbm/s. Glycerin is available at a temperature of 75°F. Determine water and glycerin outlet temperatures as a function of glycerin flow rate, which ranges from 0.01 to 20 lbm/s.

20. Repeat the previous problem using ethylene glycol as the coolant.

21. Derive Equation 7.33.

22. During one phase of the separation of crude oil into its components, the oil is to be heated by water in a 1-4 shell and tube heat exchanger. The oil has a flow rate of 110,000 lbm/hr and it enters the heat exchanger at 100°F. Water enters the exchanger at a flow rate of 66,000 lbm/hr and a temperature of 200°F. It is proposed to use an exchanger having a shell inside diameter of $23^1/_4$ in. and containing 1 in. OD tubes, 13 BWG, laid out on a $1^1/_4$ in. square pitch. The 192 tubes are 12 ft long and the exchanger contains baffles that are spaced 2 ft apart. Analyze the proposed configuration completely. Assume that oil has the following properties:

$$C_p = 2\,050 \text{ J/(kg·K)} \qquad \rho = 740 \text{ kg/m}^3$$
$$\mu = 3.4 \text{ cp} \qquad k_f = 0.132 \text{ W/(m·K)}$$

23. A 12 in. ID 1-2 shell and tube heat exchanger is used with water as a cooling medium. Kerosene [$C_p = 2\,530$ J/(kg·K), $\rho = 800$ kg/m³, $\mu = 0.4$ cp and $k_f = 0.133$ W/(m·K)] at an inlet temperature of 250°F and a flow rate of 100,000 lbm/hr is to be cooled with water available at 75°F and a flow rate of 96,000 lbm/hr. The exchanger contains forty-five 1 in. OD tubes, 6 ft long, 13 BWG, laid out on a $1^1/_4$ square pitch. Analyze the heat exchanger completely and determine if it is suitable for this service. The exchanger contains 4 evenly spaced baffles.

24. Liquid carbon dioxide at a flow rate of 110 000 kg/hr is to be heated from 0°C to 20°C in a 1-2 shell and tube heat exchanger. Water is available at a flow rate of 112 500 kg/hr and a temperature of 50°C. A 25 in. ID 1-2 shell and tube exchanger having 3/4 in., 10 BWG tubes laid out on a 1 in. triangular pitch is available. The tubes are 6 ft long and the exchanger contains baffles spaced 1 ft apart. Analyze the heat exchanger completely.

25. Kerosene at 350°F flows at a rate of 150,000 lbm/hr and must be cooled to 190°F. Crude oil is available at 70°F and flows at a rate of 150,000 lbm/hr. It is proposed to use a 21-1/4 in. ID shell and tube heat exchanger containing 1 in. OD tubes, 13 BWG laid out on a 1-1/4 in. square pitch. The 166 tubes are 15 ft long. The baffles are spaced 12 in. apart and the tube fluid will make 4 passes through the exchanger. Analyze the exchanger completely. Use the following fluid properties:

Kerosene

$C_p = 0.59$ BTU/lbm·°R $\rho - 0.73(62.4)$ lbm/ft^3
$k_f = 0.0765$ BTU/hr·ft·°R $\mu = 8.35 \times 10^{-6}$ lbf·s/ft^2

Crude Oil

$C_p = 0.49$ BTU/lbm·°R $\rho = 0.83(62.4)$ lbm/ft^3
$k_f = 0.077$ BTU/hr·ft·°R $\mu = 7.52 \times 10^{-5}$ lbf·s/ft^2

26. A sugar solution ($\rho = 1\,080$ kg/m^3, $C_p = 3\,601$ J/(kg·K), $k_f = 0.576\,4$ W/(m·K), $\mu = 1.3 \times 10^{-3}$ N·s/m^2) flows at a rate of 60 000 kg/hr and is to be heated from 25°C to 50°C. Water at 95°C is available at a flow rate of 75 000 kg/hr. It is proposed to use a 12 in. ID, 1-2 shell and tube heat exchanger containing 3/4 in. OD, 16 BWG tubes, 1 m long and laid out on a 1 in. square pitch. The exchanger contains 3 baffles spaced evenly. Will the exchanger be suitable? If not, can it be made to work by doubling the flow rate of the water? Analyze it completely.

7.8 Group Problems

A water to water system is used to test the effects of changing tube length, baffle spacing, tube pitch, pitch layout, and tube diameter. Cold water at 25°C and 100 000 kg/hr is heated by hot water at 100°C, also at 100 000 kg/hr. The exchanger has a 31 in. ID shell. Perform calculations on this 1-2 shell and tube heat exchanger for the following conditions, as outlined in the problems, and put together an overall comparison chart.

1. 3/4 in. OD tubes, 12 BWG laid out on a 1 in. triangular pitch; 3 baffles per meter of tube length. Analyze the exchanger for tube lengths of 1 m, 2 m, 3 m, 4 m, and 5 m.

2. 3/4 in. OD tubes, 12 BWG laid out on a 1 in. triangular pitch; 2 m long. Analyze for baffle placement of 1 baffle/meter of tube length, 2 baffles/meter of tube length, 3 baffles/meter of tube length, 4 baffles/meter of tube length, and 5 baffles/meter of tube length.

3. 3/4 in. OD tubes, 12 BWG; 2 m long; 4 baffles per meter of tube length. Analyze the exchanger for tube layouts of 15/16 in. triangular pitch, 1 in. triangular pitch, and 1 in. square pitch.

4. 1 in. OD tubes, 12 BWG, 4 m long; 9 baffles. Analyze the exchanger for tube layouts of 1-1/4 in. triangular pitch, and 1-1/4 in. square pitch.

5. 3/4 in. OD tubes, 12 BWG laid out on a 1 in. triangular pitch; 2 m long; 4 baffles per meter of tube length. Analyze the exchanger for 2, 4, 6, and 8 tube passes, and compare to the case of true counterflow (i.e., 1 tube pass).

CHAPTER 8

Analysis of
Heat Exchangers II

In this chapter, we continue the discussion begun in the last chapter by first considering the cross flow heat exchanger. The effectiveness-NTU method is used in the analysis although not discussed in great detail. Design methods are then presented and they include the sizing and selection of double pipe heat exchangers. An analysis for determining the optimum outlet temperature is given and it is shown how to use it to minimize the costs associated with sizing a heat exchanger.

8.1 Cross Flow Heat Exchangers

A cross flow heat exchanger brings two fluids together so that energy is transferred from the warmer to the cooler fluid. The radiator of an automobile and the condenser of an air conditioning unit are examples of such devices. The fluids flow at right angles to each other within the exchanger. In addition, each fluid stream can pass through and remain **mixed** or **unmixed**. Figure 8.1 illustrates the definitions of mixed vs unmixed flows through similar ducts. In an unmixed flow situation, the flow channel or passageway would contain internal channels (e.g., tubes or walls) that restrict lateral fluid motion. In a mixed flow passageway, internal channels are not present and the fluid particles are free to move about (and mix) over the cross section. Figure 8.2 shows examples of cross flow heat exchangers that contain mixed and unmixed flow passageways.

It is important to be able to model the energy transfer that occurs within a cross flow heat exchanger. Tests on many cross flow heat exchangers have been performed (see, for example, *Compact Heat Exchangers* by W. M. Kays and A. L. London, McGraw-Hill Book Co., 1964). Results of heat transfer and friction characteristics—overall heat transfer coefficient U and friction factor f—have been reported. Due to space limitations here, however, it is not convenient to reproduce results of such tests. We therefore assume in the problems that the overall heat transfer coefficient is known or easily found for any cross flow heat exchanger of interest.

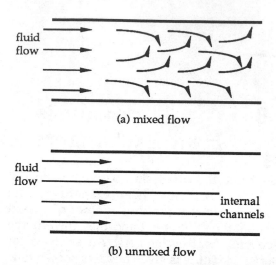

(a) mixed flow

(b) unmixed flow

FIGURE 8.1. *Mixed and unmixed flow passageways.*

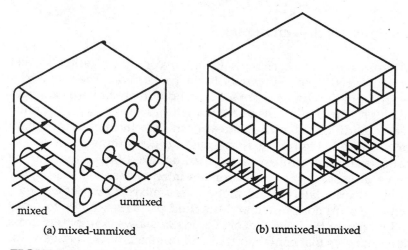

(a) mixed-unmixed   (b) unmixed-unmixed

FIGURE 8.2. *Cross flow heat exchangers with mixed and unmixed passageways.*

Once the overall heat transfer coefficient is known, the heat transfer rate is calculated with

$$q = UA\, F\, LMTD \qquad\qquad (8.1)$$

in which U is the overall heat transfer coefficient with dimensions of [F·L/(T·L²·t); BTU/(hr·ft²·°R or W/(m²·K)], A is the heat transfer area of the heat exchanger, F is a correction factor, and $LMTD$ is the log mean temperature difference for counterflow

$$LMTD = \frac{(T_1 - t_2) - (T_2 - t_1)}{\ln\left[(T_1 - t_2)/(T_2 - t_1)\right]} \qquad (8.2)$$

The heat transfer area of a cross flow heat exchanger is extremely difficult to determine because of the construction methods used and so it is not uncommon to report only the UA product. In almost all applications, it is necessary to calculate the size of an exchanger required to transfer a certain heat load. So in this case, the *frontal area* (as opposed to the heat transfer area A) is more convenient to use.

Equations for the correction factor have been derived (see *Process Heat Transfer* by D. Q. Kern, McGraw-Hill Book Co., 1950, chapter 16). As before, the correction factor is a function of the capacitance ratio R and the temperature factor S, defined in Chapter 7 as

$$R = \frac{T_1 - T_2}{t_2 - t_1} = \frac{\dot{m}_c C_{pc}}{\dot{m}_w C_{pw}} \qquad (8.3)$$

and

$$S = \frac{t_2 - t_1}{T_1 - t_1} \qquad (8.4)$$

Figure 8.3 is a sketch of the temperature variation of both fluids as they pass through a cross flow heat exchanger. As indicated, we use the lower case 't' to denote the cooler fluid and the uppercase 'T' refers to the warmer fluid. Also, the '1' subscript denotes an inlet condition and a '2' refers to an outlet condition.

When the performance of a cross flow heat exchanger is evaluated, the results show that a counterflow, double pipe heat exchanger is more efficient. The standard of comparison, then, is the counterflow heat exchanger, which is why $LMTD$ in Equation 8.1 is for counterflow.

Figure 8.4 is a graph of correction factor F as a function of S for various values of R for a mixed-unmixed cross flow heat exchanger. Figure 8.5 is a graph of correction factor F versus S for various values of R for an unmixed-unmixed cross flow heat exchanger. Figure 8.4 was graphed from the original equation while Figure 8.5 was graphed by applying a scale factor to Figure 8.4. Figure 8.5 is thus only an approximation (within 5%) of the graph obtained from the corresponding original equation for unmixed-unmixed flows.

Outlet Temperature Calculation

In many applications, only fluid inlet temperatures are known and it is necessary to calculate outlet temperatures for a given heat

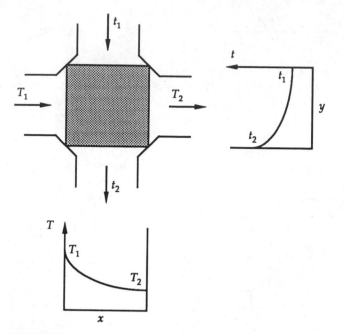

FIGURE 8.3. *Temperature variation of the cooler and warmer fluids as they flow through a cross flow heat exchanger.*

exchanger. This can be done in several ways but the **effectiveness-NTU** method presented here is most convenient. The effectiveness E is dependent upon which of the two fluids has the minimum mass flow rate x specific heat product; i.e., the minimum **capacitance**. The effectiveness E is defined as

$$E = \frac{t_2 - t_1}{T_1 - t_1} \qquad \text{if } \dot{m}_c C_{pc} < \dot{m}_w C_{pw} \qquad (8.5a)$$

$$E = \frac{T_1 - T_2}{T_1 - t_1} \qquad \text{if } \dot{m}_w C_{pw} < \dot{m}_c C_{pc} \qquad (8.5b)$$

Several important features of these definitions should be noted:

• The denominator in both is the maximum temperature difference (inlet temperatures) associated with the exchanger.

• The numerator in both is the temperature difference of the fluid having the minimum capacitance (mass flow rate x specific heat).

• When the cooler fluid has the minimum capacitance, the definition of effectiveness E equals the temperature factor S of Equation 8.4.

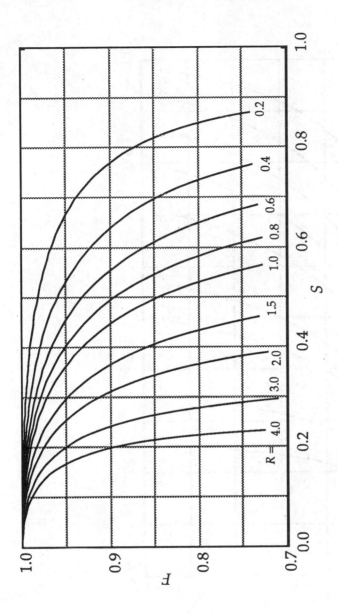

FIGURE 8.4. *Correction factor graph for a **mixed-unmixed** cross flow heat exchanger.*

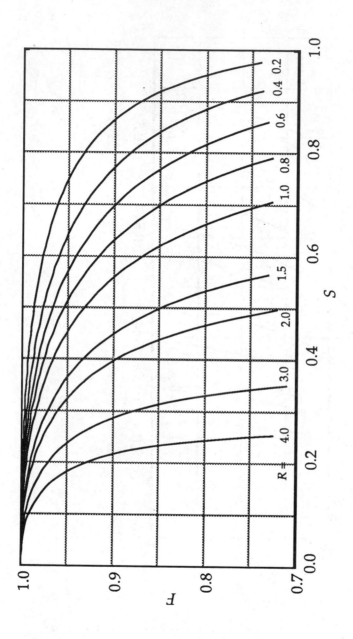

FIGURE 8.5. *Correction factor graph for an **unmixed-unmixed** cross flow heat exchanger.*

- Effectiveness E ranges from 0 to 1, similar to the correction factor F.

- The definitions above are independent of exchanger type and so can be applied to double pipe, shell and tube, cross flow heat exchangers, etc.

If effectiveness is known for the heat exchanger of interest, then the heat transfer rate can be found with

$$q = E(\dot{m}C_p)_{min}\,(T_1 - t_1) \tag{8.6}$$

All that remains in this development is an equation or graph of effectiveness E for each type of heat exchanger.

Effectiveness equations have been derived for many types of heat exchangers, and usually contain a term called the **number of transfer units**, NTU, defined as

$$NTU = \frac{UA}{(\dot{m}C_p)_{min}} \tag{8.7}$$

We also define what is known as the **ratio of capacitances** C, which is always less than 1:

$$C = \frac{(\dot{m}C_p)_{min}}{(\dot{m}C_p)_{max}} < 1 \tag{8.8}$$

This definition is similar to, but not to be confused with, the definition of R given in Equation 8.3. Equations 8.7 and 8.8 are used as conveniences in expressing complex equations for the effectiveness E. Table 8.1 lists effectiveness equations in terms of number of transfer units NTU and ratio of capacitances C for a number of heat exchanger types. Figures 8.6 and 8.7 are graphs of effectiveness E vs number of transfer units NTU for mixed-unmixed and unmixed-unmixed cross flow heat exchangers.

Like other types, cross flow heat exchangers are subject to fouling but it is difficult to accurately apply fouling factor data to them. We will therefore not consider fouling effects in such exchangers. Equally difficult to determine are pressure drops of the fluid streams as they pass through the exchanger. In cases where it is possible to calculate it, the pressure drop can be treated as a minor loss in the overall fluid flow system. The minor loss can vary tremendously for both fluid streams. For our purposes, we assume a minor loss of $K = 10$ for each stream.

It is important to know the frontal area required to transfer a required amount of energy. Figure 8.8 is an empirical graph relating frontal area to the UA product for a cross flow heat exchanger. This graph is a result of calculations made on a number of exchangers.

TABLE 8.1. *Effectiveness equations for various heat exchangers.*

Type of Exchanger	Effectiveness Equation

Double Pipe

   Parallel flow
$$E = \frac{1 - \exp[- NTU(1 + C)]}{1 + C}$$

   Counterflow
$$E = \frac{1 - \exp[- NTU(1 - C)]}{1 - C\exp[- NTU(1 - C)]}$$

Shell and Tube

   1 shell pass; 2, 4, 6, etc., tube passes

$$E = 2\left\{1 + C + \frac{1 + \exp[- NTU(1 + C^2)^{1/2}]}{1 - \exp[- NTU(1 + C^2)^{1/2}]} (1 + C^2)^{1/2}\right\}^{-1}$$

Cross flow, mixed-unmixed with   (See Figure 8.6)
   $(\dot{m}C_p)_{min}$ unmixed

$$E = C\{1 - \exp[- C(1 - \exp[- NTU])]\}$$

Cross flow, mixed-unmixed with
   $(\dot{m}C_p)_{max}$ unmixed

$$E = 1 - \exp[- C(1 - \exp[- NTU \cdot C])]$$

Cross flow, unmixed-unmixed   (See Figure 8.7)

$$E \approx 1 - \exp[C(NTU)^{0.22} \{\exp[- C(NTU)^{0.78}] - 1\}]$$

$$NTU = \frac{UA}{(\dot{m}C_p)_{min}} \qquad\qquad C = \frac{(\dot{m}C_p)_{min}}{(\dot{m}C_p)_{max}} < 1$$

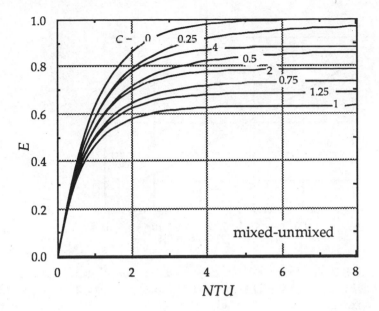

FIGURE 8.6. *Effectiveness as a function of number of transfer units for a* ***mixed-unmixed*** *cross flow heat exchanger.*

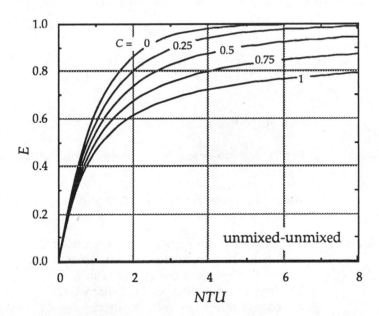

FIGURE 8.7. *Effectiveness as a function of number of transfer units for an* ***unmixed-unmixed*** *cross flow heat exchanger.*

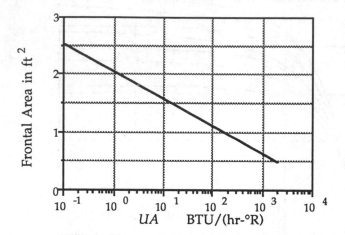

FIGURE 8.8. *Empirical relationship between frontal area (normal to airflow) and the UA product for an air-to-liquid cross flow heat exchanger.*

The method of analysis for cross flow heat exchangers has been organized into a suggested order for performing calculations, which now follows.

ANALYSIS OF CROSS FLOW HEAT EXCHANGERS
SUGGESTED ORDER OF CALCULATIONS

Problem   Complete problem statement

Discussion   Potential heat losses; other sources of difficulties

Assumptions   1. Steady state conditions exist.
                     2. Fluid properties remain constant and are evaluated at
                        a temperature of

Nomenclature   1. T refers to the temperature of the warmer fluid.
                     2. t refers to the temperature of the cooler fluid.
                     3. w subscript refers to the warmer fluid.
                     4. h subscript refers to hydraulic diameter.
                     5. c subscript refers to the cooler fluid.
                     6. 1 subscript refers to an inlet condition.
                     7. 2 subscript refers to an outlet condition.

A. Fluid Properties

$\dot{m}_w$ = T_1 =
ρ = C_p =
k_f = α =
v =   Pr =

$\dot{m}_c$ = t_1 =
ρ = C_p =
k_f = α =
v =   Pr =

B. Heat Balance

$$q_w = \dot{m}_w C_{pw}(T_1 - T_2) =$$

$$q_c = \dot{m}_c C_{pc}(t_2 - t_1) =$$

C. Log Mean Temperature Difference

$$\text{Counterflow} \qquad LMTD = \frac{(T_1 - t_2) - (T_2 - t_1)}{\ln[(T_1 - t_2)/(T_2 - t_1)]} =$$

D. Correction Factor

$$S = \frac{t_2 - t_1}{T_1 - t_1} =$$

$$R = \frac{T_1 - T_2}{t_2 - t_1} = \frac{\dot{m}_c C_{pc}}{\dot{m}_w C_{pw}} =$$

$$F =$$

$$\left(\begin{array}{c}\text{Figure 8.4 mixed-unmixed}\\ \text{Figure 8.5 unmixed-unmixed}\end{array}\right)$$

E. UA Product

$$UA = \frac{q}{F\ LMTD} =$$

F. Capacitances

Mixed-Unmixed

$$(\dot{m}C_p)_{mixed} =$$

$$(\dot{m}C_p)_{unmixed} \quad =$$

$$C = \frac{(\dot{m}C_p)_{mixed}}{(\dot{m}C_p)_{unmixed}} \quad =$$

$$Unmixed-Unmixed$$

$$(\dot{m}C_p)_w \quad =$$

$$(\dot{m}C_p)_c \quad =$$

$$C = \frac{(\dot{m}C_p)_{min}}{(\dot{m}C_p)_{max}} \quad =$$

G. Number of Transfer Units, NTU

$$NTU = \frac{UA}{(\dot{m}C_p)_{min}} =$$

H. Effectiveness, E

$$\left.\begin{array}{l} NTU = \\ \\ C = \end{array}\right\} \qquad E =$$

$$\left(\begin{array}{l}\text{Figure 8.6 mixed-unmixed} \\ \text{Figure 8.7 unmixed-unmixed}\end{array}\right)$$

I. Outlet Temperature Calculations

$$t_2 = (T_1 - t_1)E + t_1$$

$$t_2 =$$

$$T_2 = T_1 - C(t_2 - t_1)$$

$$T_2 =$$

J. Velocities in Connecting Tubes/Ducts

$$V_w =$$

$$V_c =$$

K. Pressure Drop Calculations (Use $K = 10$)

$$\Delta p_w = K \frac{\rho_w V_w^2}{2g_c} =$$

$$\Delta p_c = K \frac{\rho_c V_c^2}{2g_c} =$$

L. Frontal Area Required

$UA =$ $A =$ (Figure 8.8)

Q. Summary of Information Requested in Problem Statement

EXAMPLE 8.1. A cross flow heat exchanger (unmixed-unmixed) is used to cool water at the test stand of a diesel engine. The water enters the exchanger at 230°F and is to be cooled to 180°F. Air in the room (~70°F) is used as the cooling fluid. The fan available can move the air at 2.5 lbm/hr. Determine the frontal area (i.e., normal to the air flow) of the heat exchanger if the flow rate of the water is 0.7 lbm/hr.

Assumptions 1. Steady state conditions exist.
2. Fluid properties remain constant and are evaluated at a temperature of: Air at 540°R, Water at 212°F.

Nomenclature 1. T refers to the temperature of the warmer fluid.
2. t refers to the temperature of the cooler fluid.
3. w subscript refers to the warmer fluid.
4. h subscript refers to hydraulic diameter.
5. c subscript refers to the cooler fluid.
6. 1 subscript refers to an inlet condition.
7. 2 subscript refers to an outlet condition.

A. Fluid Properties

H_2O at 212°F		$T_1 = 230°F; T_2 = 180°F$
$\dot{m}_w = 0.7$ lbm/hr		$T_1 = 230°F; T_2 = 180°F$
$\rho = 0.960(62.4)$ lbm/ft³		$C_p = 1.007$ BTU/lbm·°R
$k_f = 0.393$ BTU/(hr·ft·°R)		$\alpha = 6.51 \times 10^{-3}$ ft²/hr
$v = 0.316 \times 10^{-5}$ ft²/s		Pr $= 1.74$

Air at 540°R		$t_1 = 70°F$
$\dot{m}_c = 2.5$ lbm/hr		$t_1 = 70°F$
$\rho = 0.0735$ lbm/ft³		$C_p = 0.240$ BTU/lbm·°R
$k_f = 0.01515$ BTU/hr·ft·°R		$\alpha = 0.859$ ft²/hr
$v = 16.88 \times 10^{-5}$ ft²/s		Pr $= 0.708$

B. Heat Balance

$H_2O \quad q_w = \dot{m}_w C_{pw}(T_1 - T_2) \quad = 0.7(1.007)(230 - 180) = 35.3 \text{ BTU/hr}$

$\text{Air} \quad q_c = \dot{m}_c C_{pc}(t_2 - t_1) \quad = 2.5(0.240)(t_2 - 70) = 35.3; \ t_2 = 129°F$

C. Log Mean Temperature Difference

$Counterflow \quad LMTD = \dfrac{(T_1 - t_2) - (T_2 - t_1)}{ln\,[(T_1 - t_2)/(T_2 - t_1)]} = 105.4°F$

D. Correction Factor

$$S = \frac{t_2 - t_1}{T_1 - t_1} = 0.369$$

$$R = \frac{T_1 - T_2}{t_2 - t_1} = \frac{\dot{m}_c C_{pc}}{\dot{m}_w C_{pw}} = 0.848$$

$\left.\begin{array}{c} \\ \\ \\ \\ \end{array}\right\} F = 0.98$

$$\left(\begin{array}{c} \text{Figure 8.4 mixed-unmixed} \\ \text{Figure 8.5 unmixed-unmixed} \end{array}\right)$$

E. UA Product

$$UA = \frac{q}{F\,LMTD} = 0.342$$

F. Capacitances

$Unmixed\text{-}Unmixed$

$H_2O \quad (\dot{m}C_p)_w \quad = 0.7(1.007) = 0.705 \text{ BTU/hr·°R}$

$\text{Air} \quad (\dot{m}C_p)_c \quad = 2.5(0.240) = 0.6 \text{ BTU/hr·°R}$

$$C = \frac{(\dot{m}C_p)_{min}}{(\dot{m}C_p)_{max}} = 0.851$$

G. Number of Transfer Units, NTU

$$NTU = \frac{UA}{(\dot{m}C_p)_{min}} = 0.569$$

H. Effectiveness, E

$$NTU = 0.569$$

$$C = 0.851$$

$$\left.\begin{array}{c}\\\\\end{array}\right\}$$

$$E = 0.39$$

$$\left(\begin{array}{c}\text{Figure 8.6 mixed-unmixed}\\\text{Figure 8.7 unmixed-unmixed}\end{array}\right)$$

I. Outlet Temperature Calculations

$$t_2 = (T_1 - t_1)E + t_1$$

Air $t_2 = 132°F$ (Step B shows $t_2 = 129°F$; not a crucial point)

$$T_2 = T_1 - C(t_2 - t_1)$$

H₂O $T_2 = 180°F$

J. Velocities in Connecting Tubes/Ducts

NA $V_w =$

NA $V_c =$

K. Pressure Drop Calculations (Use $K = 10$)

NA $\Delta p_w = K \dfrac{\rho_w V_w^2}{2g_c} =$

NA $\Delta p_c = K \dfrac{\rho_c V_c^2}{2g_c} =$

L. Frontal Area Required

$UA = 0.352$ $A = 2.25 \text{ ft}^2$ (Figure 8.8)

Q. Summary of Information Requested in Problem Statement
Frontal area (normal to the air flow direction) required
is 2.25 ft².

Discussion   The outlet temperature of the air was calculated to be
132°F which does not equal 129°F as calculated in Step B.
If the air outlet temperature were critical, we would
repeat the calculations using 132°F. The frontal area of
2.25 ft² will be sufficient to provide the necessary cooling
for the water.

8.2 Double Pipe Heat Exchanger Design Considerations

A discussion of heat transfer equipment was provided in the last chapter and in the preceding section. The discussion now is extended to include design considerations.

In many of the problems discussed thus far, only a minimum number of variables are left as unknown. The problems are therefore easy to solve and require only a few assumptions to obtain a solution. When this is not the case, we must resort to cost information to determine the particular design that will yield the lowest total cost per year. The optimum design could involve specification of tube length within an exchanger or the modification of the fluid flow rates. In addition, the apparatus is usually designed to meet certain safety codes, such as those established by ASME and by an organization called the Tubular Exchanger Manufacturers Association (TEMA).

The exchangers of interest in this chapter include the double pipe, shell and tube, and cross flow heat exchangers. Each is suitable for certain applications. A double pipe heat exchanger is used for low to moderate flow rates, and low to moderate heat transfer rates. The fluids are usually liquid to liquid or vapor to vapor or gas to gas. A shell and tube heat exchanger is used for high flow rates and high heat transfer rates (flow rates of over 10 times those in double pipe heat exchangers). The most suitable fluid combinations are liquid to liquid, gas to gas, or vapor to vapor. The cross flow heat exchanger can be sized to exchange heat at low, medium, or high flow rates, and low, medium, or high heat transfer rates. It is suited for liquid to vapor, liquid to gas, gas to gas, and vapor to vapor heat exchange. Any of the above mentioned heat exchangers can be used as condensers or evaporators, although there are devices that are better suited for such service. Many other types of heat exchangers are commercially available but they are not discussed here, however, for reasons of space. The methods of analysis are identical.

Consider a problem in which the inlet and outlet temperatures and mass flow rates are specified and it is desired to use an exchanger that will transfer the required heat load and that will minimize costs. When mass flow rates are known, it is advisable to fix the fluid velocities at the optimum values. Optimum velocity values for various fluids were presented in Table 5.4 and, for convenience, are given again in Table 8.2. It is to be remembered that when the optimum velocity is used, the total cost (first costs plus operating costs) of moving the fluid are minimized. When flow rate and velocity are known, the required cross sectional area can be easily calculated.

Consider next a double pipe heat exchanger that we wish to size for a given service. In such problem, a chart showing geometry factors for

TABLE 8.2. *Reasonable velocities for various fluids, calculated by using optimum economic diameter equations.*

Fluid	Economic Velocity Range ft/s	m/s
Acetone	4.9–9.8	1.5–3.0
Ethyl Alcohol	4.8–9.6	1.5–3.0
Methyl Alcohol	4.8–9.6	1.5–3.0
Propyl Alcohol	4.7–9.4	1.4–2.8
Benzene	4.6–9.2	1.4–2.8
Carbon Disulfide	4.2–8.4	1.3–2.6
Carbon Tetrachloride	3.9–7.8	1.2–2.4
Castor Oil	1.6–3.2	0.5–1.0
Chloroform	4.0–8.0	1.2–2.4
Decane	4.9–9.8	1.5–3.0
Ether	5.0–10.0	1.5–3.0
Ethylene glycol	3.9–7.8	1.2–2.4
R-11	4.0–8.0	1.2–2.4
Glycerine	1.4–2.8	0.43–0.86
Heptane	5.1–10.2	1.5–3.0
Hexane	5.2–10.4	1.6–3.2
Kerosene	4.7–9.4	1.4–2.8
Linseed Oil	4.9–9.8	1.5–3.0
Mercury	2.1–4.2	0.64–1.3
Octane	5.0–10.0	1.5–3.0
Propane	5.6–11.2	1.7–3.4
Propylene	5.5–11.0	1.7–3.4
Propylene Glycol	4.5–9.0	1.4–2.8
Turpentine	4.6–9.2	1.4–2.8
Water	4.4–8.8	1.4–2.8

various tubing combinations is a labor saving device. A chart of this type is given in Table 8.3.

In view of the above discussion, the problem we seek to solve is as follows: When inlet temperatures and flow rates are known, and certain outlet temperatures are desired, what size double pipe heat exchanger will transfer the required energy for a minimum cost? Without a reformulation of all the appropriate economic parameters, this problem can be solved with information already available. The method is illustrated by the next example, following a slightly modified version of the suggested order of calculations for double pipe heat exchangers given in Chapter 7.

Type M Tubing

Size	ID_a ft	ID_p ft	OD_p ft	A_p ft²	A_a ft²	D_h ft	D_e ft	\multicolumn{3}{c}{$A_o = \pi OD_p L$ in ft² for L =}
								10 ft	15 ft	20 ft
$2 \times 1\frac{1}{4}$, $1\frac{1}{4}$	0.1674	0.1076	0.1146	0.009093	0.01169	0.0528	0.1299	3.600	5.400	7.200
$2\frac{1}{2} \times 1\frac{1}{4}$	0.2079	0.1076	0.1146	0.009093	0.02363	0.0933	0.2625	3.600	5.400	7.200
3×2	0.2484	0.1674	0.1771	0.02201	0.02382	0.0713	0.1713	5.563	8.345	11.12
4×3	0.3279	0.2484	0.2604	0.04846	0.03118	0.0675	0.1524	8.180	12.27	16.36

Type M Tubing

Size	ID_a m	ID_p m	OD_p m	A_p m²	A_a m²	D_h m	D_e m	\multicolumn{3}{c}{$A_o = \pi OD_p L$ in m² for L =}
								3 m	4.5 m	6 m
$2 \times 1\frac{1}{4}$, $1\frac{1}{4}$	0.051 02	0.032 79	0.034 93	0.000 844 4	0.001 086	0.016 09	0.039 59	0.329 2	0.493 8	0.658 4
$2\frac{1}{2} \times 1\frac{1}{4}$	0.063 38	0.032 79	0.034 93	0.000 844 4	0.002 196	0.028 45	0.080 07	0.329 2	0.493 8	0.658 4
3×2	0.075 72	0.051 02	0.053 98	0.002 044	0.002 214	0.021 74	0.052 23	0.508 7	0.763 1	1.017
4×3	0.099 98	0.075 72	0.079 38	0.004 503	0.002 901	0.020 6	0.046 54	0.748 1	1.122	1.496

TABLE 8.3. *Double pipe heat exchanger tube combinations and geometry factors.*

EXAMPLE 8.2   Benzene is used in a process for the manufacture of detergent. A double pipe heat exchanger must be sized to exchange heat between benzene ($\rho = 54.3$ lbm/ft^3, $C_p = 0.425$ BTU/lbm·°R, $\mu = 50$ cp, $k_f = 0.091$ BTU/hr·ft·°R and Pr = 1.78) and water. The benzene flow rate is 10,000 lbm/hr and it is to be heated from 75°F to 125°F. The water is available at 200°F. Select an appropriate heat exchanger and determine the required water flow rate.

Discussion   For a liquid to liquid heat exchange, we can use a double pipe or a shell and tube heat exchanger. The benzene flow rate of 10,000 lbm/hr is small enough so that a double pipe heat exchanger will probably work. (If we did not know this by experience, we would begin by trying to size a double pipe heat exchanger first. If one could not be made to work, then a shell and tube exchanger would be tried next.)

The flow rate of benzene is known and the optimum velocity is found from Table 8.2. If not listed, the optimum velocity can be determined from the equations of Chapter 4. Knowing optimum velocity, it is possible to calculate the cross sectional area required for minimum cost conditions.

For benzene, the optimum velocity range is 4.6 to 9.2 ft/s, while for water, it is 4.4 to 8.8 ft/s. Moreover, the specific heat of water is greater than that of benzene and so the water will not experience as great a temperature change as will the benzene for equal flow rates.

Because a range of velocities is given, several choices need to be made. For example, it is possible to try to operate near the maximum velocity without exceeding the maximum permissible pressure drop of 10 psi (72.4 kPa). A high velocity results in a high Reynolds number, a high Nusselt number, and a high convection coefficient. On the other hand, it is prudent to remember that the velocity ranges given are merely guides and so staying at or near the median value might be better.

The fouling factors for benzene and water will influence the choice of which stream to place in the pipe and which to place in the annulus. For benzene (an organic liquid), the fouling factor from Table 7.1 is 0.001 ft^2·hr·°R/BTU, while for distilled water, the fouling factor is taken to be 0.00075 ft²·°hr·°R/BTU [= (0.0005 + 0.001)/2]. Benzene therefore has a greater tendency to cause fouling on surfaces it contacts. If placed in the annulus, benzene will form deposits on the outside surface of the inner tube *and* on the inside surface of the outer tube. So our intent at this point is to route the benzene through the pipe, or the inner tube, of the double pipe heat exchanger we select.

Assumptions

1. Steady state conditions exist.
2. Benzene properties remain constant and are given in the problem statement. Water properties are evaluated at a temperature of (200 + 75)/2 = 137.5 ≈ 140°F.
3. Exchanger length is 15 ft, and it is possible that several will be needed for this service.

Nomenclature

1. T refers to the temperature of the warmer fluid.
2. t refers to the temperature of the cooler fluid.
3. w subscript refers to the warmer fluid.
4. h subscript refers to hydraulic diameter.
5. c subscript refers to the cooler fluid.
6. a subscript refers to the annular flow area or dimension.
7. p subscript refers to the tubular flow area or dimension.
8. 1 subscript refers to an inlet condition.
9. 2 subscript refers to an outlet condition.
10. e subscript refers to equivalent diameter.

A. Fluid Properties

H_2O @ 140°F		
$\dot{m}_w$ = TO BE SELECTED		T_1 = 200°F
ρ = 0.985(62.4) lbm/ft³		C_p = 0.9994 BTU/lbm·°R
k_f = 0.376 BTU/hr·ft·°R		α = 6.02 x 10⁻³ ft²/hr
ν = 0.514 x 10⁻⁵ ft²/s		Pr = 3.02

$$4.4 \le V_{opt} \le 8.8 \text{ ft/s}$$

Benzene		
$\dot{m}_c$ =10,000 lbm/hr=2.78 lbm/s		t_1 = 75°F
ρ = 54.3 lbm/ft³		C_p = 0.425 BTU/lbm·°R
k_f = 0.091 BTU/hr·ft·°R		α = ν/Pr = 3.48 x 10⁻⁶ ft²/s
ν = $\mu g_c/\rho$ = 6.19 x 10⁻⁶ ft²/s		Pr = 1.78
$4.6 \le V_{opt} \le 9.2$ ft/s		t_2 = 125°F

Benzene   Min flow area = $\dot{m}_c / \rho V_{max}$ = 2.78/[54.3(9.2)] = 0.0056 ft²

Benzene   Max flow area = $\dot{m}_c / \rho V_{min}$ = 2.78/[54.3(4.6)] = 0.0113 ft²

Referring to Table 8.3 and assuming the given sizes are all that are available, we see that the maximum flow area corresponds roughly to A_p for a 2 x 1¹/₄ or a 2¹/₂ x 1¹/₄ double pipe heat exchanger. Arbitrarily, we select the 2 x $1^1/_4$ size and proceed with the calculations on a first trial basis.

B. Tubing Sizes

$2 \times 1^1/_4$ $ID_a = 0.1674$ ft

$ID_p = 0.1076$ ft $OD_p = 0.1146$ ft

C. Flow Areas $A_p = \pi ID_p{}^2/4 = 0.009093$ ft^2

$A_a = \pi(ID_a{}^2 - OD_p{}^2)/4 = 0.01169$ ft^2

D. Fluid Velocities [Route the benzene through the pipe or inner tube.]

Benzene $V_p = \dot{m}/\rho A = 5.63$ ft/s $G_p = \dot{m}/A = 305.7$ lbm/ft^2·s

H_2O $V_a = \dot{m}/\rho A = 6$ ft/s $G_a = \dot{m}/A = 369$ lbm/ft^2·s

Arbitrarily selected; mass flow rate is now calculated:

H_2O $\dot{m}_w = \rho A_a V_a = 0.985(62.4)(0.01169)(6) = 4.31$ lbm/s

E. Annulus Equivalent Diameters

Friction $D_h = ID_a - OD_p = 0.0528$ ft

Ht Trans $D_e = (ID_a{}^2 - OD_p{}^2)/OD_p = 0.1299$ ft

F. Reynolds Numbers

Benzene $Re_p = V_p ID_p/\nu = 9.8 \times 10^4$

H_2O $Re_a = V_a D_e/\nu = 1.52 \times 10^5$

G. Nusselt Numbers

Benzene $Nu_p = 269$

H_2O $Nu_a = 499$

H. Convection Coefficients

Benzene $h_i = Nu_p\, k_f/ID_p = 227.5$ $h_p = h_i\, ID_p/OD_p = 214$

H_2O $h_a = Nu_a\, k_f/D_e = 1444$

I. Exchanger Coefficient

$$\frac{1}{U_o} = \frac{1}{h_p} + \frac{1}{h_a}$$ $U_o = 197$

(Note: Complete step L before step J, to obtain $T_2 = 186.3°F$)

J. Outlet Temperature Calculations (Exchanger length L = to be found)

H_2O $T_2 = 186.3°F$

Benzene $t_2 = 125°F$ (given in problem statement)

K. Log Mean Temperature Difference

$$\text{Counterflow} \qquad LMTD = \frac{(T_1 - t_2) - (T_2 - t_1)}{\ln\left[(T_1 - t_2)/(T_2 - t_1)\right]} = 92.0°F$$

L. Heat Balance

Benzene $q_c = \dot{m}_c C_{pc}(t_2 - t_1)$ $= 59.1$ BTU/s $= 2.13 \times 10^5$ BTU/hr

H_2O $q_w = \dot{m}_w C_{pw}(T_1 - T_2);$ with $\dot{m}_w = 4.31$ lbm/s, we find
$T_2 = 186.3°F$

M. Fouling Factors and Design Coefficient

$R_{di} = 0.001$ $\qquad\qquad\qquad$ $R_{do} = 0.00075$ ft^2·hr/BTU

$$\frac{1}{U} = \frac{1}{U_o} + R_{di} + R_{do} \qquad\qquad U = 147$$

N. Heat Transfer Area and Tube Length (unless already known)

$$A_o = \frac{q}{U\,(LMTD)} = \frac{2.13 \times 10^5}{147(92.0)} = 15.7 \text{ ft}^2$$

$$L = \frac{A_o}{\pi\,(OD_p)} = 43.7 \text{ ft (3 heat exchangers, each 15 ft long)}$$

O. Friction Factors

$$\text{Re}_p = V_p ID_p/v = 9.79 \times 10^4$$

Benzene $\qquad\qquad \left. \dfrac{\varepsilon}{ID_p} = \text{smooth} \right\} \qquad f_p = 0.019$

$$\text{Re}_a = V_a D_h/v = 6.16 \times 10^4$$

H_2O $\qquad\qquad \left. \dfrac{\varepsilon}{D_h} = \text{smooth} \right\} \qquad f_a = 0.02$

P. Pressure Drop Calculations

$$\Delta p_p = \frac{f_p L}{ID_p} \frac{\rho_p V_p^2}{2g_c} = 206 \text{ psf} = 1.4 \text{ psi}$$

$$\Delta p_a = \left(\frac{f_a L}{D_h} + 1\right) \frac{\rho_a V_a^2}{2g_c} = 569 \text{ psf} = 4.0 \text{ psi}$$

Q. Summary of Information Requested in Problem Statement

> Use 3 of the 2 x $1^1/_4$ double pipe heat exchangers, 15 ft long. Route the benzene through the inner tube and the water through the annulus. Set the water flow rate at 4.31 lbm/s or 15,500 lbm/hr.

8.3 Shell and Tube Heat Exchanger Design Considerations

Shell and tube heat exchanger calculations performed in the last chapter were made for existing exchangers. Such problems are relatively easy to solve by following the suggested calculation procedure. There exists another class of problems in which a heat exchanger must be sized to perform a given task. For example, given inlet temperatures, flow rates, and *desired* outlet temperatures, what size heat exchanger is required to perform the task?

It is apparent that higher flow rates within a 1-2 shell and tube heat exchanger give a greater heat transfer rate (higher velocity → higher Reynolds number → higher Nusselt number → higher convection coefficient). As flow rate increases, however, so does the pressure drop. From Chapter 7, the Nusselt number equation for turbulent flow is

$$Nu_t = \frac{h_i \, ID_t}{k_f} = 0.023 \, Re_t^{4/5} \, Pr^n \tag{8.9}$$

where $Re = VD/\nu$ and $Pr = \nu/\alpha$. So the convection coefficient varies with $V^{0.8}$. The pressure drop within the system is given by

$$\Delta p_t = \frac{\rho V_t^2}{2g_c} \times (\text{geometry factors}) \tag{8.10}$$

The pressure drop varies with V^2. An increase in velocity will increase the convection coefficient which is accompanied by a greater increase in the pressure drop.

Other factors enter into the process of specifying an exchanger for a given duty. For example, if one of the fluids tends to foul surfaces more than the other, then the rapid fouling fluid should be routed through the tubes. If routed through the shell, it will foul the outside surfaces of the tubes and the inside surface of the shell. When routed through the tubes it will foul only the inside surface of the tubes and these are more readily cleaned. Larger tube sizes should be used for fluids that foul tubes rapidly. Similarly, a corrosive fluid requiring a special metal should be routed through the tubes; otherwise, special metal must be used for the shell as well. If both fluids are nonfouling, the higher pressure fluid should be routed through the tubes, avoiding the need for and expense of a thicker walled shell.

Another factor to consider when sizing an exchanger is the tube length. Tubing is available in a number of sizes but standard sizes (8, 12, or 16 ft) should be used. Exchanger size is dictated by costs associated with cleaning, by available space, and by what sizes are commonly used in other exchangers at the facility.

The problem we seek to solve regarding shell and tube heat exchanger sizing is: Given fluid properties, flow rates, and temperatures, what optimum size shell and tube heat exchanger is required to transfer the necessary heat load? The calculations for such a problem can proceed as in the last chapter by assuming a certain size exchanger and evaluating its performance. It should be noted, however, that several exchangers can do the job and analyzing via trial and error to find one can cost considerable time. A structured trial and error method can save time and is presented here.

The method is started by assuming a trial value of the overall *design* coefficient U. For water to water systems, the overall heat transfer coefficient usually ranges from 250 to 500 BTU/hr·ft²·°R [1420 to 2800 W/(m²·K)]. For other fluid combinations, the overall heat transfer coefficient ranges from 25 to 50 BTU/hr·ft²·°R [140 to 280 W/(m²·K)].

When the overall heat transfer coefficient is used with the heat transfer rate and the log mean temperature difference, a trial calculation of the heat transfer area can be made. When area is combined with tube length and pitch (arbitrarily selected), then a tube count vs shell diameter table (i.e., Table 7.3) is used to select a shell diameter. The number of tube passes is selected knowing the flow rates and the optimum velocity of the tube fluid. Current practice indicates that the best exchanger to use is the smallest one with a standard layout which fulfills dirt factor and pressure drop requirements.

Based on experience with shell and tube equipment, the baffles should be placed according to:

Maximum baffle spacing $B = ID$ of shell

$$\text{Minimum baffle spacing} \quad B = \frac{ID \text{ of shell}}{5} \left.\begin{array}{c} \\ \\ \\ \end{array}\right\} \begin{array}{l} \text{whichever} \\ \text{is larger} \end{array}$$

$$B = 2.25 \text{ in.}$$

Thus baffle spacing can be altered by a factor of five between minimum and maximum values. At wide baffle spacing, the shell fluid tends to be more axial in its flow direction rather than across the tube bundle. At closer spacing, there will exist excessive leakage between baffle and shell, and between baffle and tubes.

The number of tube fluid passes can be varied from one to eight although one tube pass is seldom used. In larger shells, the variation can range to as high as 16 passes. In 1-2 shell and tube heat exchangers, the worst performance is obtained with maximum baffle spacing and two tube passes. As indicated earlier in Equations 8.9 and 8.10, the tube fluid convection coefficient and pressure drop for turbulent flow varies according to

$$h_i \propto V_t^{0.8}$$

$$\Delta p_t \propto V_t^2 L$$

For purposes of comparison, we calculate the ratio of convection coefficients for 8 tube fluid passes to 2 fluid passes as:

$$\frac{h_{i(8 \text{ passes})}}{h_{i(2 \text{ passes})}} = \frac{[8V_t]^{0.8}}{[2V_t]^{0.8}} = 3.03$$

Similarly, for the pressure drop,

$$\frac{\Delta p_{t(8 \text{ passes})}}{\Delta p_{t(2 \text{ passes})}} = \frac{[8V_t]^2 \cdot 8}{[2V_t]^2 \cdot 2} = 64$$

Thus by increasing from 2 to 8 tube passes, the convection coefficient increases by a factor of 3 while the pressure drop increases by a factor of 64.

Consider also the variation in the shell side convection coefficient. From the equations in Chapter 7, the shell side convection coefficient and pressure drop vary according to

$$h_o \propto V_s^{0.8}$$

$$\Delta p_s \propto V_s^2 (N_b + 1)$$

where N_b is the number of baffles and $N_b + 1$ is the number of times the shell fluid crosses the tube bundle and the number of spaces between baffles from end to end within the exchanger. For variation in baffle spacing between the maximum and the minimum, we calculate

$$\frac{h_{o\ minimum}}{h_{o\ maximum}} = \frac{[5V_s]^{0.5}}{[V_s]^{0.5}} = 2.24$$

Similarly, for the pressure drop,

$$\frac{\Delta p_{s\ minimum}}{\Delta p_{s\ maximum}} = \frac{[5V_s]^2 \cdot 5}{[V_s]^2 \cdot 1} = 125$$

Thus by changing the baffle spacing from its minimum to its maximum value, the shell side pressure drop is increased by a factor of 125 while the convection coefficient is changed only by a factor of 2.24.

Based on the above discussion and on information given earlier, a format for sizing a heat exchanger for a given service has been put together. A suggested order of calculations now follows.

SUGGESTED ORDER OF CALCULATIONS FOR
SIZING
A SHELL AND TUBE HEAT EXCHANGER

Problem   Complete problem statement.

Discussion   Potential heat losses; other sources of difficulties.
                    Calculations are made remembering that:
                    • The exchanger should be as **small** as possible;
                    • The **correction factor** F should be equal to or greater
                       than 0.75;
                    • The **velocity** of the tube fluid should be within the
                       optimum range as calculated with the methods of
                       Chapter 4 or from Table 8.2; p 305
                    • The exchanger when fouled should still **deliver the
                       required energy exchange**—therefore the clean
                       overall coefficient will be greater than the design
                       value;
                    • The **overall pressure drop** for both streams should be
                       less than 10 psi.
                    Finding an exchanger that will satisfy these criteria
                    involves a trial and error procedure.

Assumptions   1. Steady state conditions exist.
                    2. Fluid properties remain constant and are evaluated at
                          a temperature of:

Nomenclature 1. T refers to the temperature of the warmer fluid.
2. t refers to the temperature of the cooler fluid.
3. w subscript refers to the warmer fluid.
4. h subscript refers to hydraulic diameter.
5. c subscript refers to the cooler fluid.
6. s subscript refers to the shell flow area or dimension.
7. t subscript refers to the tubular flow area or dimension.
8. 1 subscript refers to an inlet condition.
9. 2 subscript refers to an outlet condition.
10. e subscript refers to equivalent diameter.

A. Fluid Properties

$$\dot{m}_w = \qquad\qquad T_1 =$$
$$\rho = \qquad\qquad C_p =$$
$$k_f = \qquad\qquad \alpha =$$
$$\nu = \qquad\qquad Pr =$$
$$\qquad\qquad\qquad T_2 =$$

$$\dot{m}_c = \qquad\qquad t_1 =$$
$$\rho = \qquad\qquad C_p =$$
$$k_f = \qquad\qquad \alpha =$$
$$\nu = \qquad\qquad Pr =$$
$$\qquad\qquad\qquad t_2 =$$

B. Heat Balance for Fluids

$$q_w = \dot{m}_w C_{pw}\,(T_1 - T_2) = \quad 2.43 \times 10^5 \times 30$$

$$\frac{}{n}$$

$$q_c = \dot{m}_c C_{pc}\,(t_2 - t_1) =$$

C. Log Mean Temperature Difference

$$\text{Counterflow} \qquad LMTD = \frac{(T_1 - t_2) - (T_2 - t_1)}{\ln\,[(T_1 - t_2)/(T_2 - t_1)]} = \quad 136$$

D. Overall Heat Transfer Coefficient

(a) Assume a *design* value $U = 268$

(Better too high than too low; low flow rate stream in shell unless factors dictate otherwise)

(b) Heat Transfer area required

$$A_o = \frac{q}{U\,F\,LMTD} = \quad \frac{2.43 \times 10^5 \times 30}{268(1)(136)} = 200\ ft^2$$

$$\text{Fig. 8.10} \quad \text{Fig. 7.19} \quad \begin{cases} p_9\ 262\text{-}265 \\ (7.14)5 = \dfrac{t_2 - t_1}{T_1 - T_1} = \dfrac{2}{200 - 60} = 0.013 \end{cases}$$

$$(7.14)$$

$$R = \frac{T_1 - T_2}{t_2 - t_1} = \frac{6}{2} = 3.0$$

(c) Outside surface area of tubes (tube size assumed; dimensions from Table 7.2)

OD_t	$A'' = \pi OD_t(1 \text{ ft})$

→ 3/4 in. = 0.019 1 m	0.1963 ft²/(linear ft) = 0.0182 m²/(linear m)
1 in. = 0.021 7 m	0.2618 ft²/(linear ft) = 0.0243 m²/(linear m)

(handwritten) no 7.11 $\{ID_t = .167\,4''$ $OD_t = .1771'$ $A'' = 0.5564\,ft²/ft$.

(handwritten) $\frac{30}{\uparrow} = 29.96$

(d) Tube length (8, 12, or 16 ft) $L = 12'$

(e) Estimated number of tubes required: $N_t = \dfrac{A_o}{L\,A''} = \dfrac{200}{12(.5564)} = 29.96$

(f) Optimum Velocity for the tube fluid $V_t = V_{opt} = 13.40$ (from 7.11)

(g) Tube area $A_t = \dot{m}/\rho V_t = \dfrac{20 \times 30}{67.8(13.40)} = 0.66$

(h) Number of passes $N_p = N_t\,\pi\,(ID_t{}^2)/4A_t = \dfrac{30(\pi)(.1674)}{4(.66)} = 1.0$

(handwritten left margin) 3/4" 64 * 1.3   Table 8.3 p.251

(i) Consult tube count table and select *nearest or greater* count and shell diameter   *(handwritten)* $.1771(inch) = 2.84$ $D_s = 7.36$ $8 \times 2.84 = 22.67''$
     (handwritten) oun on   .1771(in) 3/4"
     Actual number of tubes $N_t = 32$

E. Corrected Coefficient

$$A_o = N_t L A'' = \qquad\qquad U = \dfrac{q}{A_o\,F\,LMTD} =$$

F. Tubing Sizes

$ID_t = $ $OD_t = $

$N_t = $ no. of tubes $ = $

$N_p = $ no. of passes $ = $

G. Shell Data

	D_s = shell inside diameter	=
(assumed)	B = baffle spacing	=
	N_b = number of baffles	=
	P_T = tube pitch	=
	$C = \left(\begin{array}{c}\text{clearance between}\\ \text{adjacent tubes}\end{array}\right) = P_T - OD_t$	=

H. Flow Areas

$A_t = N_t\,\pi\,(ID_t{}^2)/4N_p = $

$A_s = D_s CB/P_T = $

I. Fluid Velocities [Route the fluid with the higher flow rate through the flow cross section with the greater area.]

$$V_t = \dot{m}/\rho A_t =\qquad\qquad G_t = \dot{m}/A =$$

$$V_s = \dot{m}/\rho A_s =\qquad\qquad G_s = \dot{m}/A =$$

J. Shell Equivalent Diameter

$$\left(\begin{array}{c}\text{square}\\\text{pitch}\end{array}\right)\quad D_e = \frac{4P_T{}^2 - \pi\, OD_t{}^2}{\pi\, OD_t} =$$

$$\left(\begin{array}{c}\text{triangular}\\\text{pitch}\end{array}\right)\quad D_e = \frac{3.44P_T{}^2 - \pi\, OD_t{}^2}{\pi\, OD_t} =$$

K. Reynolds Numbers

$$Re_t = V_t\, ID_t/\nu =$$

$$Re_s = V_s\, D_e/\nu =$$

L. Nusselt Numbers

<u>Tube Side</u>

Modified Seider-Tate Equation for laminar flow:

$$Nu_t = \frac{h_i\, ID_t}{k_f} = 1.86\left(\frac{ID_t\, Re_t\, Pr}{L}\right)^{1/3}$$

$$Re_t < 2\,200 \qquad\qquad 0.48 < Pr = \nu/\alpha < 16\,700$$

Modified Dittus-Boelter Equation for turbulent flow:

$$Nu_t = \frac{h_i\, ID_t}{k_f} = 0.023\, Re_t{}^{4/5}\, Pr^n$$

$n = 0.4$ if fluid is being heated
$n = 0.3$ if fluid is being cooled

$$Re_t > 10\,000; \qquad 0.7 < Pr = \nu/\alpha < 160; \qquad L/D > 60$$

Conditions for both equations:
μ changes moderately with temperature
Properties evaluated at the average fluid
temperature [= (inlet + outlet)/2]

Shell Side

$$Nu_s = \frac{h_o D_e}{k_f} = 0.36 \, Re_s^{0.55} \, Pr^{1/3}$$

$$2 \times 10^3 < Re_s = V_s D_e/v < 1 \times 10^6 \quad Pr = v/\alpha > 0$$

μ changes moderately with temperature
Properties evaluated at the average fluid
temperature [= (inlet + outlet)/2]

$Nu_t =$

$Nu_s =$

M. Convection Coefficients

$$h_i = Nu_t \, k_f/OD_t =$$

$$h_o = Nu_s \, k_f/D_e =$$

N. Exchanger Coefficient

$$\frac{1}{U_o} = \frac{1}{h_i} + \frac{1}{h_o} \qquad\qquad U_o =$$

O. Outlet Temperatures Calculations (Exchanger length L =   )

$$R = \frac{\dot{m}_c C_{pc}}{\dot{m}_w C_{pw}} = \qquad\qquad A_o = N_t \pi \, OD_t \, L =$$

$$\frac{U_o A_o}{\dot{m}_c C_{pc}} = \qquad\qquad S = \qquad\qquad \text{(Figure 7.22)}$$

$$t_2 = S(T_1 - t_1) + t_1 =$$

$$T_2 = T_1 - R(t_2 - t_1) =$$

P. Overall Heat Balance for the Exchanger

$$F = \qquad\qquad\qquad \text{(Figure 7.19)}$$

$$q = U_o A_o F_t LMTD =$$

Q. Fouling Factors and Design Coefficient

Fluids $R_{di} =$ $R_{do} =$ $\Sigma R_d =$
(Table 7.3)

Exchanger $R_{de} = \dfrac{U_o - U}{U_o U} =$ (Must exceed ΣR_d)

R. Area Required to Transfer Heat

$$A_o = \frac{q}{UF\,LMTD} =$$

$$L = \frac{A_o}{N_t \pi\, OD_t} =$$

S. Friction Factors

<u>Tube Side</u>

Laminar flow in a tube:

$$f_t = \frac{64}{Re_t} \qquad\qquad Re_t < 2\,200 \text{ (step G above)}$$

Turbulent flow in a tube:

Chen Equation

$$\frac{1}{\sqrt{f_t}} = -2.0 \log\left\{\frac{\varepsilon}{3.7065D} - \frac{5.0452}{Re}\log\left[\frac{1}{2.8257}\left(\frac{\varepsilon}{D}\right)^{1.1098} + \frac{5.8506}{Re^{\,0.8981}}\right]\right\}$$

Churchill Equation

$$f_t = 8\left[\left(\frac{8}{Re}\right)^{12} + \frac{1}{(B+C)^{1.5}}\right]^{1/12}$$

$$\text{where } B = \left[2.457\,ln\,\frac{1}{(7/Re)^{0.9} + (0.27\varepsilon/D)}\right]^{16}$$

$$\text{and} \quad C = \left(\frac{37\,530}{Re}\right)^{16}$$

<u>Shell Side</u>

$$f_s = \exp(0.576 - 0.19\,ln\,Re_s) \qquad\qquad Re_s \text{ from step G}$$

$$Re_t = $$
$$\left.\frac{\varepsilon}{ID_t} = \right\}$$ $$f_t =$$

$$f_s =$$

T. Pressure Drop Calculations

$$\Delta p_t = \frac{\rho V_t^2}{2g_c} \left(\frac{f_t L N_p}{ID_t} + 4N_p \right) =$$

$$\Delta p_s = \frac{\rho V_s^2}{2g_c} \frac{D_s}{D_e} f_s (N_b + 1) =$$

U. Summary of Information Requested in Problem Statement

EXAMPLE 8.3 Gasoline [$\rho = 0.701(62.4)$ lbm/ft^3, $\mu = 1.07 \times 10^{-5}$ lbf·s/ft^2, $C_p = 0.5$ BTU/lbm·°R, $k_f = 0.08$ BTU/hr·hr·ft°R] flowing at 70,000 lbm/hr is to be cooled from 200°F to 100°F using city water at an inlet temperature of 80°F. Determine the specifications of a heat exchanger that will perform this service.

Discussion Calculations are made remembering that:
  • The exchanger should be as **small** as possible;
  • The **correction factor** F should be equal to or greater than 0.75;
  • The **velocity** of the tube fluid should be within the **optimum** range as calculated with the methods of Chapter 4 or from Table 8.2;
  • The exchanger when fouled should still **deliver the required energy exchange**—therefore, the clean overall coefficient will be greater than the design value;
  • The **overall pressure drop** for both streams should be less than 10 psi.

Assumptions 1. Steady state conditions exist.
2. Fluid properties remain constant and are evaluated at a temperature of: Gasoline properties given; City water at 104°F.

Nomenclature 1. T refers to the temperature of the warmer fluid.
2. t refers to the temperature of the cooler fluid.
3. w subscript refers to the warmer fluid.
4. h subscript refers to hydraulic diameter.
5. c subscript refers to the cooler fluid.
6. s subscript refers to the shell flow area or dimension.
7. t subscript refers to the tubular flow area or dimension.
8. 1 subscript refers to an inlet condition.
9. 2 subscript refers to an outlet condition.
10. e subscript refers to equivalent diameter.

A. Fluid Properties

Gasoline $\dot{m}_w = 70,000 \text{ lbm/hr}$ $T_1 = 200°F$
$\rho = 43.7 \text{ lbm/ft}^3$ $C_p = 0.5 \text{ BTU/lbm·°R}$
$k_f = 0.08 \text{ BTU/hr·ft·°R}$ $\alpha = k_f/\rho C_p = 0.00366 \text{ ft}^2/\text{hr}$
$\nu = \mu g/\rho = 7.88 \times 10^{-6} \text{ ft}^2/\text{s}$ $Pr = \nu/\alpha = 7.75$
$T_2 = 100°F$

H₂O $\dot{m}_c = $ TO BE SELECTED $t_1 = 80°F$
@ 104°F $\rho = 0.994(62.4) \text{ lbm/ft}^3$ $C_p = 0.9980 \text{ BTU/lbm·°R}$
$k_f = 0.363 \text{ BTU/hr·ft·°R}$ $\alpha = 5.86 \times 10^{-3} \text{ ft}^2/\text{hr}$
$\nu = 0.708 \times 10^{-5} \text{ ft}^2/\text{s}$ $Pr = 4.34$
$t_2 = $ TO BE FOUND

B. Heat Balance for Fluids

Gasoline $q_w = \dot{m}_w C_{pw} (T_1 - T_2) = 70,000(0.5)(200 - 100) = 3.5 \times 10^6$

H₂O $q_c = \dot{m}_c C_{pc} (t_2 - t_1) = \dot{m}_c(0.998)(t_2 - 80) = 3.5 \times 10^6 \text{ BTU/hr}$

Notes   A number of combinations of $\dot{m}_c$ and t_2 can be selected to satisfy the above equation. We rely on the correction factor to aid in finding t_2. We first calculate the ratios R and S:

$$R = \frac{\dot{m}_c C_{pc}}{\dot{m}_w C_{pw}} = \frac{T_1 - T_2}{t_2 - t_1} \qquad \text{and} \qquad S = \frac{t_2 - t_1}{T_1 - t_1}$$

Substituting,

$$R = \frac{100}{t_2 - 80} \qquad S = \frac{t_2 - 80}{200 - 80}$$

Referring to Figure 7.19 or to Equation 7.25, we can compose the following chart for various values of the outlet temperature t_2 of the cooler fluid:

t_2	R	S	F (Figure 7.19)
90	10	0.0833	~0.96
100	5	0.1667	~0.82
110	3.33	0.25	~0.725
120	2.5	0.333	< 0.7

(More exact values of F can be obtained with Equation 7.25.) For this trial, we arbitrarily use an outlet temperature of the water of $t_2 = 100°F$. The flow rate of the cooling water then becomes

$$\dot{m}_c = \frac{3.5 \times 10^6}{(0.998)(100 - 80)} = 175,400 \text{ lbm/hr}$$

We can now proceed with the calculations.

C. Log Mean Temperature Difference

$$Counterflow \qquad LMTD = \frac{(T_1 - t_2) - (T_2 - t_1)}{ln\,[(T_1 - t_2)/(T_2 - t_1)]} = 49.7°F$$

D. Overall Heat Transfer Coefficient

(a) Assume a *design* value $\qquad U = 250 \text{ BTU/hr·ft}^2\text{·°R}$

(Better too high than too low; low flow rate stream in shell unless factors dictate otherwise)

(b) Heat Transfer area required

$$A_o = \frac{q}{U\,F\,LMTD} = 320.1 \text{ ft}^2$$

(c) Outside surface area of tubes (tube size assumed; dimensions from Table 7.2)

OD_t $\qquad\qquad\qquad\qquad\qquad A'' = \pi OD_t(1 \text{ ft})$

3/4 in. = 0.019 1 m	0.1963 ft²/(linear ft) = 0.0182 m²/(linear m)
1 in. = 0.021 7 m	0.2618 ft²/(linear ft) = 0.0243 m²/(linear m)

We select the smaller size for the first trial: 3/4 in.
$ID_t = 0.04667$ ft $OD_t = 0.0625$ ft $A" = 0.1963$

(d) Tube length (8, 12, or 16 ft) $L = 8$ ft (assumed)

(e) Estimated number of tubes required: $N_t = \dfrac{A_o}{L\,A"} = \dfrac{320.1}{8(0.1963)}$

$N_t = 203.8$ use 204

(f) Optimum Velocity for the tube fluid $V_t = V_{opt} = 7.5$ ft/s

(Table 8.2, 5 to 10 ft/s; octane)

(g) Tube area $A_t = \dot{m}/\rho V_t = \dfrac{70{,}000/3600}{43.7(7.5)} = 0.05928$ ft^2

(h) Number of passes $N_p = \dfrac{N_t \pi (ID_t{}^2)}{4 A_t} = \dfrac{204\pi(0.04667)^2}{4(0.05928)} = 5.8 \sim 6$

(i) Consult tube count table and select *nearest or greater* count
     and shell diameter $D_s = 19.25$ in. $= 1.604$ ft

Actual number of tubes $N_t = 216$

E. Corrected Coefficient

$A_o = N_t L\,A" = 339.2$ ft^2 $U = \dfrac{q}{A_o\,F\,LMTD} = 236$ BTU/hr·ft^2·°R

F. Tubing Sizes, 3/4 in., 13 BWG (nothing special specified), 1 in. Δ pitch

$ID_t = 0.560$ in. $= 0.04667$ ft $OD_t = 3/4$ in. $= 0.0625$ ft
N_t = no. of tubes   = 216
N_p = no. of passes = 6

G. Shell Data

D_s	= shell inside diameter	= 1.604 ft
(assumed) N_b	= number of baffles	= 6
(assumed) B	= baffle spacing	= 8 ft/7 spaces = 1.143 ft
P_T	= tube pitch	= 1 in. = 0.08333 ft
C	$= \left(\begin{array}{c}\text{clearance between}\\ \text{adjacent tubes}\end{array}\right)$	$= P_T - OD_t = 0.02083$

H. Flow Areas $A_t = N_t \pi\,(ID_t{}^2)/4N_p = 0.06158$ ft^2
$A_s = D_s C B / P_T = 0.4583$ ft^2

I. Fluid Velocities [Route the fluid with the higher flow rate through the flow cross section with the greater area.]

Gasoline $V_t = \dot{m}/\rho A_t = 7.22$ ft/s $G_t = \dot{m}/A = 316$ lbm/(s·ft²)

H₂O $V_s = \dot{m}/\rho A_s = 1.71$ ft/s $G_s = \dot{m}/A = 106$ lbm/(s·ft²)

J. Shell Equivalent Diameter

$\left(\begin{array}{c}\text{square}\\\text{pitch}\end{array}\right)$ $D_e = \dfrac{4P_T{}^2 - \pi\, OD_t{}^2}{\pi\, OD_t} = $ NA

$\left(\begin{array}{c}\text{triangular}\\\text{pitch}\end{array}\right)$ $D_e = \dfrac{3.44P_T{}^2 - \pi\, OD_t{}^2}{\pi\, OD_t} = 0.05916$ ft

K. Reynolds Numbers

Gasoline $Re_t = V_t\, ID_t / v = 4.28 \times 10^4$

H₂O $Re_s = V_s\, D_e / v = 1.43 \times 10^4$

L. Nusselt Numbers

Gasoline $Nu_t = 215.5$

H₂O $Nu_s = 113$

M. Convection Coefficients

Gasoline $h_i = Nu_t\, k_f / OD_t = 275.8$ BTU/hr·ft²·°R

H₂O $h_o = Nu_s\, k_f / D_e = 693$ BTU/hr·ft²·°R

N. Exchanger Coefficient

$$\frac{1}{U_o} = \frac{1}{h_i} + \frac{1}{h_o} \qquad\qquad U_o = 197.0$$

We now compare the clean exchanger coefficient (197 BTU/hr·ft²·°R) to the design value calculated in step E (236 BTU/hr·ft²·°R). The clean coefficient must exceed the design value, which is not the case in this trial. So we should go back to step D, part a, and begin with an assumed coefficient. The calculations have been made (with a spreadsheet) and concluded for this problem and the results are summarized in Table 8.4.

TABLE 8.4. *Solution of Example 8.3.*

Details: $T_1 = 200°F$; $T_2 = 100°F$; $t_1 = 80°F$; $t_2 = 100°F$; $LMTD = 49.7°F$;
$R = 5$; $S = 0.166$; $F = 0.817$; $q_w = q_c = 3.5 \times 10^6$
$P_T = 0.0833$ ft $= 1$ in. Δ pitch, 13 BWG; $C = 0.0208$ ft;
$D_e = 0.0592$; $ID_t = 0.560$ in. $= 0.04667$ ft; $OD_t = {}^3/_4$ in. $= 0.0625$ ft

Trial	U BTU/hr·ft^2·°R	L ft	A_o ft^2	N_t	D_s ft
1	236	8	339.2	216	1.60
2	140	12	612.5	260	1.77
3	111	12	772.6	328	1.94
4	107	12	805.6	342	1.94
5	80	16	1074.2	342	1.94

Trial	N_b	B ft	A_s ft^2	A_t ft^2	V_t ft/s	V_s ft/s
1	6	1.143	0.458	0.0616	7.2	1.71
2	6	1.714	0.759	0.0556	8.0	1.04
3	6	1.714	0.830	0.0701	6.3	0.95
4	6	1.714	0.830	0.0975	4.6	0.95
5	8	1.714	0.861	0.0975	4.6	0.91

Trial	Re_t	Re_s	Nu_t	Nu_s	h_i BTU/hr·ft^2·°R	h_o BTU/hr·ft^2·°R
1	1.48×10^4	1.43×10^4	215	113	276	693
2	4.74×10^4	8.65×10^3	234	86	299	527
3	3.76×10^4	7.91×10^3	194	82	249	502
4	2.70×10^4	7.91×10^3	149	82	191	502
5	2.70×10^4	7.62×10^3	149	80	191	492

Trial	U_o BTU/hr·ft^2·°R	ΣR_d hr·ft^2·°R/BTU·ft2	R_{de}	Δp_t psi	Δp_s psi
1	197	0.002	-0.001[R1]	11.3[R2]	1.08[A]
2	191	0.002	0.00191[R1]	22.7[R2]	0.48[A]
3	166	0.002	0.00139[R1]	14.7[R2]	0.44[A]
4	138	0.002	0.00110[R1]	6.0[A]	0.44[A]
5	138	0.002	0.00273[A]	7.2[A]	0.53[A]

Notes: R1 = trial rejected because exchanger does not meet or exceed fouling requirements—R_{de} must exceed ΣR_d; R2 = trial rejected because exchanger exceeds 10 psi maximum; A = Acceptable.

U. Summary of Information Requested in Problem Statement

Exchanger Specifications		
Shell diameter	=	$23^1/_4$ in.
Shell fluid passes	=	1
Tube fluid passes	=	8
Number of tubes	=	342
Baffle spacing	=	1.78 ft
Number of baffles	=	8
Tubing	=	$^3/_4$ in. OD, 13 BWG
		1 in. Δ pitch
Mass flow of cooling water	=	175,400 lbm/hr

Optimum Water Outlet Temperature Analysis

Consider a conventional power plant that makes use of raw water from a nearby source as a cooling medium. This application is quite common and so it is desirable to formulate a model to optimize such a system. Although the model will be for a problem in which water is the cooling medium, the results can be applied to other fluids if cooling fluid costs are known. The optimization process becomes one of minimizing costs and determining the optimum cooling water (or fluid) temperature. This concept is illustrated in Figure 8.9.

It is possible to use a great quantity of cooling water to obtain a small temperature increase. If this is done, then less surface area is required and so a smaller heat exchanger can be used. Therefore, the original investment involves smaller cost, but we also have a greater operating cost. On the other hand, a small flow rate of cooling water will require a greater surface area and so a larger heat exchanger must be used. Therefore, the original investment is greater, but the operating costs will be reduced. In view of these two extremes, we conclude that there must be an optimum operating point that minimizes the initial plus operating costs.

The model we formulate will yield an annual cost for the exchanger. The total annual cost is the sum of the annual cost of water (cooling fluid), the fixed costs, maintenance, depreciation, amortization, etc. With regard to the discussion above, the trade-off is between surface area A_o and mass flow rate $\dot{m}_c$. It is appropriate to include these quantities as primary variables in the analysis. So for the cooling water,

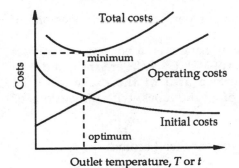

FIGURE 8.9. *Cost relationships in shell and tube equipment as a function of outlet temperature.*

$$q = \dot{m}_c C_{pc}(t_2 - t_1) = UA_o F \ LMTD \tag{8.11}$$

From the above equation,

$$\dot{m}_c = \frac{q}{C_{pc}(t_2 - t_1)}$$

and

$$A_o = \frac{q}{UF \ LMTD}$$

The cost of water per year is given as

$$C_{H_2O} = C_W \dot{m}_c \, t \tag{8.12}$$

where: C_{H_2O} is the cost of water per year with dimensions of MU/yr ($/yr);

C_W is the cost of water per unit mass with dimensions of MU/M ($/lbm or $/kg);

$\dot{m}_c$ is the mass flow rate of water (the cooling medium) with dimensions of M/T (lbm/s or kg/s); and,

t is the number of hours per year that the exchanger is in operation (hr/yr).

The annual or operating cost of the exchanger will include amortization, pumping costs, maintenance, etc. These quantities can all be included in a single term which is expressed on a per unit area basis. Thus,

$$C_A = C_F A_o \tag{8.13}$$

where: C_A is the cost of the exchanger per year with dimensions of MU/yr ($/yr);

C_F is the annual cost of the exchanger on a per unit square foot basis, with dimensions of MU/L^2 ($/ft^2 or $/m^2); and,

A is the heat transfer surface area of the exchanger (the outside surface area of all the tubes) with dimensions of L^2 (ft^2 or m^2).

The total cost of the heat exchanger per year then becomes

$$C_T = C_{H_2O} + C_A = C_W \dot{m}_c\, t + C_F A_o$$

Substituting for mass flow rate and area gives

$$C_T = \frac{C_W t q}{C_{pc}(t_2 - t_1)} + \frac{C_F q}{UF\ LMTD}$$

In terms of inlet and outlet temperatures, the above equation becomes

$$C_T = \frac{C_W t q}{C_{pc}(t_2 - t_1)} + \frac{C_F q}{UF\left(\dfrac{(T_1 - t_2) - (T_2 - t_1)}{ln\,[(T_1 - t_2)/(T_2 - t_1)]}\right)} \tag{8.14}$$

The next step involves differentiating the above expression with respect to the outlet temperature t_2 of the coolant (water) in order to minimize the total cost. Taking the partial derivative $\partial C_T / \partial t_2$ and setting the results equal to zero gives, after considerable manipulation,

$$\frac{UFtC_W}{C_F C_{pc}}\left(\frac{(T_1 - t_2) - (T_2 - t_1)}{t_2 - t_1}\right)^2 = ln\frac{T_1 - t_2}{T_2 - t_1} - \left(1 - \frac{T_2 - t_1}{T_1 - t_2}\right) \tag{8.15}$$

The above equation is graphed in Figure 8.10. The ratio $UFtC_W/C_F C_{pc}$ is plotted on the horizontal axis, which ranges from 0.1 to 10. The temperature ratio $(t_2 - t_1)/(T_2 - t_1)$ which ranges from 0.1 to 10 is plotted on the vertical axis with $(T_1 - T_2)/(T_2 - t_1)$ appearing on the graph as an independent variable. It is easier to use the graph of Figure 8.10 to find the optimum water outlet temperature than to solve for it (trial and error style) with Equation 8.15.

EXAMPLE 8.4. A shell and tube heat exchanger uses water as a cooling medium. Data on this particular heat exchanger are given below. Use the data to calculate the optimum cooling water outlet temperature. Assume

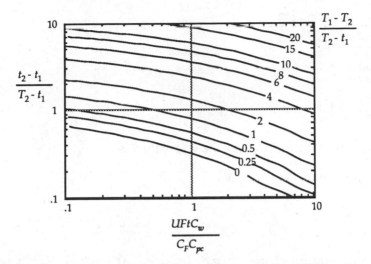

FIGURE 8.10. *Graph for finding optimum cooling fluid outlet temperature.*

that the exchanger is in operation 7800 hr/yr, that water costs $0.05/(1000 gallons) and that the annual cost of operating the exchanger amounts to $20/(ft² of surface area·yr).

A_o = 1074.2 ft² T_1 = 200°F
C_{pc} = 0.9980 BTU/lbm·°R T_2 = 100°F
U = 108 BTU/hr·ft·°R t_1 = 80°F
F = 0.817

Solution: The water cost is given as $0.05/(1000 gallons) which must be converted from monetary units per volume to monetary units per unit mass. Thus

$$C_W = \frac{\$0.05}{1000 \text{ gal}} \frac{2.831 \times 10^{-2} \text{ gal}}{3.785 \times 10^{-3} \text{ ft}^3} \frac{\text{ft}^3}{(0.994)62.4 \text{ lbm}}$$

or $C_W = \$6.03 \times 10^{-6}/\text{lbm}$

We now calculate the dimensionless ratio

$$\frac{UFtC_W}{C_F C_{pc}} = \frac{108(0.817)(7800)(6.03 \times 10^{-6})}{20(0.9980)} = 0.208$$

The temperature ratio we need is found as

$$\frac{T_1 - T_2}{T_2 - t_1} = \frac{200 - 100}{100 - 80} = 5$$

We read from Figure 8.10

$$\frac{t_2 - t_1}{T_2 - t_1} \approx 4.0$$

The optimum outlet temperature of the water then is

$$t_2 = 4(100 - 80) + 80$$

or $\quad t_2 = 160°F$

_____

The outlet temperature of the warmer fluid is 100°F which means we have a sizeable temperature cross.

_____

8.4 Show and Tell

1. Obtain the parts (from a manufacturer, perhaps) of an automobile radiator. Give a presentation on how it is made and whether the flows are mixed or unmixed.

2. Obtain the parts of a finned tube heat exchanger, like that in an air conditioning unit. Give a presentation on how it is made and whether the flows are mixed or unmixed.

3. How are shell and tube heat exchangers cleaned? Give a presentation on devices that are used for this purpose.

8.5 Problems Chapter 8

1. Suppose the cross flow heat exchanger of Example 8.1 is the mixed-unmixed type. How is the frontal area affected?

2. A cross flow heat exchanger (mixed-unmixed) is used to recover heat from the exhaust of a diesel engine. The exhaust gases enter the exchanger at 400°F. Water is available at 65°F, 1 lbm/hr, and is not of much use unless it can be heated to 90°F. Determine the size of the exchanger required if the mass flow rate of diesel exhaust is 1 lbm/hr.

3. A cross flow heat exchanger (unmixed-unmixed) is marketed as an oil cooler. Oil enters the cooler at a flow rate of 80 lbm/hr and a temperature of 170°F. The

oil should be cooled to 100°F. Air is used to cool the oil. The air inlet velocity is 800 ft/min and the frontal area through which the air flows is 16 in. x 16 in. The air inlet temperature is 70°F. Is this exchanger large enough?

4. Water enters a cross flow heat exchanger at a temperature of 85°C and a flow rate of 2.5 kg/s. Air enters the exchanger at 20°C, leaves at 40°C, and has a flow rate of 12 kg/s. Which type of cross flow heat exchanger will have the greater effectiveness—mixed-unmixed or unmixed-unmixed?

5. For the information given in Problem 4, which type of cross flow heat exchanger will need to have the larger frontal area—mixed-unmixed or unmixed-unmixed?

6. Referring to Table 8.3, suppose that a 6 x 4 double pipe heat exchanger becomes available. Make all calculations for each column in the table for this new combination.

7. Some manufacturers produce a triple pipe heat exchanger that is used to exchange heat between two fluids. This exchanger has an inner tube surrounded by two annular flow sections—see Figure P8.7. Assume that Nusselt number equations are available for this exchanger and represent them as:

inner tube	$Nu_1 = f_1(Re_1, Pr_1)$
annulus 1	$Nu_2 = f_2(Re_2, Pr_2)$
annulus 2	$Nu_3 = f_3(Re_3, Pr_3)$ $Pr_1 = Pr_3$

(same fluid)

Compose a suggested order of calculations for a triple pipe heat exchanger.

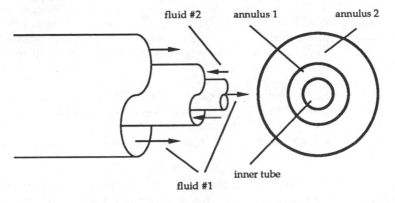

fluid #2   annulus 1   annulus 2

inner tube

fluid #1

FIGURE P8.7

8. Kerosene [$\rho = 0.73(1\ 000)$ kg/m^3, $C_p = 2\ 470$ J/(kg·K), $\mu = 0.40$ cp, $k_f = 0.132$ W/(m·K), $1.5 \le V_{optimum} \le 3$ m/s] is to be preheated in a double pipe heat exchanger before being pumped to a distillation facility. The kerosene flow rate is 8,000 lbm/hr and it is to be heated from 75 to 125°F. Water is available from

the condensed exhaust of a small steam turbine, and its flow rate can be controlled. The water is available at 200°F. Select an appropriate heat exchanger.

9. Crude oil is stored in a tank and maintained at 30°C. It is to be pumped to a distillation column for separation into usable products. Kerosene leaves the column at 200°C and it is proposed to transfer heat from the kerosene to the crude oil so that the cost associated with heating the crude oil by other means is reduced. The crude oil has a flow rate of 12 000 kg/hr and a significant savings will be realized if it can be heated at least to 75°C. Determine an appropriate heat exchanger to use and analyze it completely. Use the following properties for both fluids:

kerosene		crude oil
42	°API	34
0.73	sp. gr.	0.83
0.4 cp	viscosity	3.6 cp
2 470 J/(kg·K)	spec heat	2 050 J/(kg·K)
0.132 W/(m·K)	therm con	0.133 W/(m·K)
1.5-3 m/s	opt velocity	1.50-3 m/s

10. Ethylene glycol has a volume flow rate of 12 000 kg/hr and must be cooled from 75°C to 60°C. Select an appropriate coolant and heat exchanger. Analyze the system completely.

11. A phosphate solution [sp. gr. = 1.3, C_p = 0.757 BTU/(lbm·°R), μg_c = 2.9 lbm/(ft·hr), k_f = 0.3 BTU/(hr·ft·°R)] is used in the production of fertilizer. The solution is to be cooled from 75°C to 30°C using well water, available at 18°C. The phosphate solution flows at a rate of 7 000 kg/hr and has a dirt factor of 0.001 hr·ft·°R/BTU. Select an appropriate exchanger and analyze it completely.

12. Engine oil is used as a lubricant and due to an unforseen "hot spot" in the system, the oil temperature reaches 95°C. The oil has a flow rate of 5 000 kg/hr and it should be cooled to 25°C. Raw water is available at a temperature of 25°C. Select an appropriate heat exchanger for this service and analyze it completely.

13. Castor oil (extracted from castor oil plant seeds) is used as a fine lubricant. It is pumped to several machines for bottling which can fill 9,000 sixteen ounce bottles per hour. The oil is stored in tanks and is maintained at 45°C. A flow meter in the pipeline from tank to machine requires the oil to be at 17°C for accurate metering. City water is available at 15°C for cooling the castor oil. Select an appropriate heat exchanger and analyze it completely.

14. The process of printing an advertising brochure involves the use of 4 colors of ink. The ink itself is actually a mixture of solvent (toluene) and ink solids

(pigment). Ordinarily, the ink is stored in tanks and fed directly to printing presses, but a problem has arisen. If the ink mixture is left unattended in a tank, the solids tend to settle out which results in a nonuniform color being fed to the presses. Using a mixer in the ink tank has been ruled out as an acceptable solution to the problem. Instead, the ink is circulated about the building via a pump and piping system, and is then returned to the tank. This movement warms the ink in the tank to about 130°F. The toluene tends to vaporize too quickly, however, when the ink temperature exceeds 120°F, thus causing a health hazard. Furthermore, the printing process itself requires the ink mix to be supplied at no warmer than 80°F, and at a flow rate of 60,000 lbm/hr. Water at 65°F is available for cooling the ink. Determine the type of heat exchanger to be used to cool the ink to a desirable temperature and analyze it completely. (Should two heat exchangers be used—one to cool the ink below 120°F after it is circulated and another to cool it to 80°F on its way to the presses? Alternatively, should only one be used just to cool the ink to 80°F after it is circulated? Or is there another solution?) Take the properties of ink to be the same as those for toluene:

$$\rho = 54.3 \text{ lbm/ft}^3 \qquad\qquad C_p = 0.44 \text{ BTU/(lbm·°R)}$$
$$\mu = 0.41 \text{ cp} \qquad\qquad\quad k_f = 0.085 \text{ BTU/(hr·ft·°R)]}$$
$$4.4 \leq V_{optimum} \leq 9.5 \text{ ft/s}$$

15. Ethyl alcohol [sp. gr. = 0.78, C_p = 0.72 BTU/(lbm·°R), μg_c = 1.45 lbm/(ft·hr), k_f = 0.085 BTU/(hr·ft·°R), R_d = 0.002 hr·ft^2·°R/BTU] is heated from 80 to 150°F by water at 200°F. The ethyl alcohol flows at 115,000 lbm/hr. Select an appropriate heat exchanger for this service.

16. For turbulent flow, the development in Section 8.3 showed that the convection coefficient increases by a factor of 3.03 when increasing from 2 to 8 tube passes in a 1-2 shell and tube heat exchanger. The corresponding pressure drop increases by a factor of 64. Repeat these calculations for laminar flow conditions.

17. The details of the final results in Example 8.3 seem a little sketchy. Completely analyze the recommended exchanger to see if it will perform adequately.

18. Start with Equation 8.14 and derive Equation 8.15.

19. Calculate the optimum water outlet temperature for the conditions given below. Assume that the exchanger is in operation for 7800 hrs/yr, that water costs $0.05/(1000 gallons), and the annual fixed charges amount to $20/(ft^2·yr).

$$A_o = 616.2 \text{ ft}^2 \qquad\qquad\qquad T_1 = 130°F$$
$$C_{pc} = 0.9988 \text{ BTU/lbm·°R} \qquad T_2 = 98.4°F$$
$$U = 224.9 \text{ BTU/hr·ft·°R} \qquad\quad t_1 = 65°F$$
$$F = 0.775$$

20. If the exchanger in Example 8.4 is in operation for only 4000 hrs/yr, how does the outlet temperature of the cooler fluid t_2 change?

CHAPTER 9   The Design Process

Following in Chapter 10 are descriptions of a number of design projects. The information given thus far in this text is sufficient to provide a start for the solution of such projects. The design process (from accepting a "job" to final report) is more than merely finding a solution, however. So in this chapter, we will discuss several aspects associated with obtaining one (of many possible) solution(s) to a design project. These include: the bidding process, project management, and evaluation and assessment of performance.

9.1 Design Project Example

Consider that we are interested in working on a problem that involves the recovery of waste heat in a manufacturing facility. The problem is stated below:

Heat Recovery in a Sheetrock Plant (3 engineers)

One of the components needed in the manufacture of sheetrock is water. The process requires 70 gpm of water at a temperature of about 85°F. During summer months, the city water supply provides water whose temperature can be as high as 90°F. During other months, the average temperature of water supplied by the city is about 45°F. This water must be heated so it can be used successfully for the process. The water is heated by natural gas burners while it is in a storage tank.

One of the final phases of sheetrock production is the drying stage. Heated air is moved by a fan around the sheetrock in an oven. The air is then exhausted. It is desired to recover energy from the warm, humid exhaust air and use the energy to heat the incoming city water from 45°F (worst case) to as warm as possible. The energy recovered would reduce the need for natural gas to be used as the main heating medium. Conditions indicate that the heat recovery system will be in operation for 24 hr per day (6 day week) for 8 months.

Figure 9.1 shows the position of the drying oven and of the holding tank. As shown, the water tank is 300 ft from the oven. The volume flow rate of air in each stack is measured as 14,000 cfm and at the outlet of each stack, the air temperature is about 230 °F.

To be designed, selected, or determined:

1. Appropriate heat exchanger type, its size, location, and material of construction.
2. Pump (if necessary), size, location, and material of construction.
3. Piping, pipe fittings, size, routing, and material. (Consider that a flowmeter and/or pipe insulation may be desirable.)
4. Total cost of system including installation, operation, and maintenance.
5. Payback period on the investment.

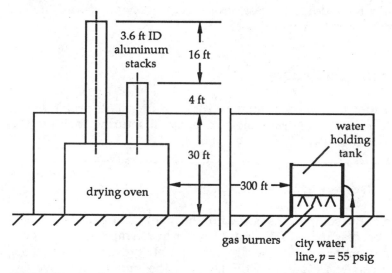

FIGURE 9.1. *Layout showing drying oven and holding tank.*

It is apparent that, to solve this problem, we have to design the piping system from the tank to one of the stacks and back. We have to specify a pipe size, determine the routing of the line itself, select a pump for the job, and size a heat exchanger to place in one of the stacks. The above list of items is the *engineering* phase of the project. The engineering part will make up only a small portion of the overall process. The other things that must be completed appear as subheadings in the discussion that follows.

The Bid Process

Suppose that the Heat Recovery Problem stated above arises in a plant where there are not enough engineers available to work on it. Management has decided to allow an outside engineering consulting company to solve the problem or at least to see if it is cost effective. Management will contact any number of consultants and invite them publicly to bid on the project. That is, each consultant is invited to submit to management (the **client**) a proposal which outlines what is to be done and how much the consulting company will charge to perform *only* the design work. Actual construction or installation might also be a part of the bid, depending on what the client requests of the bidders.

Bids are submitted in sealed envelopes and are opened with all bidders present during a **bid opening ceremony**. Bid opening is to occur at a predetermined time (known as the **closing date**) and usually (but not always) the lowest bidder is awarded the project.

The actual cost of completing the project is determined with the aid of a **budget-bid sheet**, the format of which will vary from company to company. Figure 9.2 is an example of a budget-bid sheet. An estimate of the consulting cost is obtained when the sheet is completed. Some of the items on the sheet are worth mentioning. The project title is listed first and is used on this sheet to identify the project. The item listed in part A is the number of **person-hours** required to complete the work. At best this is an estimate, but an experienced bidder can make an accurate appraisal. **Fringe benefits** for each employee working on the project are paid directly from the project budget. **Miscellaneous Costs** including materials, supplies, computer time, etc., are charged to the project. **Travel** by the group (to the facility for example) is charged to the project. It may be necessary for the design group to use the services of an expert (**Consultant Services**) and payment to the expert for his/her services is also charged to the account. The **Indirect Costs** include overhead to pay for utilities, office space, secretarial help, profit, etc., is also a part of the project cost. The total of these items is the cost that the consultant will charge the client to complete the design. Note that most of the items are tied directly to the person-hours estimated in the beginning and so an accurate estimate is highly critical.

Some companies are poorly managed and as a result tend to have a low profit. Such companies often overcharge for their services.

Project Management

Suppose next that the Heat Recovery Project has been awarded to us because our company is lowest bidder. Completion of the job requires an organized and well managed effort. The project must be divided into

			Estimated	Actual
Title of Project (number of Engineers)		Bid Amount		
Closing Date	Due Date	Hrs Req'd to Complete		
Project Director				

A. Personnel

Name	Telephone	Person-hrs	Salary	
1. _____	_____	_____	$_____	$_____
2. _____	_____	_____	$_____	$_____
3. _____	_____	_____	$_____	$_____
4. _____	_____	_____	$_____	$_____
5. _____	_____	_____	$_____	$_____
6. Subtotal			$	$

	Estimated	Actual
B. Fringe Benefits (35% of A.6)	$_____	$_____
C. Total Salaries, Wages, and Fringe Benefits (A.6 + B)	$_____	$_____

D. Miscellaneous Costs

	Estimated	Actual
1. Materials and Supplies	$_____	$_____
2. Computer Time ($0.25 per minute of on line time)	$_____	$_____
3. Other	$_____	$_____
4. Subtotal	$	$

	Estimated	Actual
E. Travel	$_____	$_____

F. Consultant Services (Name and Amount)

	Estimated	Actual
1. _____	_____	_____
2. _____	_____	_____
3. Subtotal	$_____	$_____

	Estimated	Actual
G. Total Direct Costs (C + D.4 + E + F.3)	$_____	$_____
H. Indirect Costs (50% of G)	$_____	$_____
I. Amount of this Bid (G + H)	$_____	$_____

Signature of Engineers	Date	Initials
1. _____	_____	_____
2. _____	_____	_____
3. _____	_____	_____
4. _____	_____	_____
5. _____	_____	_____

FIGURE 9.2. *Example of a budget-bid sheet.*

several smaller jobs that are finally synthesized into the overall solution. This phase involves identifying the smaller jobs, assigning the completion of each small job to an individual or individuals, and requiring each small job to be completed at a certain time. This breakdown can be done by the **Project Manager** or **Project Director**, who is ultimately responsible for ensuring that the job is finished on time and within budget.

It is convenient for the Project Director to compose a **bar chart** of project activities which outlines the tasks, or smaller jobs, to be performed in completing the projects. The bar chart is much like a graph in which time is laid out on a horizontal axis and project activities appear on the vertical axis. The advantages of such a layout are that all activities are mapped out and assigned, that the order of the activities can be readily seen and that an overall readable picture with the expected completion time is on hand. The main disadvantage of such a chart is that it will probably need updating which can require much time.

A bar chart for the Heat Recovery Project is shown in Figure 9.3. It is a "first draft" that includes all tasks that we could identify. They are listed as **activities** in order on the left. The chart shows which activities or tasks require the completion of another task beforehand. The entire project is mapped out over an 8 week period with *estimates* of how long each activity will take. Also, letters appearing in each shaded rectangle represent the initials of the engineer(s) who is (are) responsible for completing the corresponding task. The shaded rectangles are connected with lines and arrows which indicate a succession of events. Thus, before a pump is selected, for example, the line size and its route and the heat exchanger must first be specified.

Suppose that after some time has passed, we think of (or are assigned) several other tasks to perform, or some tasks were completed before their target completion date. It is advisable to rework the chart to add the new event(s), assign a responsibility to it (them), and to update the completion of the finished smaller jobs. Suppose also that it appears as if the project will be finished earlier (or later) than what was originally scheduled. This is brought out in the modified chart as well.

To illustrate, consider again the Heat Recovery Project example. Say that after much study, we find we must use an exchanger in both stacks in order to recover the required energy. Figure 9.4 shows a modified chart. Note that the new activities have been added in the appropriate positions showing their relationship to other tasks. There are two triangles on the "week number" scale. These triangles indicate time. The inverted triangle (∇) above the graph shows where we are in actual or real time (e.g., today's date). The other triangle (Δ) shows where we are as far as task performance is concerned. Therefore, when Δ is farther to the right along the time axis than ∇, then we are ahead of schedule.

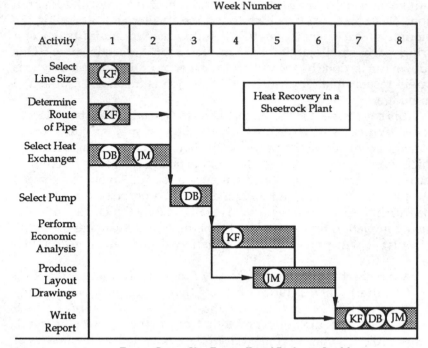

Design Group: Ken Fensin, David Birdsong, Jim Morrissey

FIGURE 9.3. *Bar chart of smaller jobs to be performed in completing the Heat Recovery Project.*

The Project Director is also responsible for handling the budget allowed for completion of the project. This would include signing all requests for payment and keeping track of how the project budget funds are expended.

The Project Director will meet frequently on an as needed basis with the group members to offer assistance if necessary. The primary job of the Project Director and the group members is:

- To always keep in mind the objective of the entire project;
- To know exactly how each group member will contribute to the overall success of the design effort (i.e., each group member will know without question exactly what his/her responsibility is);
- To identify any and all obstacles that prevent a group member from completing a task, and to remove the obstacles; and,
- To remember that all group members are "being paid" to maintain an effective working relationship with the others.

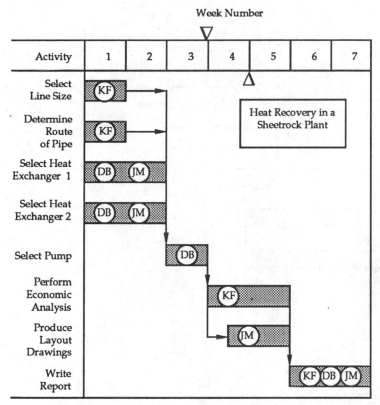

Design Group: Ken Fensin, David Birdsong, Jim Morrissey

FIGURE 9.4. *Modified bar chart of smaller jobs to be performed in completing the Heat Recovery Project.*

The Reports

Suppose next that the engineering phase of the project has been completed and it is now necessary to communicate the results. Usually a **written report** and an **oral presentation** are given by the consultant to the client. The written report will contain several items; a *suggested* example format is given in Figure 9.5. Each item is described as follows.

Letter of Transmittal—Written to the client stating that the project has been completed and the results are presented in the accompanying report.

Title Page—Lists project title, finished project due date, engineers who worked on the project, and the name of the consulting company.

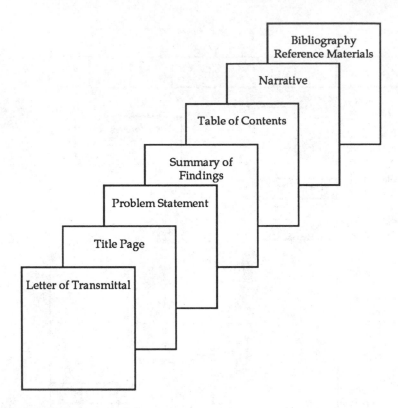

FIGURE 9.5. *Elements of the written report.*

Note that all pages of the report need to be numbered, dated, and identified somehow with the consulting company.

Problem Statement—Reiterates succinctly the problem, included so that all concerned will know what project was completed.

Summary of Findings—Summarizes the details of the solution. This section might present a list showing, for example, what pump to buy, what line size to use, where to route the pipe, what heat exchanger to use, etc., suggested suppliers, and costs for all components. Drawings of the system would also be included. The summary should be complete enough so that the client could submit it to a contractor who could complete the installation of all components.

Table of Contents—Refers the reader to any section of the report.

Narrative—Presents the details of all components specified in the summary and why each component was selected. For example, the details of how the pump was selected would be included here. Enough written detail must be included so that the reader can follow *every* step of the development. The organization of the narrative and titles of all sections will vary from writer to writer. However: *If your audience has read your report and does not understand all that you wrote, then you have not expressed yourself clearly enough.*

Bibliography/Reference Materials—Shows text titles and publications used to arrive at the specifics of the design. This section should also include information from catalogs of suppliers such as pump performance curves, if appropriate.

The written report should appear professional in every way. A well written presentation will show that the writer is meticulous and convince the client that a great deal of care went into completing the job. Graphs and drawings are done in ink or by computer with nothing drawn freehand. Text must be typewritten or produced with a word processor. The entire report should be bound and the client should be provided with more than one copy.

The oral report should be short and it need not be detailed. The oral report consists of the problem statement and a summary of the findings, to include initial and operating costs. If questions arise, the presenter can refer to details found in the narrative. Therefore, the presenter should be prepared to give details of the *entire* study but only present the problem statement and the summary.

Internal Documentation

The design process described above listed various activities that resulted in a report. The report was then provided to the client. The consulting company, however, will need to keep on file much more information about the project than is included in the written report.

Once an engineer begins work on a project, the engineer is to obtain a notebook and keep track of all things performed in association with the project (in ink) including dates and time spent. Even the most seemingly trivial contribution (such as a phone call) should be recorded. Nothing is to be erased or eradicated from the notebook. The notebook should also contain all the engineering work and calculations done on the project. The notebook is a diary. Errors are "removed" from the diary by drawing a straight line through them but they must still be readable. Each member of the group will have his/her own notebook for each project.

The copy of the final report that stays within the consulting company files should contain the budget-bid sheet originally submitted. At the time that the project is finished, the budget-bid sheet is completed to show the actual costs of items requested as part of the project. These include the person-hours expended and the profit earned on those person-hours. Remember that we are in "business" to make money and performance on the project will be evaluated in proportion to the actual profit realized.

The engineers' notebooks, final report, and completed budget-bid sheet make up the documentation that the consulting company will want to keep on file for future reference. Should the project need to be reviewed in the future, the necessary detail will be available.

Evaluation and Assessment of Results

The work performed on a project must be evaluated if possible. The two items of importance are: Will the system work as designed, and have we made a profit by delivering a good product?

In some cases, the client will not construct the system for a number of months or even years. Moreover, it is unlikely that the installed system will contain the necessary instrumentation to evaluate performance (e.g., thermocouples, flow meters, etc.). Even after installation, it may take years to determine if the system works as designed.

Whether a good product was delivered might never be assessed. Profit realized can be assessed, however, and is usually measured by an elaborate accounting system outside the scope of engineering.

It will become evident that the engineering phase of any project requires a relatively minor amount of total time expended. Equally important is the time and effort spent in documenting activities and especially in report writing. Writing is therefore extremely important to the engineer.

9.2 Questions Chapter 9

1. List all steps involved in designing a shell and tube heat exchanger and compose a bar chart of activity vs time.

2. Is it always appropriate for a job to be awarded to the lowest bidder? List exceptions and give reasons why or why not.

3. Should profit always be a motive in the consulting business?

4. Consider the question "... have we made a profit by delivering a good product?" Define a good product with regard to the Heat Recovery Problem.

5. What instrumentation is necessary in the Heat Recovery Problem if an evaluation of system performance is to be conducted?

CHAPTER 10 Project Descriptions

A Fluid Dynamometer for Bicycles (3 engineers)

A dynamometer is a device for measuring the output power of a system. Dynamometers have been used extensively for measuring power from an internal combustion engine, from a turbine, and from an automobile. Dynamometers can be of the electric type in which the output power is used to produce electric power, or of the fluid type where the output power is used to pump a liquid (usually water) or move air.

On a smaller (than IC engines) scale, it is desirable to measure the output of a person pedaling a bicycle. Results of such tests are of interest in physical education studies of human power output and endurance, and to manufacturers of bicycles. Consequently, there is a need to have a dynamometer onto which a complete bicycle can be attached and pedaled, and from which output power can be calculated.

To be designed, selected, or determined:

1. A system for supporting the weight of the bicycle and the rider while attached to the dynamometer.
2. A fluid system that is used to measure the output power of the rider over a broad range of conditions.
3. Materials of construction and design of the entire device.
4. Costs of building such a device.

Tire Pressure Readout (3 engineers)

Tractor-trailer trucks are used extensively to haul goods over long distances. A trucker may be driving for many miles before he/she stops for fuel and inspects the vehicle. Tire pressure, however, is critical because at high speed, a loss of tire pressure can be dangerous to the driver and the vehicle. Consequently, a need exists for a system that allows tire pressure to be known at times when the vehicle is moving. Such a system can be used for buses as well.

To be designed, selected, or determined:

1. A sensor that will have an output which varies with pressure in the tire.
2. A sending system that transmits the sensor signal to the tractor.
3. Total cost of the sensor and transmitter.

Notes: Assume that an electronics box in the tractor will accept the signal and display a digital readout is psig or kPa gage pressure. Your design and costs will exclude the electronics box.

   If such devices have been designed, your own design must not infringe on the patent rights of others.

Automatic Lawn Sprinkler System (3 engineers)

   The plan view of a lawn and single story home will be provided by the instructor. It is desired to install an underground automatic lawn sprinkler system. At appropriate times the system will distribute the proper amount of water or water/fertilizer mix uniformly over the grassy areas and over the shrub and flower areas. Rainwater is collected by the homeowner in two 55-gallon steel drums and provision must be made for using rainwater rather than city water. The house already exists and this is a retrofit installation. (Alternatively, the house is yet to be built and this is a new installation, to be decided by the instructor.)

To be designed, selected, or determined: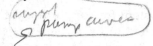

1. Location, size, and type of pump if necessary. (Take the city water pressure to be 65 psig.)
2. Routing of underground pipes or tubes and the materials of construction.
3. Provision for the user to add liquid or dissolved fertilizer to the lawn (at times specified by you) via the sprinkler system. (The grass is St. Augustine.)
4. Provision for the system to distribute collected rainwater rather than city water. When there is no rainwater the system is to use city water and must automatically make the changeover.
5. How much water to provide for the grass, shrubs, and flowers, and when to do so.
6. Sprinkler nozzle type(s), number, and location.
7. Automatic regulating device for proper system operation.
8. Total cost, including installation, expected life, and maintenance.

Note: There are manufacturers and installers of such systems. So before consulting any of those available, perform a preliminary design to express your thoughts on and to develop your intuition for designing such a system.

Domestic In-wall Vacuum Cleaning System (3 engineers)

The plan view of a single story home will be provided by the instructor. It is desired to install an in-wall vacuum system. A centrally located vacuum cleaner motor provides power for cleaning. A number of hoses connect the cleaner motor to various wall outlets throughout the dwelling. The user attaches a hose and floor nozzle (supplied by a manufacturer of portable vacuum cleaners) to any desired outlet connector and then vacuums the carpet. The house already exists and this is a retrofit installation. (Alternatively, the house is yet to be built and this is a new installation, to be decided by the instructor.)

To be designed, selected, or determined:

1. Location, size and type of vacuum motor. (Consider that the motor collects dirt into a bag and that the bag must be changed with regular use of the system.)
2. Wall outlet locations (paying strict attention to locations where accessibility is not convenient, such as where a sofa or couch might be placed).
3. Piping size, material, and routing from the motor to the wall outlets. (Must all the piping be the same diameter?)
4. Hose length and floor nozzle attachments. (Also, how will the hose and nozzles be stored?)
5. Total cost of the system including installation. Compare the cost to a portable vacuum cleaner and determine whether the in-wall system is cost effective.

Note: There are manufacturers and installers of such systems. So before consulting any of those available, perform a preliminary design to express your thoughts on and to develop your intuition for designing such a system.

Dehumidified Air Supply (3 engineers)

A 150 HP internal combustion engine is to be tested to determine its operating characteristics in a dry environment such as a desert. The main requirement is that a steady supply of dehumidified air be made available to the engine. Moreover, the dehumidified air must be supplied over a range of temperatures for a proper simulation of desert-like conditions.

To be designed, selected, or determined:

1. The air consumption of a 150 HP, 4-stroke engine.
2. A dehumidifier to remove as much moisture as possible from inlet air to simulate desert conditions.
3. A means for controlling the temperature of the dehumidified air.
4. How the air is to be made available (e.g., in a steady flow situation or from a storage tank).
5. Total cost of system including all components that make it ready for attachment to the inlet of the engine.

Signaling Funnel (3 engineers)

People who have used funnels to fill tanks often experience the problem of overfilling the container and spilling liquid over the top. This is especially common in the filling of lawn mower tanks. The user's view of the inside of the tank is obstructed by the funnel and, when the tank is full, gasoline spills out, creating a safety hazard as well as an unpleasant mess. It is proposed to design a funnel that signals the user to stop pouring when the tank is full.

To be designed, selected, or determined:

1. The proper material for a funnel designed for use with gasoline.
2. The signaling device or method.
3. An adjustment method for the signal to accommodate differing reaction times as one individual might require less time to stop than another.
4. Total cost to manufacture the signaling funnel.

Fireplace Heat Recovery (3 engineers)

Sheet metal fireplaces can be added to a room after the structure is built; that is, the fireplace need not be built when the home is. In order to enhance the usefulness of a sheet metal fireplace, it has been proposed to devise a means for recovering more of the heat that would ordinarily be discharged up the stack with the exhaust gases. It is believed that the most effective way to transfer more of the heat from combustion is by convection so that the air in the room is heated.

To be designed, selected, or determined:

1. A sheet metal fireplace for your study (with the approval of your instructor).
2. A heat exchanger for it. (If one has already been designed, your design must be substantially different. Moreover, you are also required to analyze the existing unit.)
3. The total cost of the device.

Spray Dryer for Producing Powdered Eggs (3 engineers)

Figure 10.1 is of a processing system for production of egg powders. Eggs are stored at between 5 and 15°C. They are unpacked, weighed, and candled. Dirty eggs are separated from the rest, washed, dried, and returned for candling. The eggs are then broken (which can be done by hand) and separated. An operator can break about 600–800 eggs per hour. (Automatic breaking and separating machines can process up to about 18,000 eggs per hour.)

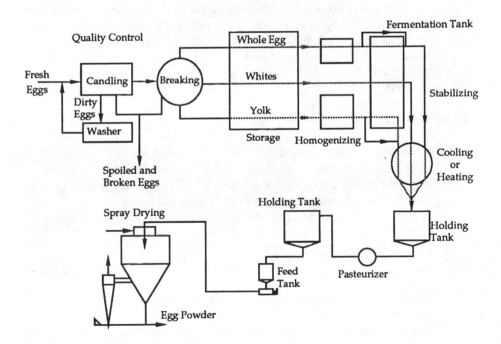

FIGURE 10.1. *Processing diagram for production of egg powder.*

For the processing of whole egg or egg-yolk powder, homogenizing and filtering are the next steps. Pasteurization then follows and, after a series of holding tanks, the liquid (whole eggs or egg yolks) is pumped to a spray dryer. It is desired to construct a dryer for the system to produce whole egg powders.

To be designed, selected, or determined:

1. The size, material of construction, and configuration of the dryer to handle 800 whole eggs per hour.
2. The nozzle(s) and types to be used.
3. The required pressure at the nozzle and liquid inlet configuration.
4. Total cost of the dryer including construction and maintenance.

A 1¹/₂ HP Centrifugal Fan (3 engineers)

Landscapers and lawn maintenance workers are responsible for (among other things) cleanup work. This often involves sweeping a sidewalk, for example, to gather and dispose of leaves and grass clippings. Recently, however, enterprising manufacturers have marketed blowers that consist of centrifugal fans and flexible hoses. Most of the fans are rated at 1 HP. It is possible to attach a hose to the inlet side of the fan and a bag to the output in order to use the fan as a vacuum cleaner. It has been found that a 1 HP fan is somewhat undersized and so a 1¹/₂ HP fan is desired.

To be designed, selected, or determined:

1. What does a sticker that says "1 HP" on a fan mean? Is the motor rated at 1 HP or does the fan receive 1 HP or does the air?
2. Design an impeller and a housing for a 1¹/₂ HP centrifugal fan. Be sure that the housing fittings mate with conventional tubes.
3. Predict the performance of the new fan using the conventional tubes.
4. Select materials of construction for the fan.
5. Determine the cost of the new fan.

Heat Recovery in a Sheetrock Plant (3 engineers)

One of the components needed in the manufacture of sheetrock is water. The process requires 70 gpm of water at a temperature of about 85°F. During summer months, the city water supply provides water whose temperature can be as high as 90°F. During other months, the average temperature of water supplied by the city is about 45°F. This water must be heated so it can be used successfully for the process. The water is heated by natural gas burners while it is in a storage tank.

One of the final phases of sheet rock production is the drying stage. Heated air is moved by a fan around the sheet rock in an oven. The air is then exhausted. It is desired to recover energy from the warm, humid exhaust air and use the energy to heat the incoming city water from 45°F (worst case) to as warm as possible. The energy recovered would reduce the need for natural gas to be used as the main heating medium. Conditions indicate that the heat recovery system will be in operation for 24 hr per day (6 day week) for 8 months.

Figure 10.2 shows the position of the drying oven and of the holding tank. As shown, the water tank is 300 ft from the oven. The volume flow rate of air in each stack is measured as 14,000 cfm and at the outlet of each stack, the air temperature is about 230°F.

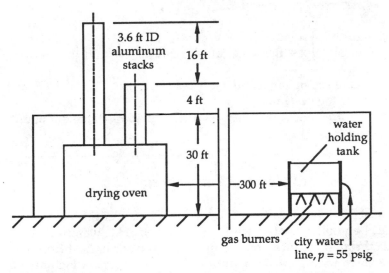

FIGURE 10.2. *Layout showing drying oven and holding tank.*

To be designed, selected, or determined:

1. Appropriate heat exchanger type, its size, location, and material of construction.
2. Pump (if necessary), size, location, and material of construction.
3. Piping, pipe fittings, size, routing, and material. (Consider that a flow meter and/or pipe insulation may be desirable.)
4. Total system cost including installation, operation, and maintenance.
5. Payback period on the investment.

Vacuum Cleaning System for Buses (3 engineers)

The local transit authority is responsible for operating and maintaining vehicles used in public transportation, specifically buses. One of the maintenance tasks includes cleaning of the buses every evening when they are returned to the garage. Currently, buses are cleaned by an individual who walks through, picks up large refuse items, and vacuums the floor. This procedure is time consuming and costly in person-hours.

It is proposed to use a huge air moving system to clean the buses. Two huge air pipes are attached to a bus. One pipe delivers high velocity, high pressure air while the other pipe takes in air exhausted from the passenger compartment. Use of such a system eliminates the need for someone to walk through the bus to clean it and would take far less time to complete the work.

To be designed, selected, or determined:

1. Total design of the system.
2. The size of the air pipes, the material of construction, and the method of attachment.
3. The size of the air moving unit and source of purchase.
4. The total cost of the system.
5. The savings in time and labor charges (if any) with a statement about cost effectiveness.

Design of a Baseboard Heater (3 engineers)

A baseboard heater consists of a fluid-tight length of copper tubing (could be another metal) with fins attached. Hot water or oil is made to pass through the tube and the fins help to transfer heat from the liquid to the surrounding air by natural convection. A manufacturer of heating devices wishes to expand their line of items by adding three of these types of heaters. Rather than circulate a fluid through the tube, however, it has been decided to fill the tube with a liquid, attach an electric heater of some sort to the tube, and seal the tube itself. Thus, a user can purchase the size heater needed, attach it to the wall at the baseboard, provide 110 V AC, and the heater will convert the electric power into heat. The manufacturer desires to market 1000, 2000, and 3000 BTU/hr size heaters as these are commonly used in conventional homes.

To be designed, selected, or determined:

1. The size of the tube needed for each heater.

2. Whether to buy finned tubes "off the shelf" or construct them.
3. The size and type of heating device needed for each heater.
4. The protective yet decorative housing available for each heater.
5. The type of heat exchange liquid (or gas) with which to fill the tube.
6. The total cost of each heater.

Ventilation System for a Spraying Booth (3 engineers)

A manufacturer of small furniture items has decided to install a spraying booth in the shop so that wood items can be given finish coats of varnish by spraying rather than by using a brush. The room allotted is 12 ft x 12 ft x 10 ft tall.

The varnish is to be applied by an operator using a spray gun. Airborne droplets of varnish cause a health problem, however, and so the booth must be adequately ventilated. Moreover, the ventilation must in no way inhibit the application process or the drying process.

To be designed, selected, or determined:

1. Whether the size of the booth is adequate.
2. The ducting system for introducing air to and exhausting air from the spray booth.
3. The fan required to move the air.
4. The total cost of the ducting, fittings, and fan.
5. The effect that the airflow through the booth has on the application and drying process.

A Sandblaster That Uses Ice (3 engineers)

Sandblasting is an operation in which fine sand particles are propelled by a jet of high velocity air at an object such as a brick wall or painted metal. The sand acts as an abrasive that removes a portion of the surface and in effect refinishes the object. Sand then permeates the air, along with the sanded material, around the workplace (unless it is enclosed), and this is objectionable from a safety viewpoint. Furthermore, sand accumulates on the ground and presents a cleanup problem.

It is proposed to use ice particles as a substitute for the sand. The ice will refinish most surfaces and, with the passage of time, the ice melts into water which then drains away.

To be designed, selected, or determined:

1. A system for producing an "ice" blaster.

2. The proper shape for the ice particles and how they are to be produced.
3. The method for propelling the ice particles. (Is there an optimum velocity?)
4. Whether or not ice can be substituted for sand in a conventional sand blaster.
5. Total cost of system.
6. A comparison to a conventional sandblaster in performance and cost.

Pneumatic Seat (3 engineers)

Elderly people and those with certain disabilities often have difficulty standing up from a seated position. It is proposed to design an inflatable device that a person would sit on, and that would help lift the person up from the seat. The device is to be inflated with a small compressor and should be comfortable to sit on. The device can be part of a chair or be something portable and placed on the chair.

To be designed, selected, or determined:

1. The inflatable seat itself.
2. The material to be used for it.
3. A suitable compressor and connecting tubes.
4. The total cost of the system.

Domestic Fresh Air Makeup System (3 engineers)

Figure 10.3 shows a plan and a profile view of a duct and fan system found in conventional residential homes. The system is located in an attic. The blower takes in air from a duct and discharges the air into a plenum chamber. Within the chamber is located a heating or cooling coil that the air passes through. At the opposite end of the plenum is a number of ducts that lead the conditioned air to various rooms in the house. The system itself is part of a cooling unit.

It is proposed to add two more ducts to the system, as indicated in Figure 10.4. These new ducts attach before and after the fan. Air would be discharged downstream of the fan and makeup air would be brought in upstream of the fan directly into the fan inlet. The makeup air would be drawn in from the attic and so this air might be colder or warmer than desired. Consequently, it is further proposed that the discharged and makeup lines pass through a heat exchanger. The makeup air temperature would then be brought close to a desired level.

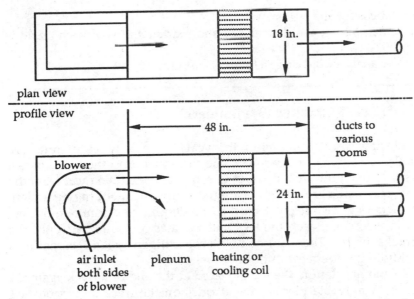

plan view

profile view

FIGURE 10.3. *Plan and profile views of a residential air handling system.*

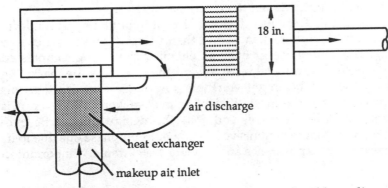

FIGURE 10.4. *Plan view showing proposed add-on lines and heat exchanger.*

To be designed, selected, or determined:

1. The air discharge and inlet line sizes and desired air flow rate. (What is the desired air flow rate based on occupancy? That is, how much fresh air should be admitted to the dwelling? How is this controlled?)
2. The material to be used for these lines.
3. The size, efficiency, and type of heat exchanger to be used. (Would this be a good application for heat pipes?)

4. The total cost of the system.
5. A statement about the desirability and marketability of such an add-on.

Perform the analysis and design for a 3 ton unit.

Pneumatic Nut Cracker (3 engineers)

During cooler months of the year, various types of nuts are harvested and sold in grocery stores. The consumer buys them and uses a nutcracker to open the shell in order to separate the edible meat from the nut. Most commercial nut crackers are manually operated. In one design, the nutcracker has two legs attached by hinge pins to a common bridge. The nut is placed between the legs and the legs are squeezed until the shell cracks open. The user then works diligently and sometimes impatiently to get the meat out of the shell.

In another design, the nut is placed at the end of a track against a metal stop. A movable piece of metal rides on the track and is pushed against the nut by a hinged lever arm. The force exerted is enough to crack the shell. Alternatively, the metal piece is accelerated toward the nut by a spring. The metal piece is moved away from the nut to compress a spring. When released, the spring pushes the metal toward the nut, and the impact force is high enough the crack the shell.

A marketing firm has decided that it would be desirable to market a pneumatic nut cracker. Rather than using the potential energy in a spring, the potential energy in an air chamber containing compressed air is used to accelerate a metal piece toward the nut. Now the marketing firm (and engineers who might work on this project) realize that this device is probably not the most efficient of nut crackers. However, it is believed that the idea is novel and that the design should be eye-catching enough so that the impulsive buyer would purchase one due to its unique aspects. Thus, appearance and novelty are extremely important in this design.

To be designed, selected, or determined:

1. A method for determining the force required to crack various nuts of the type available in the fall.
2. A pneumatic nutcracker showing design details and calculations indicating the impact forces generated.
3. A description showing the unique and attractive nature of the design.
4. Total cost of the pneumatic nut cracker.

Slab Heated Warehouse (4 engineers)

A warehouse is to be constructed, and will consist of a concrete slab with a sheet metal structure on it. The slab is 8 in. thick, and measures 50 ft x 100 ft. The metal structure bolts to the slab with anchors, and will be well insulated. The walls (12 ft tall) will have 4 in. insulation while the ceiling will have 6 in. of insulation. The interior of the warehouse is box-like—12 ft x 50 ft x 100 ft (nominally).

The warehouse is located in a region where daytime temperatures are 65°F and nighttime temperatures are 40°F. These temperatures are the same throughout the year. Paint is to be stored in the warehouse and, accordingly, it is desirable that the temperature inside be kept at 70°F at all times. It is proposed to use a heating method in the building that involves embedding tubing in the slab; that is, the tubes are laid out and the cement mix is poured directly over the tubes. Warm fluid can then be circulated through the tubing, thereby heating the slab by a convective effect, which in turn will heat the warehouse interior by convection and by radiation.

To be designed, selected, or determined:

1. The heating load for the warehouse, assuming it is 60% (by volume) occupied by 5 gallon cans of paint (accounting for aisle space, etc.).
2. An appropriate layout for the tubing in the slab. (Is an 8 in. slab thick enough? Should a serpentine flow arrangement be used or some other layout?)
3. The heat transfer fluid and how it is to be heated.
4. The method of circulating the fluid through the tubing.
5. The total cost of the installation. (What additional costs are incurred by using this method of heating over other methods?)
6. Desirability of this method over other conventional methods (e.g., radiant heaters, forced convection heaters, baseboard heaters, etc.).

Inflatable Dike (4 engineers)

Flooding in low elevation areas is disastrous and costly. Homeowners who are victims of flooding are faced with enormous cleaning and rebuilding costs. Moreover, insurance companies are hesitant to provide flood insurance policies to homeowners whose dwellings are below sea level. Consequently, some sort of protection against flooding is most desirable.

It is proposed that a properly formed inflatable material be buried around a house (that was built on a slab) and, when flooding by rising ground water occurs, the buried device is inflated. (See Figure 10.5.)

The dike must extend completely around the house and provide ample protection from flood-water that is expected to rise 10 inches above the slab, based on worst-case past experiences. In addition, when not in use, the dike must be able to be deflated and buried under the ground. The topsoil will then be landscaped and no trace of the dike is to be seen.

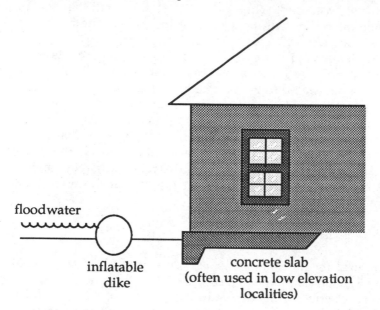

floodwater

inflatable
dike

concrete slab
(often used in low elevation
localities)

FIGURE 10.5. *View of inflatable dike concept.*

To be designed, selected, or determined:

1. The design of an inflatable dike that extends completely around the house. (Must the dike material extend under the slab? Must the material be inflated with air or will a liquid be satisfactory?)
2. The material to be used for the dike and how it is to be constructed.
3. The method for inflating the dike (or filling it with liquid).
4. Total installed cost of the entire system and its operation.

Measurement of Flipper Thrust (4 engineers)

A company that manufactures flippers is interested in modifying their traditional design and marketing a pair of flippers that provides greater thrust. It is first necessary, however, to be able to measure the thrust developed by a diver using flippers so a comparison can be made. It has been suggested that an open channel water tunnel be constructed and appropriately instrumented so that a diver can go through the motions of underwater swimming although remaining stationary and the force developed by the flippers is measured.

To be designed, selected, or determined:

1. A facility for a diver (real or mechanical) to swim underwater while not moving forward. If mechanical, then the mechanical device must also be designed.
2. The instrumentation necessary to obtain appropriate measurements.
3. The materials of construction of the facility; its design, layout, and operation.
4. Total cost of building the entire facility and cost to operate.

Hot Water Dispenser (4 engineers)

Instant coffee, hot chocolate, broth, and tea are examples of drinks that require hot water for their preparation. The hot water is prepared by heating it on a stove or in a microwave oven—both require finite amounts of time. It is proposed that an in-line heater be placed in a water line so that hot water is available from a tap at the kitchen sink. One merely places instant coffee, for example, in a cup, adds the hot water, and the drink is ready for consumption.

To be designed, selected, or determined:

1. The optimum temperature (with justification) of the hot water—if there is one.
2. An in-line heater to heat the water from an expected low temperature that ranges from 55°F to 70°F, to the proper temperature. (Should the pipe be insulated? Is a thermostat needed to control the temperature of the delivered hot water?)
3. A means for dispensing the water at a selected flow rate.
4. Safety considerations to avoid burns.
5. Total installed cost of the system and operating costs.

Vacuum Cleaner Efficiency Ratings (4 engineers)

Many devices that are manufactured for the consumer are rated. For example, refrigerators are rated by manufacturers by putting them through standard tests that reveal the cost of running them per year. Automobiles are rated in standard tests and miles per gallon figures are reported by the manufacturer. The key to developing ratings is in the establishment of standard tests. Such tests are usually developed with the aid or supervision of professional and technical societies, or alternatively, by engineers who work and have gained considerable experience in the industry.

Vacuum cleaner manufacturers have decided to allow a consulting company to design standardized methods for testing vacuum cleaners. The parameters that are deemed important are to be selected by the consultants. The procedures are to be written up in an acceptable or traditional format.

To be designed, selected, or determined:

1. Study existing methods and techniques for testing vacuum cleaners.
2. Develop a method and procedure for evaluating vacuum cleaner performance.
3. Consult ASTM standards and write your procedures in the same style.
4. Following your own procedures, select 3 different vacuum cleaners and evaluate them. Express the results in a performance factor rating that you have developed.

Paddle Wheel Design (4 engineers)

At one time, steam boats were used extensively to transport goods and passengers. Steam driven pistons provided power to rotate huge paddle wheels which propelled the ships over the water. The paddles are traditionally just flat boards bolted onto a frame. As each board was rotated under the water, thrust was developed. The shape of the boards was flat, but it is felt that a different shape or a different method of rotation under water would provide a greater transfer of power.

To be designed, selected, or determined:

1. Analysis of the traditional "blade" design including material, typical rotational speeds, thrust developed by a wheel, and efficiency.
2. Proposed new design including material, number of blades, expected thrust, and efficiency. (Rotational speed should be the same as the traditional style.)
3. Current costs of a traditional wheel.
4. Current costs of the newly designed wheel.

Bicycle Frame Finishing Process (4 engineers)

A manufacturer of bicycle frames that previously sent the frames out for finishing has decided to set up a painting facility right in the manufacturing plant. The objective of the facility is to take frames that have been treated and given a primer coat of paint and apply a finish coat. It has been decided that enamel type paints are satisfactory for finishing frames. "Orange Peel" (a finish coat that is reminiscent of the surface of an orange rind) is unacceptable.

To be designed, selected, or determined:

1. The operations required to produce an attractive, durable, rust-inhibiting finish coat on the (steel) bicycle frames.
2. Whether the finish operations should be performed in a batch type process or an assembly line type process.
3. The systems required for each operation. (For example, a chemical treatment operation performed in a tank requires design of the tank and selection of the chemicals. A paint spray booth requires design of the booth and selection of the spraying equipment.)
4. The safety requirements (adequate ventilation, for example, when a spray booth is used).
5. Total cost of the facility including operating and maintenance costs.

Personal Ice Maker (4 engineers)

People who travel for long distances by car usually like to carry coolers along in order to keep food and drinks cold. Evenings are spent in motels and motel owners often do not permit guests to have ice for their coolers. Consequently, it seems that there is a need for an alternative supply of ice. One source is in gas stations or markets. Another possible source is with the use of a portable ice maker.

The portable ice maker seems to be an attractive alternative. Upon arriving at the motel, the guest would fill the ice maker with water and provide it with electricity. By morning, the ice maker would have produced a quantity of ice sufficient for the needs of the user.

To be designed, selected, or determined:

1. A portable ice making machine to produce and store enough ice (in the form of cubes) to fill a 1 ft x 1 ft x 1 ft volume.

2. All the internal workings including (if appropriate) compressor, evaporator, and condenser.
3. The housing and mounting of all components.
4. The material used in the housing, etc.
5. The cost of producing an ice maker.
6. The cost of running the ice maker to make ice.

Ceiling Fan Blade Design (4 engineers)

Ceiling fans have been used for a number of years to cause air to circulate within a room and make the inhabitants more comfortable. The fans consist of a motor and four or more blades. The blades themselves are usually flat and wider at the tip than at the root. A performance rating of the air moving ability of conventional fans does not yet exist because a standardized method for making the needed measurements has not been devised.

To be designed, selected, or determined:

1. Compose a proposed description for a standardized test of conventional ceiling fans.
2. Select a fan and perform an analysis of the blades as marketed.
3. Determine if the blade design can be improved upon (i.e., if the same power can move more air merely by changing blade design).

Greenhouse Climatic Conditions (4 engineers)

The owners of a greenhouse desire to control the application of water (or water-fertilizer mix) and wintertime temperature within the structure to protect and feed the plants. The greenhouse is 10 m wide by 30 m long. The wall height is 2.5 m and the lengthwise sloping roof is set at a pitch of 40° with the horizontal. Tall potted plants can be set on the floor. Short potted plants are set on 3 ft high tables. The greenhouse is to produce plants sold at wholesale prices to retailers and not the general public. Thus only a few different types of plants are in the greenhouse at any time.

To be designed, selected, or determined:

1. A system for irrigating the plants, preferably automatic. (Consider that the entire floor area must be able to receive water or a water-fertilizer mix. Also, it may be desired to be able to have some areas that receive little or no water as deemed necessary by the operator.)

2. Piping (if any), piping material, and routing. (Consider that if a water-fertilizer mix is used, the fertilizer may have a corrosive effect on certain materials.)
3. Pump size (if any), type, and location. (Assume that city water is available at 55 psig.)
4. A drainage system (if necessary) which can be put into the slab before it is poured.
5. A system for heating the greenhouse in winter during times when the temperature inside might fall below 40°F.
6. Fan, ductwork, and heat source for heating purposes. (Or alternative heating technique.)
7. Total cost of all items, including installation and expected life.

Solar Heated Swimming Pool (4 engineers)

The swimming pool of a motel is currently outdoors. Management desires to erect a structure over the pool and use solar energy to heat the water. The pool dimensions are 10 m long x 7 m wide with a water depth that varies in the lengthwise direction from 1 m to 3 m. It is proposed to have flat plate collectors receive energy from the sun and use the energy to maintain the water at a comfortable temperature year round.

There exists a manufacturer of hot water heaters who markets swimming pool heaters. These heaters warm the water with natural gas burners.

To be designed, selected, or determined:

1. A comfortable water temperature for the indoor pool.
2. The number of solar collectors required. (Assume that the motel is in a city to be specified by the instructor.)
3. The placement of these collectors, their orientation, location, and number required.
4. Pump, piping, and materials of construction.
5. Total cost of system, including initial, operating, and maintenance.
6. Compare these costs to those associated with the use of a natural gas water heating system.

Economics of Fuel Tank Insulation (4 engineers)

Fuel tanks contain heated fuel oil or crude oil. The liquid is heated in order to lower its viscosity and make it less costly to pump. There are costs associated with keeping the liquid heated in a storage tank. These are investigated in this project.

It is desired to analyze heat transfer characteristics from uninsulated tanks and how these tanks would be affected if they were insulated. Three tanks are to be considered. Two types of insulation will be compared.

An uninsulated tank containing a heated liquid loses heat by conduction through its bottom to the ground, and by natural convection through its sides and roof to the ambient air, and by radiation. Referring to Figure 10.6, we see that the total heat loss is made up of:

A = heat lost from liquid through side wall
B = heat lost from vapor through side wall
C = heat lost from vapor through roof
D = heat lost to the ground

The tank wall is neglected as being a significant insulator that presents a high resistance to heat transfer.

To be designed, selected, or determined:

1. The heat loss through each tank described below, uninsulated.
2. The heat loss with a 2 in. thick layer of urethane added to the side wall.
3. The heat loss with a 2 in. thick layer of fiberglas added to the side wall.
4. Based on a 20 year life, calculate rate of return on investment including first cost, maintenance costs, and savings in fuel costs for each type of insulation. (Use a fuel cost of $2.30 for each 10^6 BTU.)
5. Make a recommendation.

Tank Descriptions

Tank #1 is 170 ft in diameter and 40 ft high. It contains #6 fuel oil maintained at 140°F. Assume that the tank on the average is 0.85 (x height of tank) full. Further assume a surface emissivity of 0.9 and a wind velocity of 10 mph. The ambient temperature is 70°F.

Tank #2 is 70 ft in diameter and 40 ft high. It contains #6 fuel oil maintained at 140°F. Assume that the tank on the average is 0.85 (x height) full of liquid, that the surface emissivity is 0.9, and that the wind velocity is 10 mph. Again the ambient temperature is 70°F.

Tank #3 is 100 ft in diameter and 40 ft high. It contains paraxylene maintained at 80°F. The tank on the average is 0.85 (x height) full of liquid and its surface emissivity is 0.9. The average wind velocity is 10 mph and the ambient air temperature is 70°F.

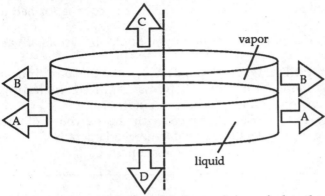

FIGURE 10.6. *Sketch of heat losses from a fuel tank.*

Insulation Information

Urethane has a thermal conductivity of 0.14 BTU/hr-ft-°R. Fiberglas has a thermal conductivity of 0.27 BTU/hr-ft-°R. Urethane costs $2.55/ft² and must be resurfaced after 10 years at a cost of 1.50/ft². The fiberglas, although it has a higher thermal conductivity, needs no resurfacing every 10 years and it has an initial cost of $3.25/ft². The urethane and fiberglas initial costs are based on a 2 in. thickness which is what your calculations should be based upon.

A Three Ton Source of Cool Air (4 engineers)

A remotely located office-shop building has been converted into a church. A one ton air conditioner originally used to cool just the offices is the only system available for making the entire building comfortable for a few hours on Sunday mornings. It is believed that finances do not permit the purchase and installation of an additional unit or units. It has been determined that 3 tons of cooling are required on Sundays to accommodate the current attendance level.

It has been suggested that the air conditioner be made to run all week to form a huge block of ice or to cool some other "storage" facility. Then on Sunday mornings, air can be moved past the ice and the cooled air used to make the occupants comfortable.

It is desired to design an appropriate "storage" facility using minimum space that can later be used to provide cool air. Ice is not necessarily the answer, although water is inexpensive.

To be designed, selected, or determined:

1. A facility (liquid, gas, solid) that can be cooled during a 6¹/₂ day

period by a one ton air conditioner to yield 3 tons of cooling for half a day.

2. The dimensions of the required facility (should be as small as possible).
3. Total cost of the facility (assuming space within the building is available). Cost should include all items required including insulation, if any, and operation of the one ton unit.
4. Compare these costs to those associated with the purchase of the proper size cooling system (including installation and operation).
5. Make a recommendation based on your calculations.

Controlled Environment for Special Metals (4 engineers)

A museum has recently inherited a number of timepieces made during the 19th century. It has been decided to display these timepieces in a glass-sided and -topped case of internal dimensions 6 ft long x 2 ft tall x 2 ft wide. It is necessary, however, to control the gaseous environment about the timepieces in order to maximize their life. For example, is dehumidifying the air sufficient for keeping the timepieces in good condition, or is maintaining a vacuum about them a better solution? The difficulty is that the composition of the metals used by the makers of the timepieces is unknown. Moreover, museum management will not permit destructive testing of any sort or magnitude to be performed on any of the timepieces to determine the composition of the metals.

To be designed, selected, or determined:

1. The probable composition of the timepieces. (Is there a nondestructive test for determining their exact composition?)
2. The most favorable environment for the timepieces.
3. A means for maintaining this environment in the 6 x 2 x 2 ft glass housing.
4. A suitable material to use as the base within the housing to set the timepieces on (felt on wood, for example).
5. Total cost of the system used to maintain the environment within the cabinet, including ductwork (if any, it should be hidden from view), fans, etc., and estimated life of this system. (If a partial or total vacuum is to be maintained within the glass housing, determine how to seal the glass pieces at the edges and corners. Also, assume that the glass is properly braced.)
6. Estimate the life of the timepieces within the cabinet and if additional care (such as cleaning at regular intervals) is required for an extended life.

Mileage Meter for a Conventional Automobile
(4 engineers)

Automobile mileage estimates are useful for comparative purposes when selecting a car. However, many drivers would like to have an inexpensive meter that gives an instantaneous reading of mileage in digital form. So while driving, the operator will know immediately how many miles (or kilometers) are obtained per gallon (cubic meter) of fuel. A driving method can then be developed to yield a savings in fuel dollars spent for the auto owner.

To be designed, selected, or determined:

1. A technique for measuring the (mass or volume) flow rate of the fuel. (An obstruction in the flow line could lead to an excessive pressure drop. Any obstructions therefore should be selected and positioned with care.)
2. A method for transmitting the measured flow rate to the "electronics box." (Assume that the design of the electronics required to process input signals will be performed elsewhere. It is sufficient to say that the microprocessor involved can add, subtract, multiply, and/or divide any two analog or digital input signals, and deliver a digital readout. Moreover, the display can read in miles/gallon or km/m^3 and this conversion is also handled electronically.)
3. A technique for measuring and transmitting to the electronics box the speed of the automobile.
4. Materials, placement, and construction of the add-on devices.
5. Total cost of all components (excluding microprocessor unit) including initial, operating, and maintenance costs.

Heated Driveway Slab (4 engineers)

It is not uncommon to see concrete driveways laid out on an incline. Snow and ice accumulate on driveway surfaces during winter months. If the snow and ice remain unattended, it becomes difficult to control an automobile on the driveway. The owner can manually clear the driveway or install a heating system within the slab when it is poured. A proposed heating system is the subject of this project.

Consider a driveway that is 20 ft wide and 40 ft long. It is desired to embed within the driveway a tubing system through which water is circulated. The water provides heat to the concrete driveway so that ice and snow melt and flow away. Embedded tubes must not detract from the strength of the driveway.

To be designed, selected, or determined:

1. A layout for the tubes over which the concrete is poured, including braces if needed.
2. The tubing material and size and joining method. (Should refrigeration tubing be used so that as few joints as possible are needed?)
3. The method used to move water through the tubing. (Is a pump preferable?)
4. The source of the water. (Should water be pumped through and discharged to a drain, or should water be pumped from and back to a sump tank?)
5. The heat source needed to warm the water, if any is necessary.
6. The appropriate fluid to use. (Water has been suggested above but would a mix of water and anti-freeze be better? Is oil a better choice?)
7. Determine the total cost of the system including installation.

Design of a Portable Car Washing System (4 engineers)

Washing a car by hand with a garden hose and drying it with perhaps a chamois is somewhat tedious and requires a certain amount of time. The method is inexpensive, however. Alternatively, car owners many times employ commercial establishments to perform the task. This too requires a certain amount of time, sometimes longer than if done by hand, and it is costly. Between these two extremes is the fund raising operation that costs little but is many times disorganized and, again, takes a certain amount of time.

It has been proposed that a portable system be set up that fund raising groups or homeowners can use. It should consist of: (a) framework of some sort with a garden hose connection for rinsing the car; (b) a method of applying a cleaning agent and washing the car; and (c) a system for drying the car that uses a high velocity output blower, the type used in cleaning sidewalks and driveways. The portable system thus uses a garden hose to connect to a water supply and a garden blower for drying, assuming that these items are readily found in most households. It is further assumed that the actual cost of using this system is comparable to that of hand washing. The objective of having this system, however, is to save time without incurring too much additional operating cost.

To be designed, selected, or determined:

1. A water distribution framework on wheels that can move around and that can be disassembled.
2. The hose connections and water application system for the frame.
3. A method of applying a cleaning agent (perhaps another frame that dispenses soap and water) and of washing the vehicle.
4. A system for attaching a blower for drying the vehicle.
5. Total cost of the system including parts, operation, and maintenance.
6. Total time involved in using the system compared to hand washing.

Massaging Air Mattress (4 engineers)

A number of people, it is believed, would like to purchase an air mattress that could be used to massage the back of the person lying on it. The mattress consists of air pockets that are attached together as indicated in Figure 10.7 (a *suggested* construction method). The air pockets extend the entire width of the mattress. The user would lie down and turn on a compressor that would inflate and deflate the air pockets in a pre-determined sequence, thus massaging the back and providing comfort.

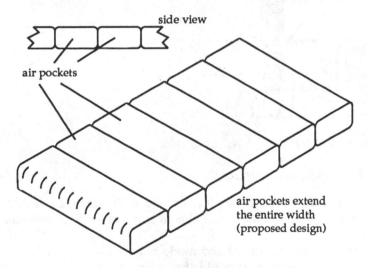

FIGURE 10.7. *A proposed design for a massaging air mattress.*

To be designed, selected, or determined:

1. The material to be used for the mattress.
2. The dimensions of the mattress and the air pockets. (Should the

mattress be the size of a conventional one?)
3. A desirable sequence for inflating and deflating the air pockets.
4. A compressor, connecting tubes, and hose attachments.
5. Total cost of the system.

Underground-stored Barricade (4 engineers)

Some city streets are used for automobile traffic during business hours and only for pedestrians in the evening. In order to keep automobiles from using such a street, posts are placed in holes in the middle of the street. To move the posts and store them elsewhere during business hours is cumbersome and requires manual labor. It is proposed that underground posts be used instead. When the post is not to restrict traffic, it "retreats" to a vertical cavity under the street surface. When it is to be up, it is raised with either an air or a hydraulic system as indicated in Figure 10.8. Air or hydraulic fluid fills and expands the expandable chamber, causing the post to move upward. Note that instead of an expanding chamber, a telescoping cylinder might be used.

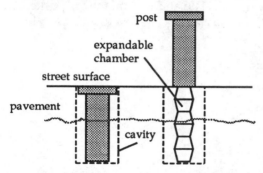

FIGURE 10.8. *An underground-stored barricade.*

To be designed, selected, or determined:

1. The post material and dimensions for it to be an effective barricade. (How much weight should the post be able to support in the stored and in the up position?)
2. The expandable chamber material and mode of operation.
3. The dimensions of the cavity. (Should the cavity be lined with a pipe or some other material?)
4. The location of the device used to inflate or pump up the expandable chamber. (Note that if an oil is used, leakage of oil from where it is stored and from the chamber into the ground is not permitted.)
5. Total cost of one set of posts for a two lane street one block long, including all parts and installation.

Ice Skating Rink (5 engineers)

An ice skating rink is to be constructed and occupy a floor area of 25 m x 15 m. The skating rink must be removable so that the floor space can be used for other purposes.

To be designed, selected, or determined:

1. The cooling system for producing an ice skating rink.
2. All components (such as piping) of the system, including materials of construction.
3. The substructure of the floor to support the rink.
4. A means for draining water when the rink is to be taken down. (Consider that a simple drain may not be sufficient and that a pump might be desirable.)
5. Total cost of the entire setup, including any special flooring required.

Analysis of "Energy Efficient" Methods in Home Construction (5 engineers)

An individual who desires to construct a new home will want to investigate the costs associated with operating it once it is built. Operating costs must be balanced with initial costs to provide an overall cost analysis for the structure. A number of "energy efficient" devices and methods have become available or recommended in recent years. In this project, cost comparisons are to be made for a number of items to determine the payback period or rate of return on investment. The comparison is between conventional construction techniques vs modifications and methods designed to conserve energy. A floor plan of a proposed home and its location will be provided by the instructor. The owner has decided to investigate the modifications listed in Table 10.1.

To be designed, selected, or determined:

1. Make a cost comparison for each of the items listed in Table 10.1.
2. Determine the additional initial cost associated with each of the proposed modifications.
3. Determine the reduction in operating cost associated with each proposed modification. ·
4. Where the modification provides a cost savings, perform an economic analysis showing the rate of return on the proposed investment.

TABLE 10.1. *Modifications to be investigated.*

Conventional		Proposed Modification
All exterior walls made with 2 x 4 wall studs spaced 16 in. center to center filled with 4 in. (nominal) of insulation.	(1)	All exterior walls made with 2 x 6 wall studs spaced with 24 in. center to center filled with 6 in. (nominal) of insulation.
Ceiling made with 2 x 10 studs spaced 24 in. center to center filled with 4 in. (nominal) of insulation.	(2)	Ceiling made with 2 x 12 studs spaced 24 in. center to center and filled with 12 in. (nominal) of insulation.
8 ft tall ceilings	(3)	9, 10, or 12 ft tall ceilings
Air cooled central air conditioner for summer, with gas heater (in one unit) for winter.	(4)	Water cooled air conditioner using lake water as a sink for summer, with heat pump (in one unit) using lake water as a source for winter.
40 gallon gas water heater	(5)	"Tankless" gas or electric water heater
—	(6)	A roof water spraying system to keep the roof from becoming excessively warm.
—	(7)	Roof vent(s)

Earth-Coupled Air Conditioning Unit (5 engineers)

Dwellings are cooled during summer months by an air conditioner. Figure 10.9 is a sketch of a conventional air conditioning unit. The fluid within the system is refrigerant. It leaves the compressor as a superheated vapor and enters the condenser. A fan moves outside air past the condenser where the air absorbs energy from the refrigerant causing the refrigerant to condense. The pressure of the refrigerant drops only slightly (due to wall friction). The refrigerant passes through an expansion valve, or capillary tube (throttling devices), after which the refrigerant pressure is greatly decreased. At this point, the liquid refrigerant is quite cold (35°F is not unusual). The refrigerant next passes through the evaporator and absorbs energy from inside air. The refrigerant vaporizes due to this energy (or heat) addition. It also experiences only a slight decrease in pressure (because of friction). The low pressure, vaporized refrigerant now returns to the compressor, and completes another cycle. While we have followed the path taken by a finite mass of refrigerant, it should be remembered that the process is continuous.

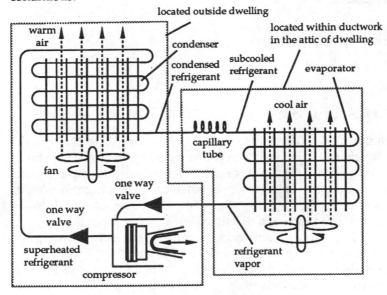

FIGURE 10.9. *A conventional air conditioning system.*

In many conventional units, the condenser is actually a refrigerant-to-air cross flow heat exchanger. Recently, it has been proposed to replace the condenser with a double pipe heat exchanger in which the refrigerant flows in the inner tube and water flows in the annulus. The water would absorb energy (heat) from the refrigerant. The

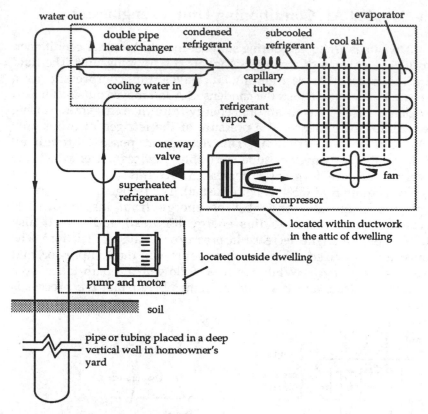

FIGURE 10.10. *An earth coupled air conditioning system.*

water is then pumped underground and, by heat transfer to the soil, the water loses the energy it gained from the refrigerant. Figure 10.10 shows the water cooled system. A number of manufacturers produce and market such systems, commonly referred to as earth-coupled air conditioners. They can operate in reverse (by changing direction of the flow of refrigerant) and be used to heat a dwelling during winter months.

Figure 10.11 shows an alternative configuration for burying the water pipe. The advantage of the setup in Figure 10.11 (over that of Figure 10.10) is that the well depth theoretically need not be so great. It is desired to perform an analysis of the system displayed in Figure 10.10.

To be designed, selected, or determined:

1. Pump, size, and material of construction.
2. Pipe or tubing size and material. (Allow at least 25 ft of pipe for each water line from the double pipe heat exchanger to the ground surface.)
3. Depth of well required for the system of Figure 10.10.

4. Diameter and depth of well required for the system of Figure 10.11 is used.
5. Double pipe heat exchanger size and material of construction. (Use a commercially available one if possible.)
6. Total cost of system, including exchanger, pump, piping, maintenance, and operation. Compare costs to those encountered in Figure 10.10.
7. Perform the analysis for a 3 ton air conditioner.

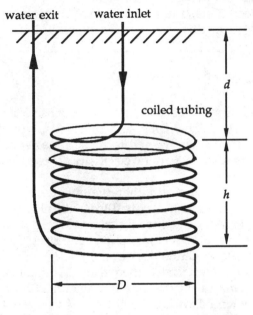

FIGURE 10.11. *Alternative configuration to the deep vertical well.*

Amusement Park Water Slide (5 engineers)

It is desired to construct a water slide in an amusement park. People of all ages and body shapes will climb stairs or a ramp to the top of the slide and ride down on a rubber or rubber substitute mat. One acre of land can be devoted to the slide and its associated operation (e.g., ticket selling booth, steps or ramp, etc.).

To be designed, selected, or determined:

1. Slide height, width, path, and material of construction.
2. Support structure and material of construction, including ground support.

3. "Catch" basin or method for individuals coming off the slide at the bottom.
4. Pump size, type, and location.
5. Piping, piping material, and routing.
6. Additives required to maintain the water (consider also that water may freeze during winter months).
7. Mat size, material, color, and number to be ordered.
8. Layout of the slide and its mode of operation on the acre of land.
9. Total cost of the slide, its expected life, and an estimate of the portion of the cost of a ticket that will cover this cost over the expected life.

A Heating System for Tank Car (5 engineers)

Conventional railroad tank cars are used to transport glucose from a supplier to a company that makes jellies, jams, and preserves. Upon arrival, the glucose in the tank car must be heated to make it easier to pump out. Heating can be done by using steam or electricity. These two methods are compared in this project.

If the glucose is to be heated with steam, then the tank car must be manufactured with tubes inside. When the car is filled with glucose, the tubes will be completely submerged. An external steam source is then connected to the tube inlet at the car, and steam is pumped inside. The tubes must be completely sealed so steam cannot leak inside.

If the glucose is to be heated electrically, then electric resistance heaters are to be installed in the tank car at the time it is manufactured. The resistance heaters will be submerged within the glucose. An external power supply is then connected to the heaters and the glucose is heated.

To be designed, selected, or determined:

1. A layout for the steam tubes within the car.
2. A tube diameter.
3. Materials to be used for the tubing.
4. Steam flow rate and inlet temperature. (After passing through, should the steam be discharged to atmosphere, or collected and reheated?)
5. Cost of the steam heating system (tubing cost, labor associated with fluid tight installation, and operating costs).
6. A layout or positioning of the immersion heater(s).
7. Electric power required.
8. Cost of electric heating system (heating, labor associated with installation, and operating costs).
9. The optimum temperature at which glucose should be maintained, as dictated by minimum cost analysis.

10. Perform all calculations for a conventional 40 foot tank car. Should the tank be insulated? What are its exact dimensions?

Viscometer for Bingham Fluids (5 engineers)

There are a number of commercial devices available for measuring viscosity. A Saybolt viscometer is used in the petroleum industry for oils. A Stormer viscometer is used in the paint industry. A rotating cup viscometer and a cone and plate viscometer are used for other fluids.

A capillary tube viscometer is commonly used to measure the viscosity of some Newtonian fluids. Figure 10.12 shows a capillary tube viscometer in which fluid is made to flow from one leg of the system to the other. As the fluid flows, it passes through a capillary tube at such a low velocity that the flow is laminar.

As indicated in the figure, this device consists of a glass tube of small diameter etched at three locations. By applying a vacuum to the right side, the liquid level is raised until it just reaches the uppermost etched line. At this point the liquid is released and allowed to flow under the action of gravity. The time required for the liquid level to fall from the middle to the lowest etched line is measured. Obviously, this problem is unsteady, but if average values are used, acceptable results are obtained. The volume of liquid involved is called the *efflux volume* which is carefully measured for the experiment. The average flow rate through the capillary tube is

$$Q = \frac{V}{t}$$

where V is the efflux volume and t is the time recorded. For laminar flow through a tube,

$$Q = \left(-\frac{dp}{dz} \right) \frac{\pi R^4}{8\mu}$$

where R is the tube radius and $(-dp/dz)$ is a positive pressure drop

$$\left(-\frac{dp}{dz} \right) = \frac{p_2 - p_1}{L}$$

The pressure drop equals the available hydrostatic head, which contains the gravity term (gravity is the driving force):

$$p_2 - p_1 = \frac{\rho g}{g_c} z$$

Substituting into the volume flow equation and solving for kinematic viscosity, we get

$$\frac{\mu g_c}{\rho} = v = \left(\frac{z \pi R^4 g}{8 L \not{V}}\right) t$$

For a given viscometer, the quantity in parentheses (a geometric quantity) is a constant. So kinematic viscosity is proportional to the time.

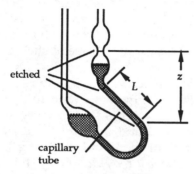

FIGURE 10.12. A capillary
tube viscometer.

To summarize how this viscometer works for a Newtonian fluid:
(1) We set up a laminar flow within the tube for which an exact flow rate equation is known.
(2) We related terms in this equation to flow parameters.
(3) We solved for viscosity in terms of the geometry of the system and the elapsed time for conducting the experiment.

It is desired to make up an analogous apparatus and analysis for measuring the viscosity of a Bingham fluid, using peanut butter as the test fluid. Figure 10.13 is a schematic of a proposed system. The test fluid is in a pressurized container on the left. The fluid is forced through a capillary tube to another container on the right. For a Bingham fluid flowing under laminar conditions, the Buckingham-Reiner Equation applies:

$$Q = \frac{\pi R^4}{8 \mu_o} \left(-\frac{dp}{dz}\right)\left[1 - \frac{4}{3}\left(\frac{\tau_o}{\tau_w}\right) + \frac{1}{3}\left(\frac{\tau_o}{\tau_w}\right)^4\right]$$

where μ_o is the apparent viscosity, τ_o is the initial yield stress of the fluid, and τ_w is the momentum flux at the wall $[= (- dp/dz)(R/2)]$. Thus by making certain measurements on the apparatus, the viscosity μ_o and the initial yield stress τ_o of a Bingham fluid can be calculated.

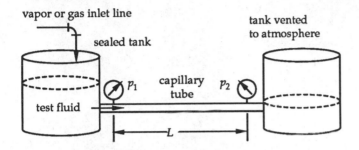

FIGURE 10.13. *Suggested design for a Bingham fluid capillary tube viscometer.*

To be designed, selected, or determined:

1. The design of an apparatus to force a Bingham fluid through a capillary tube. (The sketch in Figure 10.13 shows a horizontal tube. Would a vertically oriented tube be better to use?)
2. The capillary tube diameter. (What is the maximum Reynolds number for laminar flow conditions to exist?)
3. A method for assuring that laminar flow does indeed exist within the tube.
4. A method for determining the appropriate parameters of the flow rate equation.
5. A cost for the viscometer.

Appendix Tables

TABLE A.1. *Prefixes applicable to units in SI.*

Factor by which unit is multiplied	Prefix	Symbol
10^{12}	tera	T
10^9	giga	G
10^6	mega	M
10^3	kilo	k
10^2	hecto	h
10	deka	da
10^{-1}	deci	d
10^{-2}	centi	c
10^{-3}	milli	m
10^{-6}	micro	μ
10^{-9}	nano	n
10^{-12}	pico	p
10^{-15}	femto	f
10^{-18}	atto	a

Source: *E. A. Mechtly, NASA SP 7012, 1973.*

TABLE A.2. *Factors for converting to SI units; listed by physical quantity.*

To convert from	To	Multiply by
Acceleration (L/T^2)		
foot/second^2	*meter/second^2	3.048×10^{-1}
inch/second^2	*meter/second^2	2.54×10^{-2}
Area (L^2)		
foot^2	*meter^2	$9.290\,304 \times 10^{-2}$
inch^2	*meter^2	6.4516×10^{-4}
mile^2 (U.S. statute)	*meter^2	$2.589\,988\,110\,336 \times 10^6$
yard^2	*meter^2	$8.361\,273\,6 \times 10^{-1}$
Convection Coefficient or Film Conductance $[F{\cdot}L/(T{\cdot}L^2{\cdot}t)]$		
BTU/hour·ft^2·°R	W/(m^2·K)	5.678×10^0
Density (M/L^3)		
gram/centimeter^3	*kilogram/meter^3	1.00×10^3
lbm/inch^3	kilogram/meter^3	$2.767\,990\,5 \times 10^4$
lbm/foot^3	kilogram/meter^3	$1.601\,846\,3 \times 10^1$
slug/inch^3	kilogram/meter^3	$8.912\,929\,4 \times 10^5$
slug/foot^3	kilogram/meter^3	$5.153\,79 \times 10^2$

* Indicates an exact conversion. Otherwise, conversions are approximate representations of definitions or are results of physical measurements. Source: *E. A. Mechtly, NASA SP 7012, 1973.*

TABLE A.2. *Factors for converting to SI units, continued*

To convert from	To	Multiply by
Energy (F·L)		
BTU	joule	$1.054\ 350 \times 10^3$
calorie	*joule	4.184×10^0
foot–pound force	joule	$1.355\ 817\ 9 \times 10^0$
kilocalorie	*joule	4.184×10^3
kilowatt hour	*joule	3.60×10^6
watt hour	*joule	3.60×10^3
Energy/(area·time) or Heat Flux [F·L/(L²·T)]		
BTU/(ft² second)	watt/meter²	$1.134\ 893\ 1 \times 10^4$
BTU/(foot² minute)	watt/meter²	$1.891\ 488\ 5 \times 10^2$
BTU/(foot² hour)	watt/meter²	$3.152\ 480\ 8 \times 10^0$
BTU/(inch² second)	watt/meter²	$1.634\ 246\ 2 \times 10^6$
calorie/(cm² minute)	watt/meter²	$6.973\ 333\ 3 \times 10^2$
watt/centimeter²	*watt/meter²	1.00×10^4
Force (F)		
lbf	*newton	$4.448\ 221\ 615\ 260\ 5 \times 10^0$
Length (L)		
foot	*meter	3.048×10^{-1}
inch	*meter	2.54×10^{-2}
mile (U.S. statute)	*meter	$1.609\ 344 \times 10^3$
mile (U.S. nautical)	*meter	1.852×10^3
yard	*meter	9.144×10^{-1}
Mass (M)		
gram	*kilogram	1.00×10^{-3}
lbm (pound mass)	*kilogram	$4.535\ 923\ 7 \times 10^{-1}$
slug	kilogram	$1.459\ 390\ 29 \times 10^1$
Power or Heat Flow Rate (F·L/T)		
BTU/second	watt	$1.054\ 350\ 264\ 488 \times 10^3$
BTU/hour	watt	$2.928\ 750\ 7 \times 10^{-1}$
calorie/second	*watt	4.184×10^0
calorie/minute	watt	$6.973\ 333\ 3 \times 10^{-2}$
foot lbf/hour	watt	$3.766\ 161\ 0 \times 10^{-4}$
foot lbf/minute	watt	$2.259\ 696\ 6 \times 10^{-2}$
foot lbf/second	watt	$1.355\ 817\ 9 \times 10^0$
horsepower	watt	$7.456\ 998\ 7 \times 10^2$
kilocalorie/minute	watt	$6.973\ 333\ 3 \times 10^1$

* Indicates an exact conversion. Otherwise, conversions are approximate representations of definitions or are results of physical measurements. Source: *E. A. Mechtly, NASA SP 7012, 1973.*

TABLE A.2. *Factors for converting to SI units, continued*

To convert from	To	Multiply by
Pressure (F/L^2)		
atmosphere	*newton/meter^2	$1.013\ 25 \times 10^5$
bar	*newton/meter^2	1.00×10^5
centimeter Hg (0°C)	*newton/meter^2	$1.333\ 22 \times 10^3$ *
centimeter H$_2$O (4°C)	newton/meter^2	$9.806\ 38 \times 10^1$
dyne/centimeter^2	*newton/meter^2	1.00×10^1
foot of H$_2$O (39.2°F)	newton/meter^2	$2.988\ 98 \times 10^3$
inch of Hg (32°F)	newton/meter^2	$3.386\ 389 \times 10^3$
inch of Hg (60°F)	newton/meter^2	$3.376\ 85 \times 10^3$
inch of water (39.2°F)	newton/meter^2	$2.490\ 82 \times 10^2$
inch of water (60°F)	newton/meter^2	$2.488\ 4 \times 10^2$
lbf/foot^2	newton/meter^2	$4.788\ 025\ 8 \times 10^1$
lbf/inch^2 (psi)	newton/meter^2	$6.894\ 757\ 2 \times 10^3$
pascal	*newton/meter^2	1.00×10^0
torr (0°C)	newton/meter^2	$1.333\ 22 \times 10^2$
Specific Heat Capacity [F·L/(M·t)]		
BTU/(lbm·°F)	joule/(kilogram·K)	4.187×10^3
Speed or Velocity (L/T)		
foot/hour	meter/second	$8.466\ 666\ 6 \times 10^{-5}$
foot/minute	*meter/second	5.08×10^{-3}
foot/second	*meter/second	3.048×10^{-1}
inch/second	*meter/second	2.54×10^{-1}
kilometer/hour	meter/second	$2.777\ 777\ 8 \times 10^{-1}$
knot (international)	meter/second	$5.144\ 444\ 444 \times 10^{-1}$
mile/hour	*meter/second	$4.470\ 4 \times 10^{-1}$
mile/minute	*meter/second	$2.682\ 24 \times 10^1$
mile/second	*meter/second	$1.609\ 344 \times 10^3$
Temperature (t)		
Celsius	Kelvin	$t_K = t_C + 273.15$
Fahrenheit	Kelvin	$t_K = (5/9)(t_F + 459.67)$
Fahrenheit	Celsius	$t_C = (5/9)(t_F - 32)$
Rankine	Kelvin	$t_K = (5/9)t_R$
Thermal Conductivity [F·L/(T·L·t)]		
BTU/(hour·foot·°R)	watt/(meter·K)	1.731×10^0
Thermal Diffusivity (L^2/T)		
foot^2/second	meter^2/second	9.29×10^{-2}
foot^2/hour	meter^2/second	2.581×10^{-5}

* Indicates an exact conversion. Otherwise, conversions are approximate representations of definitions or are results of physical measurements. Source: *E. A. Mechtly, NASA SP 7012, 1973.*

TABLE A.2. *Factors for converting to SI units, continued*

To convert from	To	Multiply by
Thermal Resistance [T·t/(F·L)]		
hour·°R/BTU	K/watt	$1.895\ 8 \times 10^0$
Time (T)		
day	*second	8.64×10^4
hour	*second	3.60×10^3
minute	*second	6.00×10^1
month	*second	2.628×10^6
year	*second	$3.153\ 6 \times 10^7$
Viscosity–kinematic (L^2/T)		
centistoke	*meter^2/second	1.00×10^{-6}
stoke	*meter^2/second	1.00×10^{-4}
foot^2/second	*meter^2/second	$9.290\ 304 \times 10^{-2}$
Viscosity–absolute $(F·T/L^2)$		
centipoise	*newton·second/meter^2	1.00×10^{-3}
lbm/(foot·second)	newton·second/meter^2	$1.488\ 163\ 9 \times 10^0$
lbf·second/foot^2	newton·second/meter^2	$4.788\ 025\ 8 \times 10^1$
poise	*newton·second/meter^2	1.00×10^{-1}
Volume (L^3)		
acre·foot	meter^3	$1.233\ 481\ 9 \times 10^{-3}$
barrel (42 gallons)	meter^3	$1.589\ 873 \times 10^{-1}$
cup	*meter^3	$2.365\ 882\ 365 \times 10^{-4}$
foot^3	*meter^3	$2.831\ 684\ 659\ 2 \times 10^{-2}$
gallon (U.S. dry)	*meter^3	$4.404\ 883\ 770\ 86 \times 10^{-3}$
gallon (U.S. liquid)	*meter^3	$3.785\ 411\ 784 \times 10^{-3}$
inch^3	*meter^3	$1.638\ 706\ 4 \times 10^{-5}$
liter	*meter^3	1.00×10^{-3}
ounce (U.S. fluid)	*meter^3	$2.957\ 352\ 956\ 25 \times 10^{-5}$
pint (U.S. dry)	*meter^3	$5.506\ 104\ 713\ 575 \times 10^{-4}$
pint (U.S. liquid)	*meter^3	$4.731\ 764\ 73 \times 10^{-4}$
quart (U.S. dry)	*meter^3	$1.101\ 220\ 942\ 715 \times 10^{-3}$
quart (U.S. liquid)	meter^3	$9.463\ 529\ 5 \times 10^{-4}$
tablespoon	*meter^3	$1.478\ 676\ 478\ 125 \times 10^{-5}$
teaspoon	*meter^3	$4.928\ 921\ 593\ 75 \times 10^{-6}$
yard^3	*meter^3	$7.645\ 548\ 579\ 84 \times 10^{-1}$

* Indicates an exact conversion. Otherwise, conversions are approximate representations of definitions or are results of physical measurements. Source: *E. A. Mechtly, NASA SP 7012, 1973.*

TABLE A.3. *Temperature conversions.*

K	°C	°R	°F
250	-23.15	450	-10
260	-13.15	468	8
270	-3.15	486	26
280	6.85	504	44
290	16.85	522	62
295	21.85	531	71
300	26.85	540	80
310	36.85	558	98
320	46.85	576	116
330	56.85	594	134
340	66.85	612	152
350	76.85	630	170
360	86.85	648	188
370	96.85	666	206
380	106.85	684	224
390	116.85	702	242
400	126.85	720	260
410	136.85	738	278
420	146.85	756	296

TABLE A.4. *The Greek Alphabet.*

English Spelling	Greek Capital Letters	Greek Lowercase Letters	English Spelling	Greek Capital Letters	Greek Lowercase Letters
Alpha	A	α	Nu	N	ν
Beta	B	β	Xi	Ξ	ξ
Gamma	Γ	γ	Omicron	O	o
Delta	Δ	δ	Pi	Π	π
Epsilon	E	ε	Rho	P	ρ
Zeta	Z	ζ	Sigma	Σ	σ
Eta	H	η	Tau	T	τ
Theta	Θ	θ	Upsilon	Y	υ
Iota	I	ι	Phi	ϑ	φ
Kappa	K	κ	Chi	X	χ
Lambda	Λ	λ	Psi	Ψ	ψ
Mu	M	μ	Omega	Ω	ω

TABLE B.1. *Properties of liquids at room temperature and pressure.*

Fluid	Specific Gravity	Absolute Viscosity $(N \cdot s/m^2) \times 10^3$	$(lbf \cdot s/ft^2) \times 10^5$	Surface Tension $(N/m) \times 10^3$
Acetone	0.787	0.316	0.659	23.1
Benzene	0.876	0.601	1.26	28.18
Carbon Disulfide	1.265	0.36	0.752	32.33
Carbon Tetrachloride	1.59	0.91	1.90	26.3
Castor Oil	0.96	650	1356	–
Chloroform	1.47	0.53	1.11	27.14
Decane	0.728	0.859	1.79	23.43
Ether	0.715	0.223	0.466	16.42
Ethyl Alcohol	0.787	1.095	2.29	22.33
Ethylene Glycol	1.100	16.2	33.8	48.2
Glycerine	1.263	950	1983	63.0
Heptane	0.681	0.376	0.786	19.9
Hexane	0.657	0.297	0.622	18.0
Kerosene	0.823	1.64	3.42	–
Linseed Oil	0.93	33.1	69.0	–
Mercury	13.6	1.53	3.20	484
Methyl Alcohol	0.789	0.56	1.17	22.2
Octane	0.701	0.51	1.07	21.14
Propane	0.495	0.11	0.23	6.6
Propyl Alcohol	0.802	1.92	4.01	23.5
Propylene	0.516	0.09	0.19	7.0
Propylene Glycol	0.968	42	88	36.3
Turpentine	0.87	1.375	2.87	–
Water	1.00	0.89	1.9	71.97

Source: *Reprinted with permission from* CRC Handbook of Tables for Applied Engineering Science *(2nd ed), pages 90 and 92, Tables 1-44 and 1-46, 1973. Copyright CRC Press, Inc., Boca Raton, FL.*

Notes on reading the table:

Density $\rho = (sp.gr. \times 1.94)$ slug/ft^3 = (sp.gr. $\times$ 1 000) kg/m^3 = (sp.gr. $\times$ 62.4) lbm/ft^3

Viscosity of Acetone: $\mu \times 10^3 = 0.316 \, N \cdot s/m^2$; $\mu = 0.316 \times 10^{-3} \, N \cdot s/m^2$

TABLE B.2. *Properties of saturated liquids: unused engine oil.*

temp °C	°F	specific gravity	specific heat C_p $\frac{J}{kg \cdot K}$	$\frac{BTU}{lbm \cdot °R}$	kinematic viscosity ν $m^2/s \times 10^4$	$ft^2/s \times 10^3$	thermal conductivity k $\frac{W}{m \cdot K}$	$\frac{BTU}{hr \cdot ft \cdot °R}$	thermal diffusivity α $m^2/s \times 10^8$	$ft^2/hr \times 10^3$	Prandtl number Pr
0	32	0.899	1796	0.429	42.8	46.1	0.147	0.085	9.11	3.53	47 100
20	68	0.888	1880	0.449	9.0	9.7	0.145	0.084	8.72	3.38	10 400
40	104	0.876	1964	0.469	2.4	2.6	0.144	0.083	8.34	3.23	2 870
60	140	0.864	2047	0.489	0.839	0.903	0.140	0.081	8.00	3.10	1 050
80	176	0.852	2131	0.509	0.375	0.404	0.138	0.080	7.69	2.98	490
100	212	0.840	2219	0.530	0.203	0.219	0.137	0.079	7.38	2.86	276
120	248	0.828	2307	0.551	0.124	0.133	0.135	0.078	7.10	2.75	175
140	284	0.816	2395	0.572	0.080	0.086	0.133	0.077	6.86	2.66	116
160	320	0.805	2483	0.593	0.056	0.060	0.132	0.076	6.63	2.57	84

$\beta = 0.70 \times 10^{-3}/K = 0.39 \times 10^{-3}/°R$ Example of reading values: kinematic viscosity at 0°C is $\nu = 42.8 \times 10^{-4}$ m^2/s

Source: *Data taken from Analysis of Heat and Mass Transfer, by E. R. G. Eckert and R. M. Drake, Jr., Taylor & Francis Group—Hemisphere Publishing Corp., New York, © 1987. Used with permission of the publisher.*

TABLE B.3. *Properties of saturated liquids: ethylene glycol* $C_2H_4(OH_2)$.

| temp | | specific gravity | specific heat C_p | | kinematic viscosity v | | thermal conductivity k | | thermal diffusivity α | | Prandtl number Pr |
°C	°F		$\dfrac{J}{kg \cdot K}$	$\dfrac{BTU}{lbm \cdot °R}$	$m^2/s \times 10^6$	$ft^2/s \times 10^5$	$\dfrac{W}{m \cdot K}$	$\dfrac{BTU}{hr \cdot ft \cdot °R}$	$m^2/s \times 10^8$	$ft^2/hr \times 10^3$	
0	32	1.130	2 294	0.548	57.53	61.92	0.242	0.140	9.34	3.62	615
20	68	1.116	2 382	0.569	19.18	20.64	0.249	0.144	9.39	3.64	204
40	104	1.101	2 474	0.591	8.69	9.35	0.256	0.148	9.39	3.64	93
60	140	1.087	2 562	0.612	4.75	5.11	0.260	0.150	9.32	3.61	51
80	176	1.077	2 650	0.633	2.98	3.21	0.261	0.151	9.21	3.57	32.4
100	212	1.058	2 742	0.655	2.03	2.18	0.263	0.152	9.08	3.52	22.4

$\beta = 0.65 \times 10^{-3} /K = 0.36 \times 10^{-3} /°R$ Example of reading values: kinematic viscosity at 0°C is $v = 57.53 \times 10^{-6} \ m^2/s$

Source: *Data taken from Analysis of Heat and Mass Transfer, by E. R. G. Eckert and R. M. Drake, Jr., Taylor & Francis Group— Hemisphere Publishing Corp., New York,* © *1987. Used with permission of the publisher.*

TABLE B.4. *Properties of saturated liquids: glycerin $C_3H_5(OH)_3$.*

temp °C	°F	specific gravity	specific heat C_p $\dfrac{J}{kg \cdot K}$	$\dfrac{BTU}{lbm \cdot °R}$	kinematic viscosity ν $m^2/s \times 10^3$	$ft^2/s \times 10^2$	thermal conductivity k $\dfrac{W}{m \cdot K}$	$\dfrac{BTU}{hr \cdot ft \cdot °R}$	thermal diffusivity α $m^2/s \times 10^8$	$ft^2/hr \times 10^3$	Prandtl number Pr
0	32	1.276	2 261	0.540	8.31	8.95	0.282	0.163	9.83	3.81	84 700
10	50	1.270	2 319	0.554	3.00	3.23	0.284	0.164	9.65	3.47	31 000
20	68	1.264	2 386	0.570	1.18	1.27	0.286	0.156	9.47	3.67	12 500
30	86	1.258	2 445	0.584	0.50	0.54	0.286	0.165	9.29	3.60	5 380
40	104	1.252	2 512	0.600	0.22	0.24	0.286	0.165	9.14	3.54	2 450
50	122	1.244	2 583	0.617	0.15	0.16	0.287	0.166	8.93	3.46	1 630

$\beta = 0.50 \times 10^{-3}/K = 0.28 \times 10^{-3}/°R$ Example of reading values: kinematic viscosity at 0°C is $\nu = 8.31 \times 10^{-3}$ m²/s

Source: *Data taken from Analysis of Heat and Mass Transfer, by E. R. G. Eckert and R. M. Drake, Jr., Taylor & Francis Group— Hemisphere Publishing Corp., New York, © 1987. Used with permission of the publisher.*

TABLE B.5. *Properties of saturated liquids: water H_2O.*

temp °C	°F	specific gravity	specific heat C_p $\frac{J}{kg \cdot K}$	$\frac{BTU}{lbm \cdot °R}$	kinematic viscosity ν $m^2/s \times 10^7$	$ft^2/s \times 10^6$	thermal conductivity k $\frac{W}{m \cdot K}$	$\frac{BTU}{hr \cdot ft \cdot °R}$	thermal diffusivity α $m^2/s \times 10^7$	$ft^2/hr \times 10^3$	Prandtl number Pr
0	32	1.002	4 217	1.0074	17.88	19.25	0.552	0.319	1.308	5.07	13.6
20	68	1.000	4 181	0.9988	10.06	10.83	0.597	0.345	1.430	5.54	7.02
40	104	0.994	4 178	0.9980	6.58	7.08	0.628	0.363	1.512	5.86	4.34
60	140	0.985	4 184	0.9994	4.78	5.14	0.651	0.376	1.554	6.02	3.02
80	176	0.974	4 196	1.0023	3.64	3.92	0.668	0.386	1.636	6.34	2.22
100	212	0.960	4 216	1.0070	2.94	3.16	0.680	0.393	1.680	6.51	1.74
120	248	0.945	4 250	1.015	2.47	2.66	0.685	0.396	1.708	6.62	1.446
140	284	0.928	4 283	1.023	2.14	2.30	0.684	0.395	1.724	6.68	1.241
160	320	0.909	4 342	1.037	1.90	2.04	0.670	0.393	1.729	6.70	1.099
180	356	0.889	4 417	1.055	1.73	1.86	0.675	0.390	1.724	6.68	1.004
200	392	0.866	4 505	1.076	1.60	1.72	0.665	0.384	1.706	6.61	0.937
220	428	0.842	4 610	1.101	1.50	1.61	0.572	0.377	1.680	6.51	0.891
240	464	0.815	4 756	1.136	1.43	1.54	0.635	0.367	1.639	6.35	0.871
260	500	0.785	4 949	1.182	1.37	1.48	0.611	0.353	1.577	6.11	0.874
280	537	0.752	5 208	1.244	1.35	1.45	0.580	0.335	1.481	5.74	0.910
300	572	0.714	5 728	1.368	1.35	1.45	0.540	0.312	1.324	5.13	1.109

$\beta = 0.18 \times 10^{-3}/K = 0.10 \times 10^{-3}/°R$ Example of reading values: kinematic viscosity at 0°C is $\nu = 17.88 \times 10^{-7} \ m^2/s$

Source: *Data taken from Analysis of Heat and Mass Transfer, by E. R. G. Eckert and R. M. Drake, Jr., Taylor & Francis Group— Hemisphere Publishing Corp., New York, © 1987. Used with permission of the publisher.*

TABLE C.1. *Physical properties of gases at room temperature and pressure.*

Gas	density ρ		specific heat C_p		kinematic viscosity ν		gas constant R		$\gamma = C_p/C_v$
	kg/m³	lbm/ft³	$\frac{1}{kg \cdot K}$	$\frac{BTU}{lbm \cdot °R}$	m²/s×10⁶	ft²/s×10⁵	$\frac{1}{kg \cdot K}$	$\frac{ft \cdot lbf}{lbm \cdot °R}$	
Air	1.177	0.0735	1 005.7	0.240	15.68	16.88	0.026 24	0.01516	1.4
Carbon dioxide	1.797	0.1122	871	0.208	8.321	8.957	0.016 572	0.009575	1.3
Nitrogen	1.142	0.0713	1 040.8	0.2486	15.63	16.82	0.026 20	0.01514	1.4
Oxygen	1.300	0.0812	920.3	0.2198	15.86	17.07	0.026 76	0.01546	1.4

Example of reading values: kinematic viscosity of air is $\nu = 15.68 \times 10^{-6}$ m²/s

Gas Constant $R = R_u/MW$; $R_u = 1545$ ft·lbf/lbmol·°R = 49,700 ft·lbf/slugmol·°R = 8 312 N·m/(mol·K)

MW = molecular weight (Engineering units) = molecular mass (SI units)

1 BTU = 778 ft·lbf

Source: Data taken from a number of sources; see references at end of text.

TABLE C.2. *Properties of gases at atmospheric pressure (101.3 kPa = 14.7 psia).*

Air (Gas Constant = 286.8 J/(kg·K) = 53.3 ft·lbf/(lbm·°R); $\gamma = C_p/C_v = 1.4$)

temp		density ρ		specific heat C_p		kinematic viscosity ν		thermal conductivity k		thermal diffusivity α		Prandtl number Pr
K	°R	kg/m³	lbm/ft³	J/kg·K	BTU/lbm·°R	m²/s×10⁶	ft²/s×10⁵	W/m·K	BTU/hr·ft·°R	m²/s×10⁴	ft²/hr	
100	180	3.601	0.225	1026.6	0.245	1.923	2.070	0.009 246	0.005342	0.025 01	0.0969	0.770
150	270	3.268	0.148	1009.9	0.241	4.343	4.674	0.013 735	0.007936	0.057 45	0.223	0.753
200	360	1.768	0.110	1006.1	0.240	7.490	8.062	0.018 09	0.01045	0.101 65	0.394	0.739
250	450	1.413	0.0882	1005.3	0.240	9.49	10.2	0.022 27	0.01287	0.131 61	0.510	0.722
300	540	1.177	0.0735	1005.7	0.240	15.68	16.88	0.026 24	0.01516	0.221 60	0.859	0.708
350	630	0.998	0.0623	1009.0	0.241	20.76	22.35	0.030 03	0.01735	0.298 3	1.156	0.697
400	720	0.883	0.0551	1 104.0	0.242	25.90	27.88	0.033 65	0.01944	0.376 0	1.457	0.689
450	810	0.783	0.0489	1 020.7	0.244	28.86	31.06	0.037 07	0.02142	0.422 2	1.636	0.683
500	900	0.705	0.0440	1 029.5	0.245	37.90	40.80	0.040 38	0.02333	0.556 4	2.156	0.680
550	990	0.642	0.0401	1 039.2	0.248	44.34	47.73	0.043 60	0.02519	0.653 2	2.531	0.680
600	1080	0.589	0.0367	1 055.1	0.252	51.34	55.26	0.046 59	0.02692	0.751 2	2.911	0.680
650	1170	0.543	0.0339	1 063.5	0.254	58.51	62.98	0.049 53	0.02862	0.857 8	3.324	0.682
700	1260	0.503	0.0314	1 075.2	0.257	66.25	71.31	0.052 30	0.03022	0.967 2	3.748	0.684
750	1350	0.471	0.0294	1 085.6	0.259	73.91	79.56	0.055 09	0.03183	1.077 4	4.175	0.686
800	1440	0.441	0.0275	1 097.8	0.262	82.29	88.58	0.057 79	0.03339	1.195 1	4.631	0.689
850	1530	0.415	0.0259	1 109.5	0.265	90.75	97.68	0.060 28	0.03483	1.309 7	5.075	0.692
900	1620	0.393	0.0245	1 121.2	0.268	99.3	107.0	0.062 79	0.03628	1.427 1	5.530	0.696
950	1710	0.372	0.0232	1 132.1	0.270	108.2	116.5	0.065 25	0.03770	1.551 0	6.010	0.699
1 000	1800	0.352	0.0220	1 141.7	0.273	117.8	126.8	0.067 52	0.03901	1.677 9	6.502	0.702
1 100	1980	0.320	0.0120	1 160	0.277	138.6	149.2	0.073 2	0.0423	1.969	7.630	0.704

Example of reading values: kinematic viscosity at 100 K is $\nu = 1.923 \times 10^{-6}$ m²/s

TABLE C.2. *Properties of gases at atmospheric pressure (101.3 kPa = 14.7 psia), continued.*

Air (Gas Constant = 286.8 J/(kg·K) = 53.3 ft·lbf/(lbm·°R); $\gamma = C_p/C_v = 1.4$)

| temp | | density ρ | | specific heat C_p | | kinematic viscosity ν | | thermal conductivity k | | thermal diffusivity α | | Prandtl |
K	°R	kg/m³	lbm/ft³	J kg·K	BTU lbm·°R	m²/s×10⁶	ft²/s×10⁵	W m·K	BTU hr·ft·°R	m²/s×10⁴	ft²/hr	number Pr
1 200	2160	0.295	0.0184	1 179	0.282	159.1	171.3	0.0782	0.0452	2.251	8.723	0.707
1 300	2340	0.271	0.0169	1 197	0.286	182.1	196.0	0.0837	0.0484	2.583	10.01	0.705
1 400	2520	0.252	0.0157	1 214	0.290	205.5	221.2	0.0891	0.0515	2.920	11.32	0.705
1 500	2700	0.236	0.0147	1 230	0.294	229.1	246.6	0.0946	0.0547	3.262	12.64	0.705
1 600	2880	0.221	0.0138	1 248	0.298	254.5	273.9	0.100	0.0578	3.609	13.98	0.705
1 700	3060	0.208	0.0130	1 267	0.303	280.5	301.9	0.105	0.0607	3.977	15.41	0.705
1 800	3240	0.197	0.0123	1 287	0.307	308.1	331.6	0.111	0.0641	4.379	16.97	0.704
1 900	3420	0.186	0.0115	1 309	0.313	338.5	364.4	0.117	0.0676	4.811	18.64	0.704
2 000	3600	0.176	0.0110	1 338	0.320	369.0	397.2	0.124	0.0716	5.260	20.38	0.702
2 100	3780	0.168	0.0105	1·372	0.328	399.6	430.1	0.131	0.757	5.715	22.15	0.700
2 200	3960	0.160	0.0100	1 419	0.339	432.6	465.6	0.139	0.0803	6.120	23.72	0.707
2 300	4140	0.154	0.00955	1 482	0.354	464.0	499.4	0.149	0.0861	6.540	25.34	0.710
2 400	4320	0.146	0.00905	1 574	0.376	504.0	542.5	0.161	0.0930	7.020	27.20	0.718
2 500	4500	0.139	0.00868	1 688	0.403	543.5	585.0	0.175	0.101	7.441	28.83	0.730

Example of reading values: kinematic viscosity at 100 K is $\nu = 159.1 \times 10^{-6}$ m²/s

Source: Data taken from Analysis of Heat and Mass Transfer, by E. R. G. Eckert and R. M. Drake, Jr., Taylor & Francis Group—Hemisphere Publishing Corp., New York, © 1987. Used with permission of the publisher.

TABLE C.3. *Properties of gases at atmospheric pressure (101.3 kPa = 14.7 psia).*

Carbon Dioxide (Gas Constant = 188.9 J/(kg·K) = 35.11 ft·lbf/(lbm·°R); $\gamma = C_p/C_v = 1.3$)

temp		density ρ		specific heat C_p		kinematic viscosity ν		thermal conductivity k		thermal diffusivity α		Prandtl
K	°R	kg/m³	lbm/ft³	J kg·K	BTU lbm·°R	m²/s×10⁶	ft²/s×10⁵	W m·K	BTU hr·ft·°R	m²/s×10⁴	ft²/hr	number Pr
220	396	2.473	0.1544	783	0.187	4.490	4.833	0.010 805	0.006243	0.059 20	0.2294	0.818
250	450	2.165	0.1352	804	0.192	5.813	6.257	0.012 884	0.007444	0.074 01	0.2868	0.793
300	540	1.797	0.1122	871	0.208	8.321	8.957	0.016 572	0.009575	0.105 88	0.4103	0.770
350	630	1.536	0.0959	900	0.215	11.19	12.05	0.020 47	0.01183	0.148 08	0.5738	0.755
400	720	1.342	0.0838	942	0.225	14.39	15.49	0.024 61	0.01422	0.194 63	0.7542	0.738
450	810	1.191	0.0744	980	0.234	17.90	19.27	0.028 97	0.01674	0.248 13	0.9615	0.721
500	900	1.073	0.067	1013	0.242	21.67	23.33	0.033 52	0.01937	0.308 4	1.195	0.702
550	990	0.973	0.0608	1047	0.250	25.74	27.72	0.038 21	0.02208	0.375 0	1.453	0.685
600	1080	0.893	0.0558	1076	0.257	30.02	32.31	0.043 11	0.02491	0.448 3	1.737	0.668

Example of reading values: kinematic viscosity at 100 K is $\nu = 4.490 \times 10^{-6}$ m²/s

Source: *Data taken from Analysis of Heat and Mass Transfer, by E. R. G. Eckert and R. M. Drake, Jr., Taylor & Francis Group— Hemisphere Publishing Corp., New York, © 1987. Used with permission of the publisher.*

TABLE C.4. Properties of gases at atmospheric pressure (101.3 kPa = 14.7 psia).

Nitrogen (Gas Constant = 296.8 J/(kg·K) = 55.16 ft·lbf/(lbm·°R); $\gamma = C_p/C_v = 1.4$)

temp		density ρ		specific heat C_p		kinematic viscosity ν		thermal conductivity k		thermal diffusivity α		Prandtl number Pr
K	°R	kg/m³	lbm/ft³	J kg·K	BTU lbm·°R	m²/s ×10⁶	ft²/s ×10⁵	W m·K	BTU hr·ft·°R	m²/s ×10⁴	ft²/hr ×10⁴	
100	180	3.480	0.2173	1 072.2	0.2571	1.971	2.122	0.009 450	0.005460	0.025 319	0.09811	0.786
200	360	1.710	0.1068	1 042.9	0.2491	7.568	8.146	0.018 24	0.01054	0.102 24	0.3962	0.747
300	540	1.142	0.0713	1 040.8	0.2486	15.63	16.82	0.026 20	0.01514	0.220 44	0.8542	0.713
400	720	0.853	0.0533	1 045.9	0.2498	25.74	27.71	0.033 35	0.01927	0.373 4	1.447	0.691
500	900	0.682	0.0426	1 055.5	0.2521	37.66	40.54	0.039 84	0.02302	0.553 0	2.143	0.684
600	1080	0.568	0.0355	1 075.6	0.2569	51.19	55.10	0.045 80	0.02646	0.748 6	2.901	0.686
700	1260	0.493	0.0308	1 096.9	0.2620	65.13	70.10	0.051 23	0.02960	0.946 6	3.668	0.691
800	1440	0.427	0.0267	1 122.5	0.2681	81.46	87.68	0.056 09	0.03241	1.168 5	4.528	0.700
900	1620	0.379	0.0237	1 146.4	0.2738	91.06	98.02	0.060 70	0.03507	1.394 6	5.404	0.711
1 000	1800	0.341	0.0213	1 167.7	0.2789	117.2	126.2	0.064 75	0.03741	1.625 0	6.297	0.724
1 100	1980	0.310	0.0194	1 185.7	0.2382	136.0	146.4	0.068 50	0.03958	1.859 1	7.204	0.736
1 200	2160	0.285	0.0178	1 203.7	0.2875	156.1	168.0	0.071 84	0.04151	2.093 2	8.111	0.748

Example of reading values: kinematic viscosity at 100 K is $\nu = 1.971 \times 10^{-6}$ m²/s

Source: Data taken from *Analysis of Heat and Mass Transfer*, by E. R. G. Eckert and R. M. Drake, Jr., Taylor & Francis Group—Hemisphere Publishing Corp., New York, © 1987. Used with permission of the publisher.

TABLE C.5. *Properties of gases at atmospheric pressure (101.3 kPa = 14.7 psia).*

Oxygen (Gas Constant = 260 J/(kg·K) = 48.3 ft·lbf/(lbm·°R); $\gamma = C_p/C_v = 1.4$)

temp		density ρ		specific heat C_p		kinematic viscosity ν		thermal conductivity k		thermal diffusivity α		Prandtl number Pr
K	°R	kg/m³	lbm/ft³	J kg·K	BTU lbm·°R	m²/s×10⁶	ft²/s×10⁵	W m·K	BTU hr·ft·°R	m²/s×10⁴	ft²/hr	
100	180	3.992	0.2492	947.9	0.2264	1.946	2.095	0.009 03	0.00522	0.023 876	0.09252	0.815
150	270	2.619	0.1635	917.8	0.2192	4.387	4.722	0.013 67	0.00790	0.056 88	0.2204	0.773
200	360	1.955	0.1221	913.1	0.2181	7.593	8.173	0.018 24	0.01054	0.102 14	0.3958	0.745
250	450	1.561	0.0975	915.7	0.2187	11.45	12.32	0.022 59	0.01305	0.157 94	0.6120	0.725
300	540	1.300	0.0812	920.3	0.2198	15.86	17.07	0.026 76	0.01546	0.223 53	0.8662	0.709
350	630	1.113	0.0695	929.1	0.2219	20.80	22.39	0.030 70	0.01774	0.296 8	1.150	0.702
400	720	0.975	0.0609	942.0	0.2250	26.18	28.18	0.034 61	0.02000	0.376 8	1.460	0.695
450	810	0.868	0.0542	956.7	0.2285	31.99	34.43	0.038 28	0.02212	0.460 9	1.786	0.694
500	900	0.780	0.0487	972.2	0.2322	38.34	41.27	0.041 73	0.02411	0.550 2	2.132	0.697
550	990	0.709	0.0443	988.1	0.2360	45.05	48.49	0.045 17	0.02610	0.644 1	2.496	0.700
600	1080	0.650	0.0406	1 004.4	0.2399	52.15	56.13	0.048 82	0.02792	0.739 9	2.867	0.704

Example of reading values: kinematic viscosity at 100 K is $\nu = 1.946 \times 10^{-6}$ m²/s

Source: Data taken from *Analysis of Heat and Mass Transfer*, by E. R. G. Eckert and R. M. Drake, Jr., Taylor & Francis Group—Hemisphere Publishing Corp., New York, © 1987. Used with permission of the publisher.

TABLE C.6. *Properties of gases at atmospheric pressure (101.3 kPa = 14.7 psia).*

Water Vapor or Steam (Gas Constant = 461.5 J/(kg·K) = 85.78 ft·lbf/(lbm·°R); $\gamma = C_p/C_v = 1.333$)

temp K	°R	density ρ kg/m³	lbm/ft³	specific heat C_p J kg·K	BTU lbm·°R	kinematic viscosity ν m²/s×10⁵	ft²/s×10⁴	thermal conductivity k W m·K	BTU hr·ft·°R	thermal diffusivity α m²/s×10⁴	ft²/hr	Prandtl number Pr
380	684	0.586	0.0366	2 060	0.492	2.16	2.33	0.024 6	0.0142	0.293 6	0.789	1.060
400	720	0.554	0.0346	2 014	0.481	2.42	2.61	0.026 1	0.0151	0.233 8	0.906	1.040
450	810	0.490	0.0306	1 980	0.473	3.11	3.35	0.029 9	0.0173	0.307	1.19	1.010
500	900	0.440	0.0275	1 985	0.474	3.86	4.16	0.033 9	0.0196	0.386	1.50	0.996
550	990	0.400	0.0250	1 997	0.477	4.70	5.06	0.037 9	0.0219	0.475	1.84	0.991
600	1080	0.0365	0.0228	2 026	0.484	5.66	6.09	0.042 2	0.0244	0.573	2.22	0.986
650	1170	0.338	0.0211	2 056	0.491	6.64	7.15	0.046 4	0.0268	0.666	2.58	0.995
700	1260	0.314	0.0196	2 085	0.498	7.72	8.31	0.050 5	0.0292	0.772	2.99	1.000
750	1350	0.293	0.0183	2 119	0.506	8.88	9.56	0.054 9	0.0317	0.883	3.42	1.005
800	1440	0.274	0.0171	2 152	0.514	10.20	10.98	0.059 2	0.0342	1.001	3.88	1.010
850	1530	0.258	0.0161	2 186	0.522	11.52	12.40	0.063 7	0.0368	1.130	4.38	1.019

Example of reading values: kinematic viscosity at 380 K is $\nu = 2.16 \times 10^{-5}$ m²/s

Source: Data taken from a number of references.

TABLE D.1. *Pipe dimensions.*

Nominal Diameter	Outside Diameter in. (ft)	cm	Schedule	Inside Diameter ft	cm	Flow Area ft²	cm²
1/8	0.405 (0.03375)	1.029	40 (std)	0.02242	0.683	0.0003947	0.366 4
			80 (xs)	0.01792	0.547	0.0002522	0.235 0
1/4	0.540 (0.045)	1.372	40 (std)	0.03033	0.924	0.0007227	0.670 6
			80 (xs)	0.02517	0.768	0.0004974	0.463 2
3/8	0.675 (0.05625)	1.714	40 (std)	0.04108	1.252	0.001326	1.233
			80 (xs)	0.03525	1.074	0.0009759	0.905 9
1/2	0.840 (0.070)	2.134	40 (std)	0.05183	1.580	0.002110	1.961
			80 (xs)	0.04550	1.386	0.001626	1.508
			160	0.03867	1.178	0.001174	1.090
			(xxs)	0.02100	0.640	0.0003464	3.217
3/4	1.050 (0.0875)	2.667	40 (std)	0.06867	2.093	0.003703	3.441
			80 (xs)	0.06183	1.883	0.003003	2.785
			160	0.05100	1.555	0.002043	1.898
			(xxs)	0.03617	1.103	0.001027	9.555
1	1.315 (0.1095)	3.340	40 (std)	0.08742	2.664	0.006002	5.574
			80 (xs)	0.07975	2.430	0.004995	5.083
			160	0.06792	2.070	0.003623	3.365
			(xxs)	0.04992	1.522	0.001957	1.815
1¹/₄	1.660 (0.1383)	4.216	40 (std)	0.1150	3.504	0.01039	9.643
			80 (xs)	0.1065	3.246	0.008908	8.275
			160	0.09667	2.946	0.007339	6.816
			(xxs)	0.07467	2.276	0.004379	4.069
1¹/₂	1.900 (0.1583)	4.826	40 (std)	0.1342	4.090	0.01414	13.13
			80 (xs)	0.1250	3.810	0.01227	11.40
			160	0.1115	3.398	0.009764	9.068
			(xxs)	0.09167	2.794	0.007700	6.131

Notes: std implies standard; xs is extra strong; xxs is double extra strong.

TABLE D.1. *Pipe dimensions, continued.*

Nominal Diameter	Outside Diameter in. (ft)	cm	Schedule		Inside Diameter ft	cm	Flow Area ft^2	cm^2
2	2.375	6.034	40	(std)	0.1723	5.252	0.02330	21.66
	(0.1979)		80	(xs)	0.1616	4.926	0.02051	19.06
			160		0.1406	4.286	0.01552	14.43
				(xxs)	0.1253	3.820	0.01232	11.46
2^1/$_2$	2.875	7.303	40	(std)	0.2058	6.271	0.03325	30.89
	(0.2396)		80	(xs)	0.1936	5.901	0.02943	27.35
			160		0.1771	5.397	0.02463	22.88
				(xxs)	0.1476	4.499	0.01711	15.90
3	3.500	8.890	40	(std)	0.2557	7.792	0.05134	47.69
	(0.2917)		80	(xs)	0.2417	7.366	0.04587	42.61
			160		0.2187	6.664	0.03755	34.88
				(xxs)	0.1917	5.842	0.02885	26.80
3^1/$_2$	4.000	10.16	40	(std)	0.2957	9.012	0.06866	63.79
	(0.3333)		80	(xs)	0.2803	8.544	0.06172	57.33
4	4.500	11.43	40	(std)	0.3355	10.23	0.08841	82.19
	(0.375)		80	(xs)	0.3198	9.718	0.07984	74.17
			120		0.3020	9.204	0.07163	66.54
			160		0.2865	8.732	0.06447	59.88
				(xxs)	0.2626	8.006	0.05419	50.34
5	5.563	14.13	40	(std)	0.4206	12.82	0.1389	129.10
	(0.4636)		80	(xs)	0.4011	12.22	0.1263	117.30
			120		0.3803	11.59	0.1136	105.50
			160		0.3594	10.95	0.1015	94.17
				(xxs)	0.3386	10.32	0.09004	83.65
6	6.625	16.83	40	(std)	0.5054	15.41	0.2006	186.50
	(0.5521)		80	(xs)	0.4801	14.64	0.1810	168.30
			120		0.4584	13.98	0.1650	153.50
			160		0.4823	13.18	0.1467	136.40
				(xxs)	0.4081	12.44	0.1308	121.50

Notes: std implies standard; xs is extra strong; xxs is double extra strong.

TABLE D.1. *Pipe dimensions, continued.*

Nominal Diameter	Outside Diameter in. (ft)	cm	Schedule	Inside Diameter ft	cm	Flow Area ft²	cm²
8	8.625	21.91	20	0.6771	20.64	0.3601	334.60
	(0.7188)		30	0.6726	20.50	0.3553	330.10
			40 (std)	0.6651	20.27	0.3474	322.70
			60	0.6511	19.85	0.3329	309.50
			80 (xs)	0.6354	19.37	0.3171	294.70
			100	0.6198	18.89	0.3017	280.30
			120	0.5989	18.26	0.2817	261.90
			140	0.5834	17.79	0.2673	248.60
			(xxs)	0.5729	17.46	0.2578	239.40
			160	0.5678	17.31	0.2532	235.30
10	10.750	27.31	20	0.8542	26.04	0.5730	332.60
	(0.8958)		30	0.8446	25.75	0.5604	520.80
			40 (std)	0.8350	25.46	0.5476	509.10
			60 (xs)	0.8125	24.77	0.5185	481.90
			80	0.7968	24.29	0.4987	463.40
			100	0.7760	23.66	0.4730	439.70
			120	0.7552	23.02	0.4470	416.20
			140 (xxs)	0.7292	22.23	0.4176	388.10
			160	0.7083	21.59	0.3941	366.10
12	12.750	32.39	20	1.021	31.12	0.8185	760.60
	(1.058)		30	1.008	30.71	0.7972	740.71
			(std)	1.000	30.48	0.7854	729.70
			40	0.9948	30.33	0.773	722.50
			(xs)	0.9792	29.85	0.7530	699.80
			60	0.9688	29.53	0.7372	684.90
			80	0.9478	28.89	0.7056	655.50
			100	0.9218	28.10	0.6674	620.20
			120 (xxs)	0.8958	27.31	0.6303	585.80
			140	0.8750	26.67	0.6013	558.60
			160	0.8438	25.72	0.5592	519.60
14	14.000	35.57	30 (std)	1.104	33.65	0.9575	889.30
	(1.1667)		160	0.9323	28.42	0.6827	634.40

Notes: std implies standard; xs is extra strong; xxs is double extra strong.

TABLE D.1. *Pipe dimensions, continued.*

Nominal Diameter	Outside Diameter in. (ft)	cm	Schedule	Inside Diameter ft	cm	Flow Area ft²	cm²
16	16.000	40.64	30 (std)	1.271	38.73	1.268	1 178.00
	(1.333)		160	1.068	32.54	0.8953	831.60
18	18.000	45.72	(std)	1.438	43.81	1.623	1 507.00
	(1.500)		160	1.203	36.67	1.137	1 056.00
20	20.000	50.80	20 (std)	1.604	48.89	2.021	1 877.00
	(1.6667)		160	1.339	40.80	1.407	1 307.00
22	22.000	55.88	20 (std)	1.771	53.97	2.463	2 288.00
	(1.8333)		160	1.479	45.08	1.718	1 596.00
24	24.000	60.96	20 (std)	1.938	59.05	2.948	2 739.00
	(2.00)		160	1.609	49.05	2.034	1 890.00
26	26.000 (2.167)	66.04	(std)	2.104	64.13	3.477	3 230.00
28	28.000 (2.333)	71.12	(std)	2.271	69.21	4.050	3 762.00
30	30.000 (2.500)	76.20	(std)	2.438	74.29	4.666	4 335.00
32	32.000 (2.667)	81.28	(std)	2.604	79.34	5.326	4 944.00
34	34.000 (2.833)	86.36	(std)	2.771	84.45	6.030	5 601.00
36	36.000 (3.000)	91.44	(std)	2.938	89.53	6.777	6 295.00
38	38.000 (3.167)	96.52	—	3.104	94.61	7.568	7 030.00
40	40.000 (3.333)	101.6	—	3.271	99.69	8.403	7 805.00

Notes: std implies standard; xs is extra strong; xxs is double extra strong.

Source: *Dimensions in English units obtained from* ANSI B36.10–79, American National Standard Wrought Steel and Wrought Iron Pipe. *Reprinted by permission of the publisher, The American Society of Mechanical Engineers.*

TABLE D.2. *Dimensions of seamless copper tubing.*

Standard Size	Outside Diameter in. (ft)	cm	Type	Inside Diameter ft	cm	Flow Area ft^2	cm^2
1/4	0.375 (0.03125)	0.953	K	0.02542	0.775	0.0005074	0.471 7
			L	0.02625	0.801	0.0005412	0.503 9
3/8	0.500 (0.04167)	1.270	K	0.03350	1.022	0.0008814	0.820 3
			L	0.03583	1.092	0.001008	0.936 6
			M	0.03750	1.142	0.001104	1.024
1/2	0.625 (0.05208)	1.588	K	0.04392	1.340	0.001515	1.410
			L	0.04542	1.384	0.001620	1.505
			M	0.04742	1.446	0.001766	1.642
5/8	0.750 (0.0625)	1.905	K	0.05433	1.657	0.002319	2.156
			L	0.05550	1.691	0.002419	2.246
3/4	0.875 (0.0729)	2.222	K	0.06208	1.892	0.003027	2.811
			L	0.06542	1.994	0.003361	3.123
			M	0.06758	2.060	0.003587	3.333
1	1.125 (0.09375)	2.858	K	0.08292	2.528	0.005400	5.019
			L	0.08542	2.604	0.005730	5.326
			M	0.08792	2.680	0.006071	5.641
1^1/$_4$	1.375 (0.1146)	3.493	K	0.1038	3.163	0.008454	7.858
			L	0.1054	3.213	0.008728	8.108
			M	0.1076	3.279	0.009090	8.444
1^1/$_2$	1.625 (0.1354)	4.128	K	0.1234	3.762	0.01196	11.12
			L	0.1254	3.824	0.01235	11.48
			M	0.1273	3.880	0.01272	11.82
2	2.125 (0.1771)	5.398	K	0.1633	4.976	0.02093	11.95
			L	0.1654	5.042	0.02149	19.97
			M	0.1674	5.102	0.02201	20.44
2^1/$_2$	2.625 (0.21875)	6.668	K	0.2029	6.186	0.03234	30.05
			L	0.2054	6.262	0.03314	30.80
			M	0.2079	6.338	0.03395	40.17

TABLE D.2. *Dimensions of seamless copper tubing, continued.*

Standard Size	Outside Diameter in. (ft)	cm	Type	Inside Diameter ft	cm	Flow Area ft^2	cm^2
3	3.125 (0.2604)	7.938	K	0.2423	7.384	0.04609	42.82
			L	0.2454	7.480	0.04730	43.94
			M	0.2484	7.572	0.04847	45.03
3$^1/_2$	3.625 (0.3021)	9.208	K	0.2821	8.598	0.06249	58.06
			L	0.2854	8.700	0.06398	59.45
			M	0.2883	8.786	0.06523	60.63
4	4.125 (0.34375)	10.48	K	0.3214	9.800	0.08114	75.43
			L	0.3254	9.922	0.08317	77.32
			M	0.3279	9.998	0.08445	78.51
5	5.125 (0.4271)	13.02	K	0.4004	12.21	0.1259	117.10
			L	0.4063	12.38	0.1296	120.50
			M	0.4089	12.47	0.1313	112.10
6	6.125 (0.5104)	15.56	K	0.4784	14.58	0.1798	167.00
			L	0.4871	14.85	0.1863	173.20
			M	0.4901	14.93	0.1886	175.30
8	8.125 (0.6771)	20.64	K	0.6319	19.26	0.3136	291.50
			L	0.6438	19.62	0.3255	302.50
			M	0.6488	19.78	0.3306	307.20
10	10.125 (0.84375)	25.72	K	0.7874	24.00	0.4870	452.50
			L	0.8021	24.45	0.5053	469.50
			M	0.8084	24.64	0.5133	476.80
12	12.125 (1.010)	30.80	K	0.9429	28.74	0.6983	648.80
			L	0.9638	29.38	0.7295	677.90
			M	0.9681	29.51	0.7361	684.00

Note: *Type K is for underground service and general plumbing; type L is for interior plumbing; type M is for use only with soldered fittings.*

Source: *Dimensions in English units obtained from ANSI/ASTM B88–78,* Standard Specifications for Seamless Copper Water Tube. *Copyright ASTM. Reprinted with permission.*

Bibliography

General References

Transport Phenomena by R. B. Bird, W. E. Stewart and E. N. Lightfoot, John Wiley and Sons, Inc., New York, 1960.

Standard Mathematical Tables edited by S. M. Selby, 16th ed., CRC Press, Inc., Boca Raton FL, 1968.

Standard for Metric Practice, American Society for Testing and Materials, Philadelphia PA, Designation: E 380-76, 1976.

ASME Orientation and Guide for Use of SI (Metric) Units, 8th edition, published by ASME, New York, ASME Guide SI-1.

NASA SP-7012 by E. A. Mechtly, 1969. (Contains conversion factors from most units to SI units.)

1992 ASHRAE Handbook—HVAC Systems and Equipment, American Society of Heating, Refrigerating, and Air Conditioning Engineers, Atlanta GA, 1992. Other handbooks by ASHRAE that are useful include: *HVAC Applications, Refrigeration*, and *Fundamentals*.

Fluid Mechanics References

Introduction to Fluid Mechanics by W. S. Janna, PWS-KENT Publishing Company, Boston MA, 1987.

Fluid Flow by R. H. Sabersky, A. J. Acosta and E. G. Hauptmann, Macmillan Publishing Co., New York, 1989.

Introduction to Fluid Mechanics by J. E. A. John and W. Haberman, 2nd ed., Prentice-Hall, Inc., Englewood Cliffs, NJ, 1980.

Essentials of Engineering Fluid Mechanics by R. M. Olson, 4th ed., Harper and Row, Publishers, New York, 1980.

"An Explicit Equation for Friction Factor in Pipe," by N. H. Chen, American Chemical Society Communication, 1979.

Engineering Data Book, Hydraulic Institute, Cleveland OH, 1979.

"Friction Factors for Pipe Flow," by L. F. Moody, *Trans ASME*, v. 66, 1944, p. 671.

Heat Transfer References

Engineering Heat Transfer by W. S. Janna, PWS-KENT Publishing Company, Boston MA, 1983.

Fundamentals of Heat Transfer by F. P. Incropera and D. P. DeWitt, John Wiley and Sons, Inc., New York, 1981.

Heat Transfer by Helmut Wolf, Harper and Row Publishers, New York, 1983.

Process Heat Transfer by Don Q. Kern, McGraw-Hill Publishing Co., New York, 1990.

Properties of Substances

Handbook of Tables for Applied Engineering Science, edited by R. E. Bolz and G. L. Tuve, CRC Press, Inc., Boca Raton FL, 1973. (Highly recommended.

Handbook of Fundamentals published by American Society of Heating, Refrigerating, and Air Conditioning Engineers (ASHRAE), Chapters 17 and 31, 1972.

Transport Processes and Unit Operations by Christie J. Geankoplis, Allyn and Bacon, Inc., Boston MA, 1978.

Heat and Mass Transfer by E. R. G. Eckert and R. M. Drake, 2nd ed., McGraw-Hill Book Co., New York, 1958.

U. S. National Bureau of Standards Circular 564, 1955.

Heat Exchanger References

A Working Guide to Shell-and-Tube Heat Exchangers by S. Yokell, McGraw-Hill Publishing Co., New York, 1990.

Process Heat Transfer by Don Q. Kern, McGraw-Hill Publishing Co., New York, 1990.

Heat Exchanger Design by A. P. Fraas, John Wiley and Sons, Inc., New York, 1989.

Heat Exchangers—Selection, Design and Construction by E. A. D. Saunders, Longman Scientific and Technical Publishers; co-published with John Wiley and Sons, Inc., New York, 1988.

Design Analysis of Thermal Systems by R. F. Boehm, John Wiley and Sons, Inc., New York, 1987.

Compact Heat Exchangers by W. M. Kays and A. L. London, McGraw-Hill Book Co., New York, 1964.

Pipe and Tube Specifications

ANSI B36.10-1979, *American National Standard Wrought Steel and Wrought Iron Pipe,* published by ASME, New York, 1979.

ANSI/ASTM B88-78, *Standard Specifications for Seamless Copper Water Tube,* published by ASTM, New York, 1978.

Index